Special Research Project

The Ocean Characteristics and Their Changes

Newsletter Nos. 1–17

SPECIAL RESEARCH PROJECT

# The Ocean Characteristics and Their Changes

NEWSLETTER
NOS. 1–17

Translated from Japanese

AMERIND PUBLISHING CO. PVT. LTD.
New Delhi Bombay Calcutta New York

Translator: P.R. Kanade
General Editor: Dr. V.S. Kothekar

Published by Gulab Primlani, Amerind Publishing Co. Pvt. Ltd.,
66 Janpath, New Delhi. Typeset by Baba Barkhanath Printers, New Delhi and
printed at Pauls Press, New Delhi

# CONTENTS

Special Research Project

# The Ocean Characteristics and Their Changes

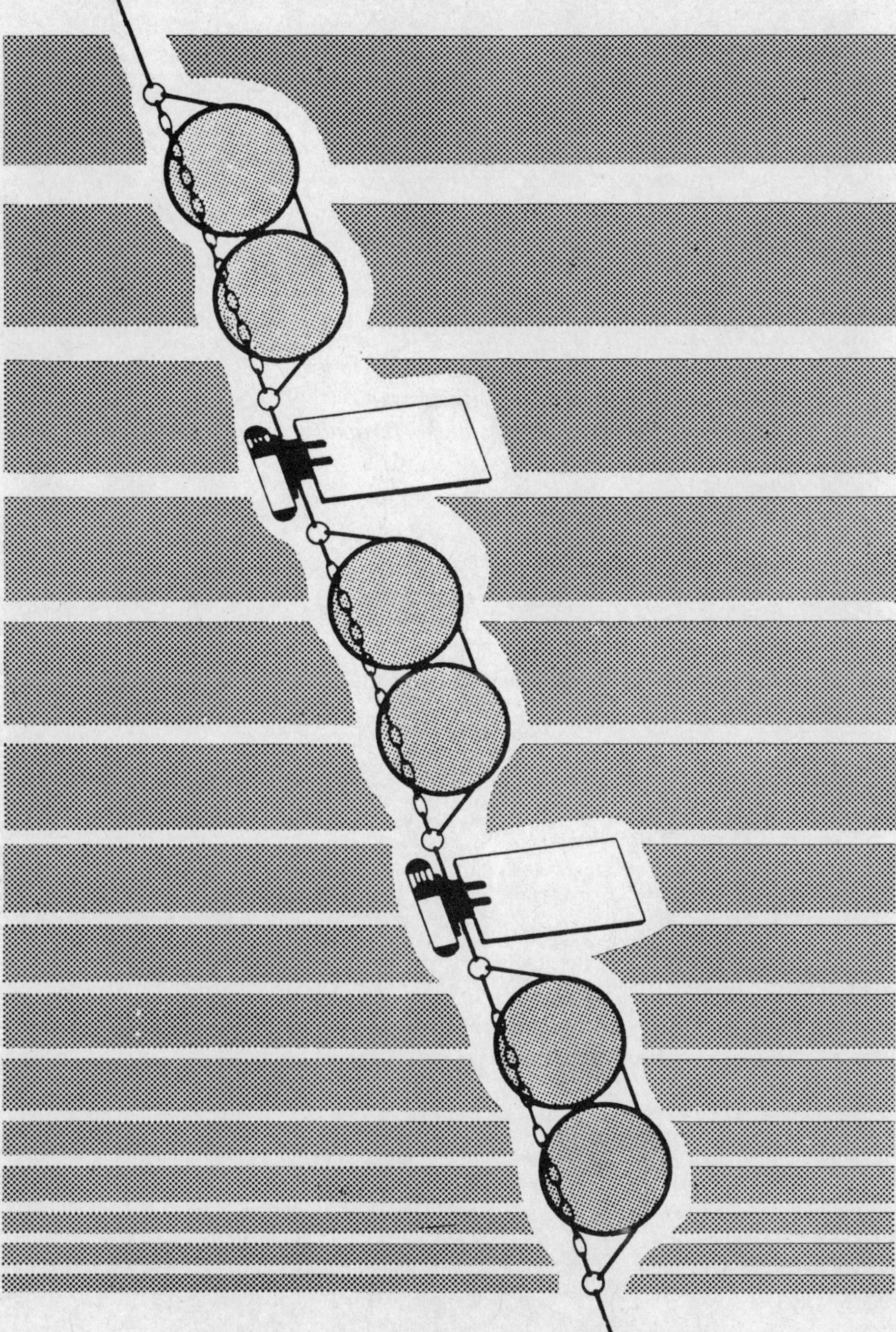

NEWSLETTER NO. 1, SEPTEMBER, 1981

# Contents

# On the Publication of First Issue of Newsletter

*Kinjiro Kajiura**

Oceanographic studies today have refocused from static phenomena to dynamic phenomena. Rapid progress in physical and chemical data collection techniques of very high precision and developments in high-speed computers through which we can collect and process voluminous data collection or numerically analyze complicated theoretical models enable us to "observe" even a rough sea where measurements have so far been very difficult to determine. With new technology we can examine the structure of the ocean including the meso-scale variations, circulation of deep sea water and other phenomena including oceanic circulation and atmospheric circulation.

Modern methods of analysis are vital here since without them we cannot utilize the data already available and important observations might be overlooked. On the other hand if we have only theories but no data, research cannot progress. It is, therefore, necessary to combine theories, a research scheme which can test them and appropriate technology to support the scheme.

During the 1970s, exciting results were obtained from such oceanographic studies as the ocean dynamics experiment project (MODE) sponsored by the International Decade of Ocean Exploration (IDOE), the Geochemical Ocean Sections Study (GEOSECS) and others. These results, however, have helped only to formulate theories based on knowledge accumulated in the 1960s. The research cycle, consisting of the discovery of new facts through new measuring techniques and methods, then new questions arising from this information followed by further detailed research schemes to answer them, is in motion today covering various phases of oceanographic studies, including the mid-sea or ocean floor.

The emphasis on oceanographic research the world over is the result of the demand for both the development of the ocean and the preservation of ocean environments as the existence of humanity in the future greatly depends on the conditions of the ocean. The United Nations Committee for Oceanographic Regulations symbolizes this need. The need to understand long-term meteorological variations, including the possibility of global in-

*Member, Special Study Group, Seismic Research Laboratory, Tokyo University.

crease in temperature because of increase in $CO_2$ in the atmosphere as a result of human activities, is vital in terms of food and energy availability in the 21st century. At the same time, the role of the ocean in meteorological variations also needs clarification.

Based on these trends, the Technological Study Committee released a proposal entitled, "On Progress in Oceanographic Studies" in 1978 and the Ocean Development Study Committee released their report, "The basic concepts and promotional measures for ocean development and its long-term prospects," in 1979-80.

Both these papers have stressed that our factual knowledge of the ocean is extremely limited and we have only recently begun to understand the complexity of time and space phenomena and their intercorrelation.

The current three-year Special Study Project beginning this year follows the guidelines of a special research project, "Biological Processes in the Ocean and Their Application," started in 1980 and the proposals of the Technological Study Committee. This may become the basis for systematic studies to increase our understanding of ocean dynamics.

The aim of this Newsletter is to publicize the research activities of various scientific groups from a wide range of fields, to further the understanding of interested individuals in the activities of various special research projects and their objectives as a medium of information exchange and to create opportunities for the promotion of joint studies by different scientists. The co-operation of all concerned is earnestly solicited.

## HIGHLIGHTS

# Long-term Measurements of Deep Ocean Currents on the Eastern Side of the Izu-Ogasawara Trench

*Keisuke Taira and Toshihiko Teramoto**

Deep ocean currents is one of the least understood phenomena of the ocean. Earlier, these currents were gaged by the distribution of materials in sea water or the condition of the water. However, to fully understand ocean currents it is necessary to actually measure the flow.

Measurement of the flow is very difficult and little was known about deep ocean currents until 1950 when J. Swallow first measured the flow by the Lagrangian method, using a neutrally buoyant (or a balanced) float. It was believed that even when they exist, deep ocean currents are very weak. In measurements with the neutrally buoyant float, the depth of measurement (that is, the depth of the buoy) varied by a few hundred meters during the series of measurements between the drifts of the float. This measurement, however, indicated that the velocity of these currents is much greater than imagined and that the flow varies considerably with time.

In 1960, a new technique was developed at Woods Hole Oceanographic Institution, after several efforts to measure the current over a long period by anchoring current meters (at various depths). This type of measurement was also tried in Japan and the importance of this technique was emphasized by the Special Study Project for Ocean Preservation (Representative: Mr. Horibe) which began its research in 1975; working group (WG) was created to acquire this technology. As planned, the first three years of the Special Study Committee were spent in acquiring the technology. Subsequently the measurement of flow was made for a long period around the eastern side of the trench (142°-146°E, 29°-31°N) by the Special Study Group on Environmental Science (Fiscal 1978) and the Special Study Group on Fundamentals of Oceanography (Fiscal 1979, Fiscal 1980). For this purpose a total of nine voyages were made of which eight voyages were on the *Hakucho-maru* (Oceanographic Laboratory, Tokyo University) and one on the *Keiten-maru* (Marine Research Laboratory, Kogoshima University). Today's measurements are taken using three systems installed in April 1981. Some of the measuring systems were repeatedly lowered and withdrawn at the same point (depth of

*Institute of Oceanography, Tokyo University.

water about 5900 m). Thus, at one point (TA point), the flow was measured for 1007 days but consecutive observations were made on 929 days. In the Woods Hole Oceanographic Institution Experiment, measurements were made from January 1965 to December 1974, over ten years, at off-shore site D (39°10'N, 70°W, 2660 m). However, there were many interruptions. During 1972, measurements were made for only two months, April and May. The consecutive observations made from March 1973, for 670 days are the longest in this experiment. Since 1973, the Woods Hole Oceanographic Institution has collected data for 731 consecutive days at site 28°N, 70°W (depth of water 5500 m), regarding Mode 0, Mode 1, Post Polymode and others. The record 929 consecutive days and our measurements for 1063 consecutive days made at Hachijojima are the longest consecutive measurements of this type in the world. Such long-term records are very useful in understanding deep ocean currents as long-term changes (in flow) are apparently very significant.

Our measurements of flow have been reported in many places [Horibe (Ed.), 1979; Taira and Teramoto, 1979; Taira, Imawaki and Teramoto, 1979; Kimura, 1979; Taira, 1980; Horibe, 1981; Nagata, 1981; Teramoto and Taira, 1981; Taira and Teramoto, 1981) and a technical paper is in preparation. In this paper, the authors who actually conducted the systematic measurements have discussed the broad outline of their results. This is good reference material for the study of deep ocean currents. The measurement of deep ocean currents using the anchored buoy technique is a principal method considered in this study and the results of this method are awaited.

## Measurement of Currents

We describe here in simple terms the method of measurement. A schematic diagram of the moored buoy system at a depth of 5900 m below sea is shown in Fig. 1. The total length (of the float) is 1900 m and the top of the float is 4000 m below sea level. A sonic sounding system is fixed at the bottom. This sonic sounding system can detect an underwater sound signal from a distance of 15 km. It releases and confirms the release of messenger upon receipt of a sound signal and makes hauling as well as recovery (of the float) possible. A 17 inch diameter pressure resistant glass sphere is used as the float and 6 mm Kevlar rope is used as the chain. Six Sanderae current meters are fixed to the rope as shown in the figure. The velocity sensor of the current meter is a rotor which revolves with a current exceeding 1.5 cm/sec. The direction of the current is monitored by a vane and its course is measured with a magnetic compass. The measured data are recorded on the built-in magnetic tape in digital form. The tape can accommodate data of 416 days if it is recorded at intervals of one hour. The recorded data is retrieved and processed on the computer. We analyzed the 24 hour mean values.

We will now explain the measurement conditions. Fig. 2 shows the topography of the ocean floor in and around the contour lines of 5000 m and 6000 m depth. Symbols A, B and C are fixed to the submarine mountain range in the east-west direction at 30° ± 20′N. The positions of moors at TA, TB, TD, TF and TG which lie in this region are also shown in the figures. Thus, TG, TF, TA lie on the 30°N line whereas TD lies on the 29°35′N line. A cross-section of the topography of the ocean floor along the 30°N line is shown in Fig. 3. The depth of the water increases continuously from the eastern bank of the trench to about 150°E. (The slope as shown here is 500 m per 300 km). The black dots in this figure indicate the depth at which the current is measured.

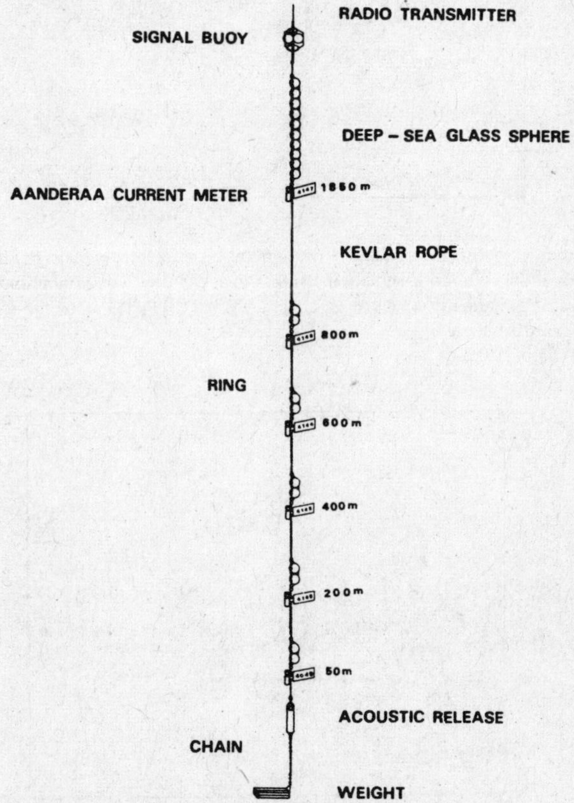


Fig. 1. Anchored buoy set at a depth of 5900 m at 30°N, 145°45′E. Six current meters were fixed at intervals of 50 m–1850 m from the sea bed.

## Variations in the Velocity of Current at Point TA and the Mean Current

The longest records were obtained at point TA as shown in Figs. 4 to 7. The current was measured at a place where ocean floor was 5900 m deep at depth of 50 m, 200 m, 800 m and 1850 m from the ocean floor. In addition, the current was measured with a moor interval of 25 m, 100 m, 600 m and

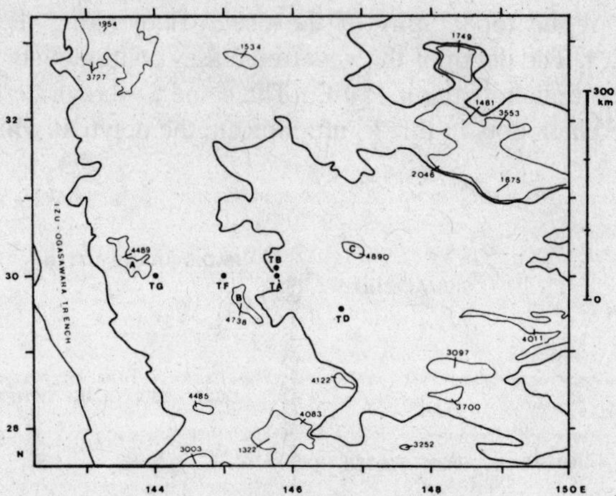


Fig. 2. Topography of ocean floor. Thick lines—6000 m, thin lines—5000 m. The depth of the water at the peak of the submarine mountains is less than 5000 m. The mountain range extending east-west around 30°N is identified as A, B, C. TA, TB, TD, TF, TG show the position of mooring points to be discussed here.

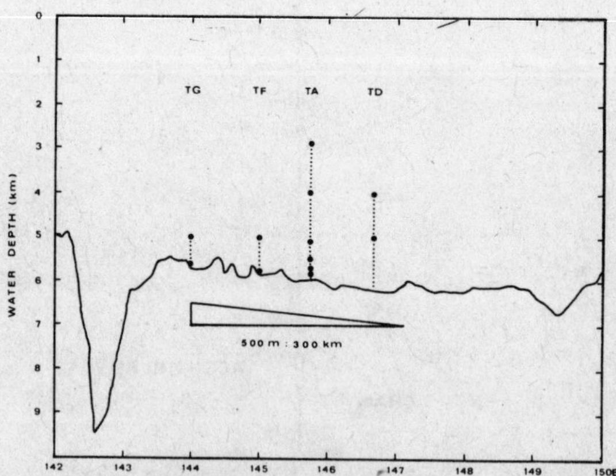


Fig. 3. Cross-section of the topography of the ocean floor at 30°N, 142–150°E. The depth of water increases eastward from the eastern bank of Izu-Ogasawara trench. TG, TF, TA, TD are mooring positions and the black dots indicate the depth of the current meter.

2750 m. If we consider Sept. 30, 1978 as the beginning of continuous measurement, then the last record was made on Apr. 16, 1981, the 929th day. The current velocity is expressed as a vector with East component as U and North component as V (both U and V were to the same scale).

The flow of the deep ocean currents is much greater than supposed and reached a value of 18.1 cm/sec on the 384th day, as seen in Fig. 4. Similarly, the maximum velocity was 20.1 cm/sec during the 85-day long observations beginning on Dec. 4, 1977, although this is not shown in this figure (depth 25.5 m above the ocean floor). Our observations reveal that strong currents exist deep in the oceans of the world. However, we still do not clearly understand the mechanism which causes them. What does create the strong currents we observed? Analysis of our data may answer this question.

The variation in currents is irregular but, as is clear from Figs. 4 to 7, variation in currents at all layers is in the range of 10 cm/sec. The velocity spectrum of the north component of the current is shown in Fig. 8.

The energy peak is around the frequency of $0.75\times10^{-2}$ CPD (for 133

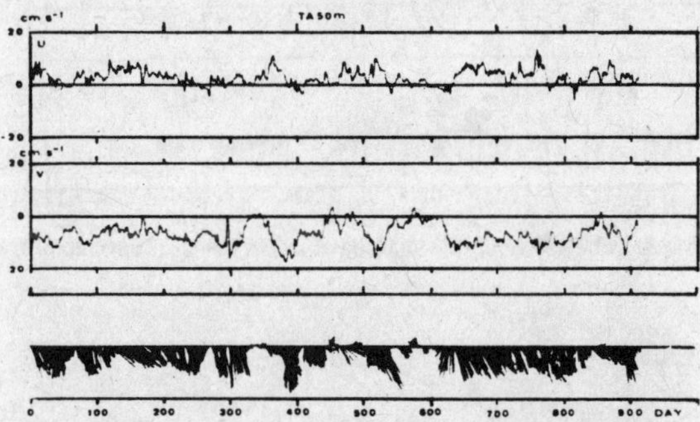


Fig. 4. Record showing daily average velocity 50 m above sea bed at point TA (ref. Figs. 2 and 3) starting Sept. 30, 1978 and 929th day is Apr. 14, 1981.

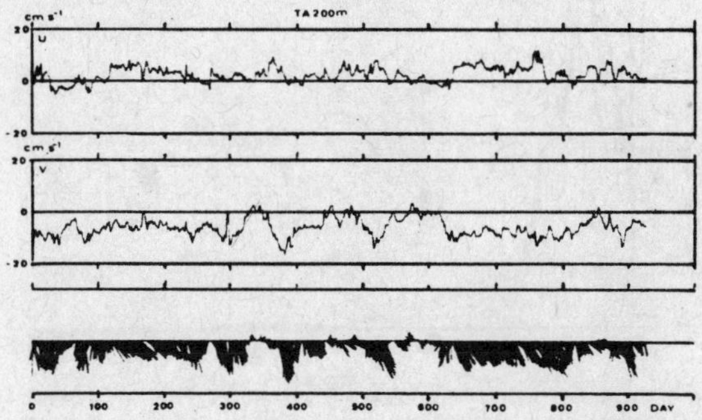


Fig. 5. Record of velocity at 200 m above sea bed at point TA (ref. explanations of Fig. 4).

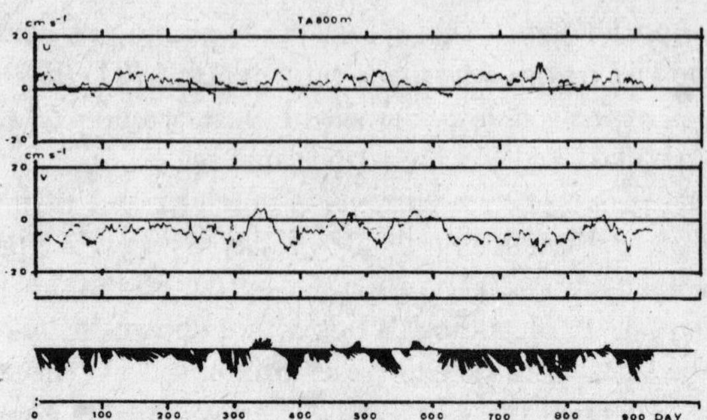


Fig. 6. Record of velocity at 800 m above sea bed at point TA (ref. explanations of Fig. 4).

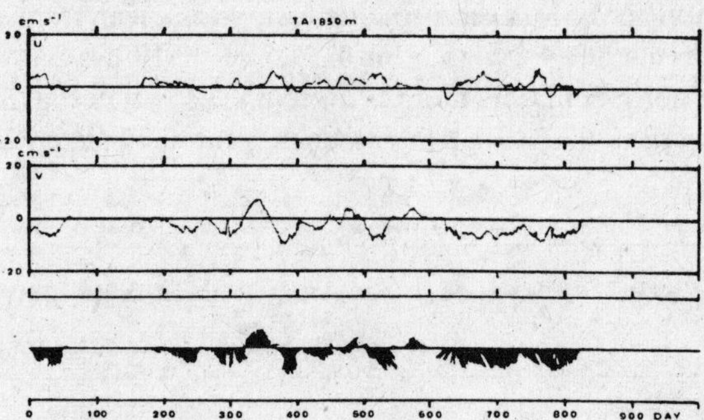


Fig. 7. Record of velocity at 1850 m above sea bed at point TA (ref. explanations of Fig. 4).

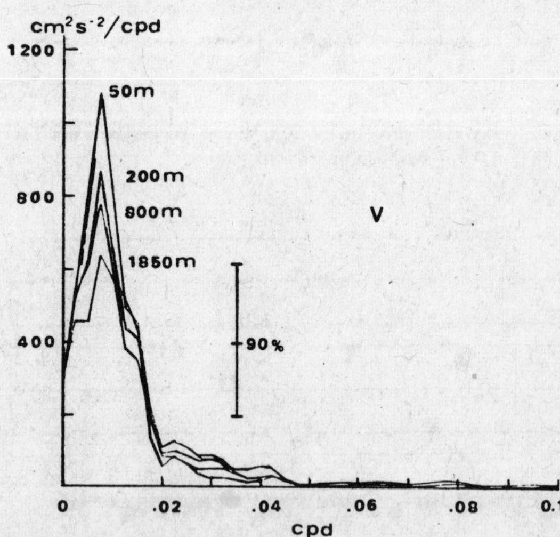


Fig. 8. North component spectrum of the daily average velocity at point TA, 50 m, 200 m, 800 m and 1850 m above the ocean floor. The period of spectral peak is about 133 days.

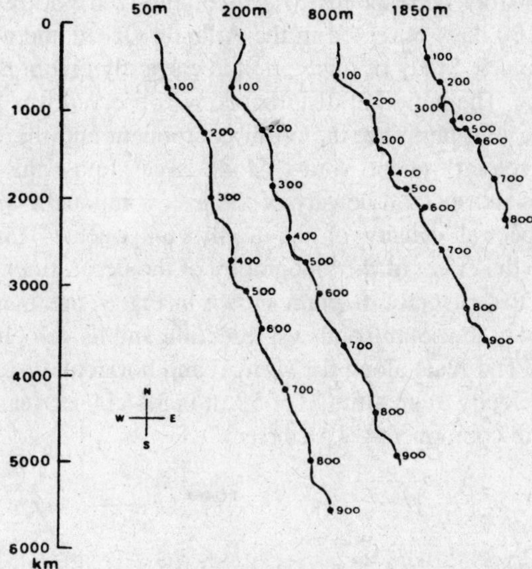


Fig. 9. Progressive vector diagram of daily average velocity, 50 m, 200 m, 800 m and 1850 m above sea bed at point TA. A strong south-southeast average is found near the sea bed (at 50 m above sea bed it is 7.6 cm/sec).

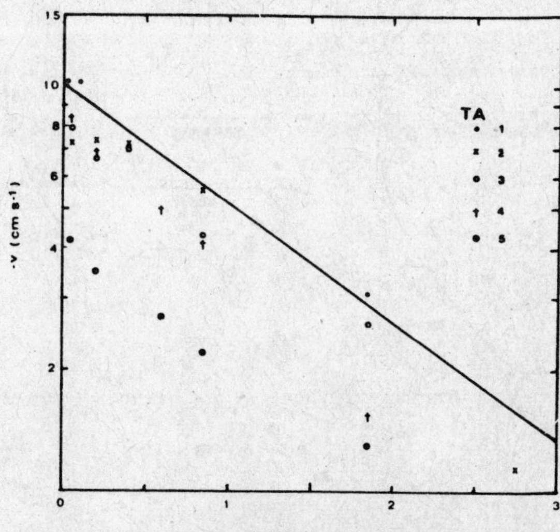


Fig. 10. Vertical distribution of average southward current during each period of observation at point TA. TA (1): Dec. 4, 1977 to Feb. 27, 1978: 85 days; TA (2): Sept. 30, 1978 to Mar. 16, 1979: 167 days; TA (3): Mar. 20, 1979 to Jul. 23, 1979: 125 days; TA (4): Jul. 26, 1979 to Oct. 31, 1979: 97 days; TA (5): Oct. 31, 1979 to Jun. 14, 1980: 227 days.

days). This probably corresponds to a medium sized vortex (Mode Vortex) period off 50–100 days observed in the Atlantic Ocean and occurring due to stormy disturbances. Study of this vortex is currently one of the main themes of ocean physics. The period of disturbance we observed was longer than that of Vortex Mode. Furthermore, the north component and the east component of the current velocity under Vortex Mode have almost the same spectrum density whereas the spectral density of the east component we observed was one-half the spectral density of the north component. This anisotropy is probably due to the effect of the topography of the ocean floor.

According to the vector diagram shown in Fig. 9, the average current for each larger flow in the south-southeast direction and its velocity increase near the ocean floor. The scale along the vertical and horizontal axes is in km. The mean current velocity at an altitude of 50 m is 8.1 cm/sec (east component—2.7 cm/sec, north component—7.1 cm/sec).

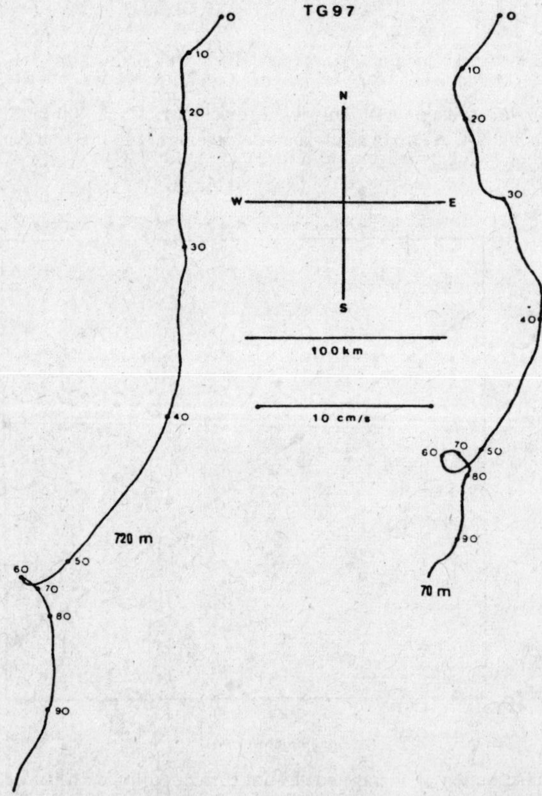


Fig. 11. Progressive vector diagram of the velocity records at point TG (depth of water 5730 m) during the 100 days between Jul. 22, 1979 and Oct. 30, 1979.

The strength of the mean current is a function of the height above the ocean floor. In Fig. 10, the logarithmic of the mean current velocity is plotted as a function of the height of each layer above the ocean floor for five moors. The strength of the mean current decreases exponentially with height. The straight line in this figure shows the relationship, assuming that the strength of the current becomes 1/e at an altitude of 1500 m.

## Variation in Deep Ocean Currents in the East-West Direction

We have explained above that the average current at point TA flows in the south-southeast direction. Point TA lies above a small submarine ridge with a diameter of 5 km and an altitude difference of 50 m. Point TB was set 10 km north of point TA to determine the effect of this topography. This point was 95 m deeper than point TA. The record (of ocean currents) obtained at point TB was similar to that of TA (figure not shown). We now examine conditions at points TG, TF and TD which are east and west of TA. Although the period of observation was relatively shorter, 100 days (TG), 95 days (TF) and 169 days (TD), we can still determine the approximate value of the mean current during prevailing current variations. The latter greatly affect the evaluation of the mean current in 133 days.

Figure 11 shows the progressive vector diagram of the current velocity at point TG (depth of water 5730 m) at altitudes of 70 m and 720 m. The period of mooring was the same as the fourth test at point TA. Here the mean current was southward. Unlike point TA, the velocity of the current was higher in the upper layers.

Figure 12 shows the progressive vector diagram of the current velocity at point TF (depth of water 5930 m) at altitudes of 110 m and 930 m. The period of mooring was the same as the fourth test at point TA. The mean current is northward and velocity is higher for the lower layer.

Figure 13 shows the progressive vector diagram velocity at point TD (depth of water 6040 m) at altitudes of 1040 m and 2040 m. The period of measurement is the same as the second test at point TA. Here, again, the velocity of the upper layer is higher.

Thus, the mean current over a long period at 30°N is mainly northward or southward. The vertical velocity distribution of the current at all points of measurement along 30°N can be assumed to follow the distribution at point TA. If all these velocity values are converted to an altitude of 200 m above the ocean floor and plotted against the latitude, the result is the lowest plotting on Fig. 14 (shown by an arrow mark). The northward current and southward current exist alternately. The distance between the two adjacent points TA and TG, at which the southward flow was observed, is 160 km. If we consider a

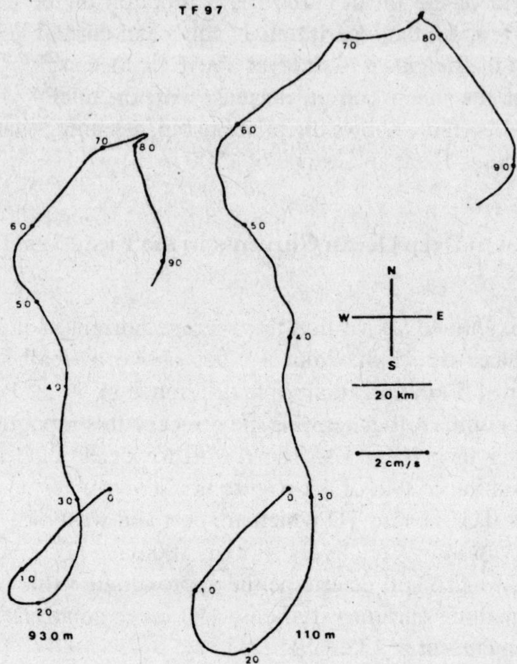


Fig. 12. Progressive vector diagram of the velocity records at point TF (depth of water 5930 m) during the 95 days between Jul. 27, 1979 and Oct. 30, 1979.

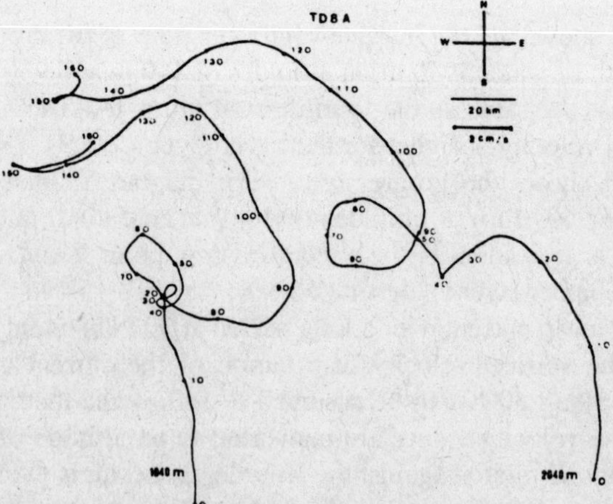


Fig. 13. Progressive vector diagram of the velocity records at point TD (depth of water 6040 m) during the 169 days between Oct. 1, 1978 and Mar. 19, 1979.

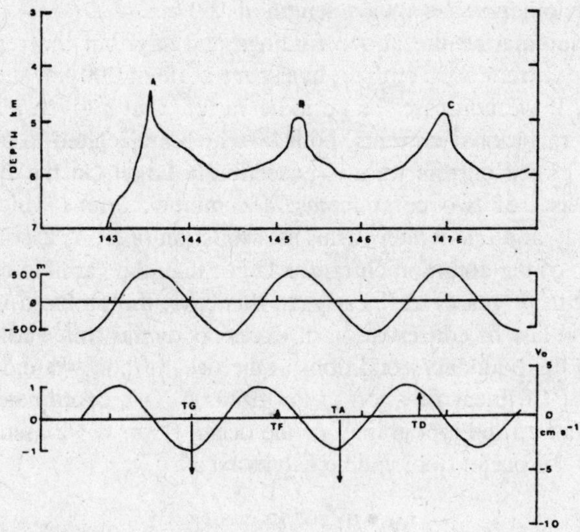


Fig. 14. The east-west topography of the submarine ridge around 30°N (top). The sinusoidal wave representing this topography (middle) and its derivative (lower). The arrow mark in the lower part indicates amplitude of the southward current at each point of observation.

sinusoidal wave with this wave length, then a northward current is expected at points TF and TG. The results of measurement support these hypotheses (Fig. 14).

What then is the range over which this flow pattern extends in the north-south direction? We can confirm that the current at and up to 10 km north of point TA (point TB) flows in the same direction as at TA. Similarly, point TD (northward current) lies 46 km south of TA. Therefore, the north-south direction of the current flow can be assumed to extend over a range of 100 km at least.

## Discussion

Our observations in the Izu-Ogasawara trench revealed for the first time that an unusual pattern exists in the deep ocean currents in the east-west direction where the current alternately flows northward and southward. Our results can be summarized as follows:

1) Neither the directions nor the velocity of this current change during a period of over 1000 days.

2) The velocity of this current increases nearer the ocean floor whereas it decreases exponentially at higher altitudes.

3) Both north and south currents exist on the 30°N line and this

sinusoidal phenomenon has a wave length of 160 km.

We can summarize the above findings and say that the representative velocity of the current is 10 cm/sec, in a space scale of 200 km and time scale of 2000 days. If we compare the periodic variation at a given place and the amplitude of transitional currents, both of which are related to acceleration, then the transitional current term is exceedingly large. On the spatial scale, the flow consists of two components: a common current which does not change spatially and one which varies as a function of X, Y, Z coordinates. If the proportion of the common current is larger than the variable current, then the question can be resolved linearly. In this case, the problem is solved by considering the law of conservation of excess potential (ref. Pedlosky, 1979, § 6.13). From the boundary conditions at the ocean floor, we understand that the dependence of linear function of the flow on X–Y coordinates has to be the same as that on the topography of the ocean floor. If we assume that the topography of the ocean floor can be expressed as:

$$\eta_B = \eta_0 \cos kx \cos Iy,$$

where $k$ is the wave number in the east-west direction and $I$ is wave number in the north-south direction, then the north component of flow varying spatially can be expressed as follows (standard wave motion based on topography):

$$v(x, y, z) = \frac{S\eta_0}{m \sin hm} \cos m(z-I) \sin kx \cdot \cos Iy.$$

Here $z$ is normalized by depth of water, $D$;

$$S = N^2D^2/f_0{}^2L^2, \; m^2 = S\left\{(k^2+I^2)-\beta/u_0\right\},$$

$u_0$ is the common flow in the east direction.

When the topography at the ocean floor in the east-west direction varies according to cos $kx$, then the velocity of the current becomes proportional to sin $kx$. The topographical cross section along 30°N (Fig. 3) does not have such sinusoidal variation. However, if we look at the east-west region in the belt defined as 30°±20′N, we find that it consists of a range of ridges in the east-west direction as mentioned earlier (A, B, C, Fig. 2).

The upper part of Fig. 14 shows the topography of this ridge by connecting their cross sectional view. If this is represented as a sinusoidal wave with the amplitude of 600 m and wave length of 164 km, the result is the central portion of Fig. 14. The lower part of this figure is the wave form of the north component of velocity with the normal topography.

The arrows indicate the mean value of the current at the points of measurement mentioned above. The theoretically estimated current which follows

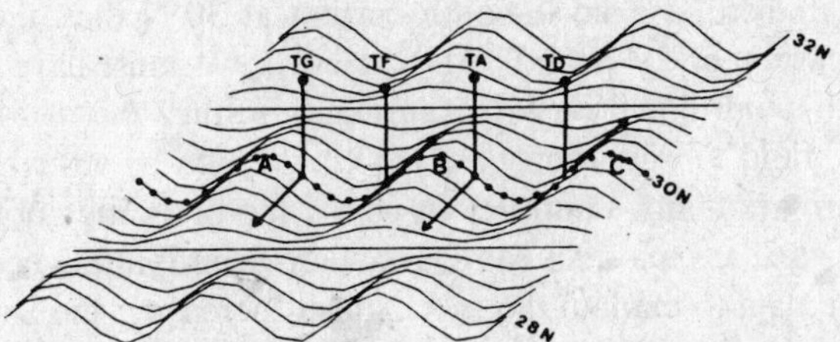


Fig. 15. A model topographical layout of the ocean bed considering a normal topographical wave motion. 30°N is indicated as a dotted line and the direction of the current expected at each point is also shown. The currents due to the normal topographical wave motion circulate clockwise around the ocean mountains.

the wave form caused by the topography of the ocean floor matches the measured values of the current.

We shall now evaluate the wave length of the topography of the ocean floor in the north-south direction. Fig. 2 shows that the east-west distribution of ridges at the sea bottom consists of three belts, 28°N, 30°N and 32°N.

Here the north-south wave length can be assumed to be 400 km. If the east-west wave length is assumed to be 164 km, then the topography model at the bottom of the sea can be represented as Fig. 15. The crest at 28°N becomes the trough at 30°N and again becomes a crest at 32°N. The crests at 30°N were identified as A, B, C. We have shown schematically in this figure the position of the moor and the direction of the current during the normal topographical wave motions at those points.

Once the wave numbers *k* and *l* for the topography of the ocean floor are known the transverse variation in the velocity of flow due to topographical motion can be determined. (Variation according to latitude in Coriolis coefficient, β, is ignored.) The velocity of the flow is greater near the bottom and decreases exponentially upward. The value of N according to CTD observations (Horibe and Urase, 1980) at 30°N, 145°40′E is $1.2\times10^{-3}$ sec and the attenuation in the velocity at this time is 1/e over a distance of 1500 m. Thus, the expected value of the attenuation coefficient is also closer to the observed value.

We have indicated that when the transitional current term is larger than the time dependent term (a similar result can be obtained even for a north bound general current if $\beta = 0$ is assumed) then an infinitesimal amplitude wave normal topographical wave motion—which is a linear function of topography of the sea bottom, is supported within the area under observation. This wave motion has common properties with the mean flow we have observed so far.

However, if the above model is assumed, two problems arise. Firstly, a drifting field exists in a clockwise direction around the sea mountains. Accordingly, if there is a north-south current at 30°N then there has to be an east-west current at 29°N and 131°N. Hence, we must have a mooring point in the north-south direction for examination. Since the drifting field follows the density field, observation of CTD would also be valuable. Secondly, although there are many examples in which the properties of the infinitesimal amplitude wave are maintained (even when the amplitude becomes finite) we do not have data to explain the mechanism increasing the amplitude of wave motion. Comparison of Figs. 4 to 7 indicates that the short period variations are prominent in the lower layers whereas at the upper layers the amplitude variations are smooth and attenuated. The length of these variations is about ten days and there may be some correlation with atmospheric changes.

The variation over a representative period of 133 days may act as a source of energy. Although it is not clear from the observation, the general current in the east or the south direction may also be considered a source of energy.

This discussion explores only one possibility. If joint studies are undertaken for one problem at a time with observations and discussions, more effective results can be expected. It is, therefore, desirable to organize joint ventures for oceanographic studies.

## Acknowledgment

We express our sincere appreciation to Prof. Horibe who rendered continuous support during the study and to the crew of *Hakucho-maru* and *Keiten-maru* who actually conducted the mooring operation and to other scientists. This study was conducted with the help of Mombusho (Ministry of Education) Science Research Fellowships, Special Study Project Preservation of the Ocean (1975–77), Special Study Environmental Science (1978) and Special Study Fundamentals of Oceanography (1979–80).

## REPORTS

First Executive Committee meeting held on May 8, 1981 at the Marine Research Laboratory of Tokyo University.

The outline for the development of the special study was prepared. The functions of each group were studied and a proposal for submission to the Managing Committee was prepared.

First Managing Committee meeting held on June 18, 1981 at Aoi Kaikan, Minato-ku, Tokyo.

The organization of the special study was discussed.

The general organizational chart prepared for the development of this special study is explained in the following section. The ultimate authority is the Managing Committee itself which consists of all the group leaders. The Executive Committee consists of two members from each of the four study groups and is entrusted with the task of day to day management. The following eight members form the Executive Committee:

| | | |
|---|---|---|
| General | Kajiura, Kinjiro | Seismic Research Laboratory, Tokyo University |
| Physics | Kunishi, Hideaki | Physics Department, Kyoto University |
| | <u>Teramoto, Toshihiko</u> | Institute of Oceanography, Tokyo University |
| Chemistry | Kitano, Yasushi | Hydrospheric Research Laboratory, Nagyo University |
| | <u>Horibe, Yoshio</u> | Institute of Oceanography, Tokyo University |
| Geology | <u>Takayanagi, Yokichi</u> | Physics Department, Tohoku University |
| | Nasu, Noriyuki | Institute of Oceanography, Tokyo University |
| Measurements | <u>Mochizuki, Hitoshi</u> | Tokyo Institute of Electrical Communication |
| | Teramoto, Toshihiko | Institute of Oceanography, Tokyo University |

The underlining indicates group leaders.

A Planning and Public Relations Group is also included in the above groups. Each study will be undertaken under the guidance of the group leaders. Preparation of an agenda for the meeting of the Managing Committee

## Organization Chart

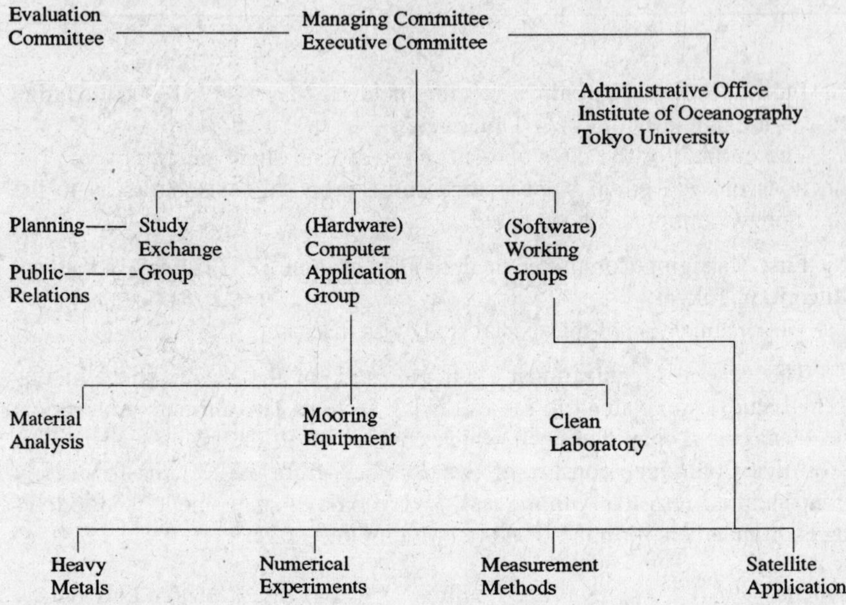


or the Executive Committee or intimation of such meetings to members will be made in the same order. The administrative responsibility of the entire project rests with the Institute of Oceanography, Tokyo University. Those responsible for general affairs, accounts, publicity and other matters are Teramoto, Administrative Officer, Taira and Kitano.

Within the Managing Committee, a study exchange group, a computer application group and other working groups have been constituted so that coordination among the various study groups is strengthened and the progress of work as well as the actual management are facilitated. Special committees can be formed as necessary from the computer application group and other working groups.

It has been proposed that an independent evaluation committee be established to monitor the progress and efficiency of this special study.

In this fiscal year, it is proposed that the Managing Committee meet three times (October and February in addition to the meeting held already) whereas the Executive Committee should meet at least once a month. The computer application group and working groups organize their meetings, and commence their activities in consultation with the Managing Committee.

To publicize the project, the publication of, "An Outline of Research Organization and its Plan", as well as the Newsletter have been undertaken this year. This will facilitate the exchange of information within the special study project and promote the dissemination of the results of the studies to domestic and foreign readers.

*The first meeting of the Study Exchange Group:* held on June 18–19, 1981, at Aoi Kaitan, Minato-ku, Tokyo.

The working plans of each group including the planning group, public relations group and others, were introduced (the details are mentioned in, "An Outline of Research Organization and its Plan during Fiscal 1981"). On the second day, Prof. Kajiura, as a representative, explained the Organization or the Management of the Study as discussed in the first Managing Committee meeting held prior to this meeting.

*The second meeting of Executive Committee* was held on July 6, 1981 at the Institute of Oceanography, Tokyo University.

M/s. Saito, Shisuka and Koyo were elected members of the Evaluation Committee.

The work of the computer application group and other working groups was planned and allocated under the leadership of following members:

| | |
|---|---|
| Material analysis | Horibe, Yoshio |
| Mooring equipment | Teramoto, Toshihiko |
| Clean laboratory | Horibe, Yoshio |
| Heavy metals | Kanamori, Satoru |
| Numerical experiments | Suginohara, Nobuo |
| Methods of measurements | Mochizuki, Hitoshi |
| Satellite application | Teramoto, Toshihiko |

*The first meeting of the Executive Committee* in the capital was held on July 18, 1981 at the Institute of Oceanography, Tokyo University.

The name, frequency and contents of the proposed Newsletter were discussed.

*The third meeting of Executive Committee* was held on August 8, 1981 at the Institute of Oceanography, Tokyo University.

Informal consent of the three nominees was obtained for appointment as members of the Evaluation Committee and this was announced.

The essentials of public relation activity for the next year were discussed and it was decided to continue these activities in the same manner as this year.

The budget for next year is to be prepared according to the report submitted last year.

For the training of Iwate researchers, scholarships were awarded by the group attending the meeting. The leader must plan for the support of each group to the extent of 150,000.

Printing of "An Outline of Research Organization and its Plans" for the fiscal year 1981 is complete and distribution has started.

## Guidelines for Authors

Newsletter welcomes any enterprising article.

The contents are generally classified as:

| | |
|---|---|
| Highlights: | Quick (intermediate) results of this Special Research Project |
| Bulletin: | Domestic or international news reports closely related to this Special Research Project |
| Reports: | Reports regarding the activities of each section or working group |
| Announcements: | Announcements regarding the proposed meetings or conferences of each group |
| Bibliography: | Listing of papers published by the members on this Special Research Project or other reports. |

Manuscripts should be written on standard 400 character size paper (handwritten material also accepted). Figures should be drawn in such a way that they can be reproduced as they are.

Manuscripts may be sent to this office through the respective group leader.

Manuscripts should reach this office at least 20 days prior to the scheduled date of publication.

The future publication schedule is January 30, 1982 and March 10, 1982.

**Note:**

When reproducing or citing material published in this Newsletter, due acknowledgment should be given. Authors' names, wherever mentioned, should also be quoted. A copy of the publication containing such a reproduction may be sent to the Editorial Section of the Administrative Office.


Published by: Administrative Office, Special Research Project "The Ocean Characteristics and Their Changes"

Published at: Administrative Office,
Physical Oceanography Department,
Institute of Oceanography,
Tokyo University,
1-15-1 Minami-dai, Nakano-ku,
Tokyo 164
Tel: 03-376-1251, Ext. 251.

Special Research Project

# The Ocean Characteristics and Their Changes

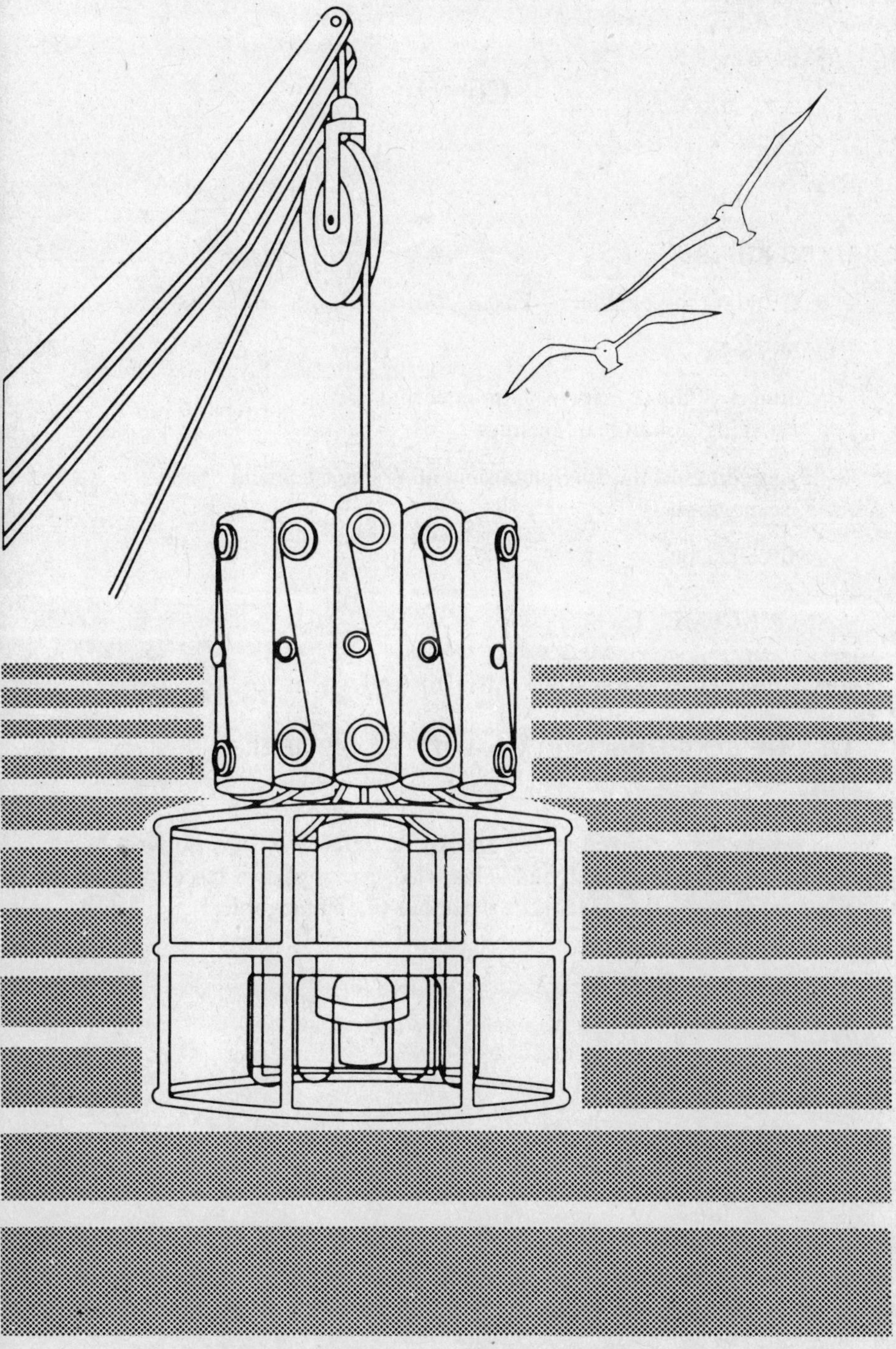

NEWSLETTER NO. 2, NOVEMBER, 1981

## Contents



Researchers engaged in this Special Research Project, who wish to submit their papers in English and acknowledge receipt of a research grant for this special study, should indicate the above title of the project.

HIGHLIGHTS

# $^{14}C$ and $^{3}H$ in Sea Water

*Yoshio Horibe**

The *Hakucho-maru* reached 30°34′N, 170°36′E (point CY 11) on June 6, 1980. This point lies in the northwest Pacific region and is connected to equatorial sea on the south by the Wakey Channel. According to CTD observations the temperature of sea water at 6,000 m depth along 30°N latitude is the lowest in this region representing the effect of the deep waters of the South Pacific quite prominently.

At this point we collected sea-water samples at 15 levels using 2 large water tanks alternately, each with 2 barrels of 230 l capacity. One of the objectives of the experiment was to measure the concentration of $^{14}C$ in the form of carbon dioxide in the sea water. According to the present methods of measurement, it is necessary to use 10 l of carbon dioxide for an accurate measurement of $^{14}C$; however, 1 l of sea water contains only 55 ml of carbon dioxide and hence it is necessary to use as much as 200 l of sea water.

The CY 11 point was also the study area (G 226) of the *Melville,* a research vessel of the Scripps Institute of Oceanography on November 9, 1972. During the 1970s an experiment entitled Geochemical Ocean Sections Study (GEOSECS) was planned as part of the International Decade of Ocean Exploration (IDOE). This plain aimed to particularly study ocean cycles using chemical parameters. Not only a large number of chemical oceanographers but also eminent physical oceanographers like Stommel, Veronis, and others participated in it actively. They observed 147 points in the Pacific Ocean and in addition to standard oceanographic observations (temperature, salinity, mineral nutrients, etc.) the data was also used to analyze radioactive elements like $^{3}H$, $^{14}C$, $^{210}Pb$, $^{226}Ra$, $^{228}Ra$, $^{228}Th$, $^{90}Sr$, $^{137}Cs$, Pu, etc. as well as the particle structure and Ba gas.

As a result of a reaction between atmosphere and ionosphere $^{3}H$ or $^{14}C$ are formed. According to experiments conducted after 1950, they were found to be released in large quantities in the atmosphere. Their temperature in the atmosphere was highest around 1963 with a decreasing tendency thereafter. In

*Institute of Oceanography, Tokyo University.

the atmosphere $^{14}C$ exists in the form of carbon dioxide and it remains totally oxidized even in the ocean. Tracing the path of $^{14}C$, released in spurts in large quantities in the atmosphere, can help us prevent the spread of carbon dioxide in the atmosphere.

As mentioned by Broecker, the distribution of $^{14}C$ in sea water does not change considerably with time. Fig. 1a and b show the vertical distribution at G 226 (November 9, 1972) and CY 11 (June 6, 1980) respectively.

Thus the vertical distribution is almost the same to 6,000 m. However, if we observe the readings to 1,200 m carefully, as shown in Fig. 2a, b, the recent CY 11 point shows a higher concentration in the first few hundred meters.

Figure 3 shows a comparison of $^{14}C$ ($mol \cdot m^{-3}$) concentrations calculated from the above figures for G 226 and CY 11 as a function of depth up to

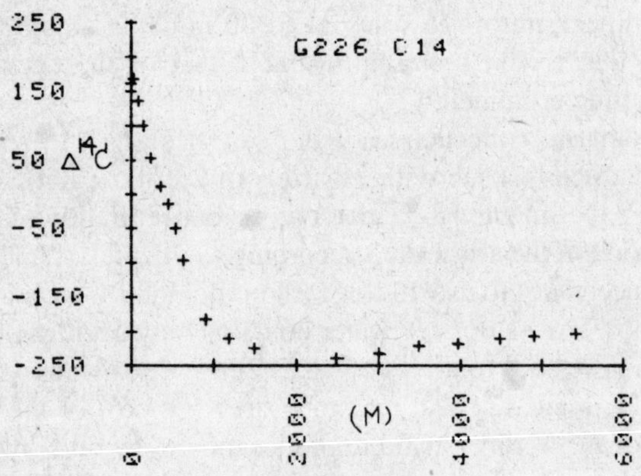


Fig. 1a. Vertical distribution of $^{14}C$ during November 1972.

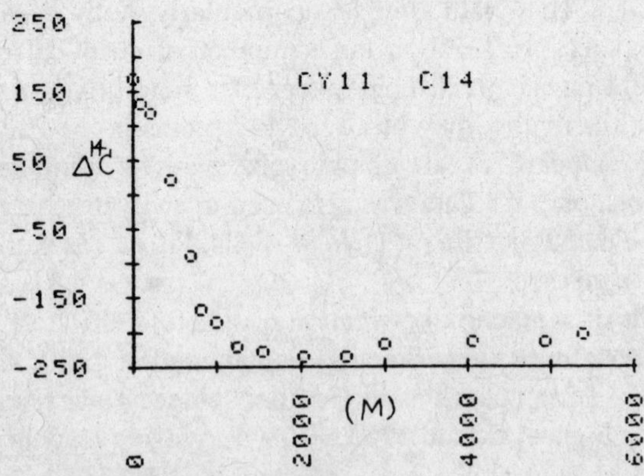


Fig. 1b. Vertical distribution of $^{14}C$ during June 1980.

1,200 m. The total amount of $^{14}C$ in the hatched region is the quantity of $^{14}C$ dissolved in this water column over a period of 7.5 years. From elementary considerations, this quantity of $^{14}C$ can be assumed to have come from the atmosphere. This quantity is $5.5\times10^{-11}$ mol · $m^{-2}$ and accordingly the flow of carbon dioxide from the atmosphere to the ocean would be $0.8\times10^{-11}$ mol · $m^{-2}$, $yr^{-11}$. This value is larger than those obtained by other methods.

During GEOSECS, the $^{3}H$ distribution was measured by Östrund of Miami University. In our group, it was not possible to measure $^{3}H$ during the experiments in 1980. We, therefore, sent the samples collected at CY 11 in glass ampules to Östrund for analysis. The result was available by October, 1981. Fig. 4 shows the concentration of $^{3}H$ (Tritium Unit) at G 226 and CY 11 with + and 0 marks respectively.

It is interesting to note that the concentration of $^{3}H$ at the surface is much

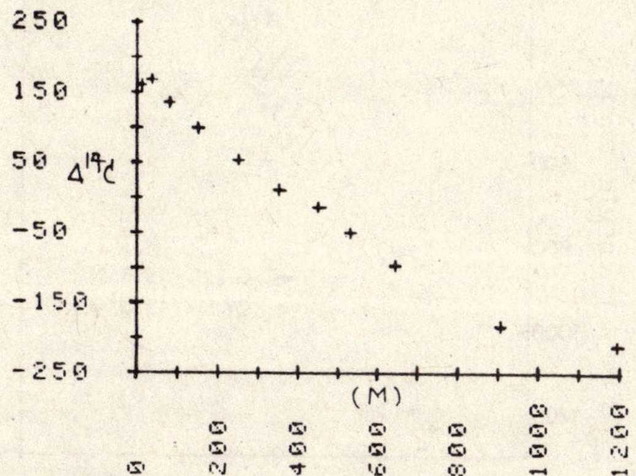


Fig. 2a. Vertical distribution of $^{14}C$ to first 1,200 m in November 1972.

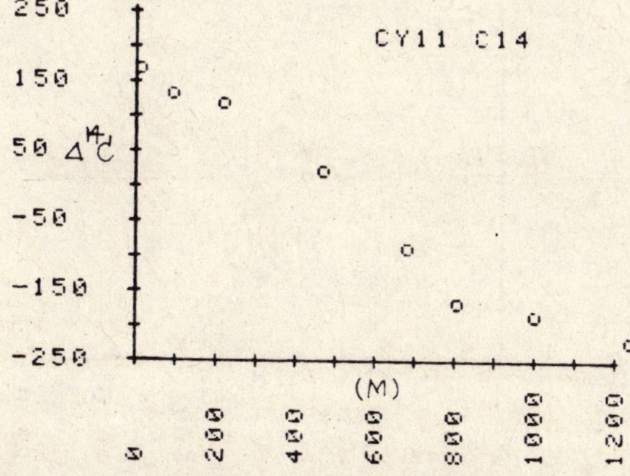


Fig. 2b. Vertical distribution of $^{14}C$ to first 1,200 m in June, 1980.

lower than the reduction expected according to its half-life, whereas at around 800 m, there is no apparent change in concentration. This may mean that $^3$H has penetrated down to 800 m. Similarly, the $^3$H concentration at 600 m remained unchanged in 1980. Now, during 1972 observations, concentration of $^3$H seems to have dropped at 200 m but this only shows the existence of sinking motion. Here, in order to explain the $^3$H distribution along with $^{14}$C as well as the $^3$H without any contradiction, it may be better if we could also measure $^3$H formed by the disintegration of $^3$He. Also, we should know the periodic change in the distribution at a given depth.

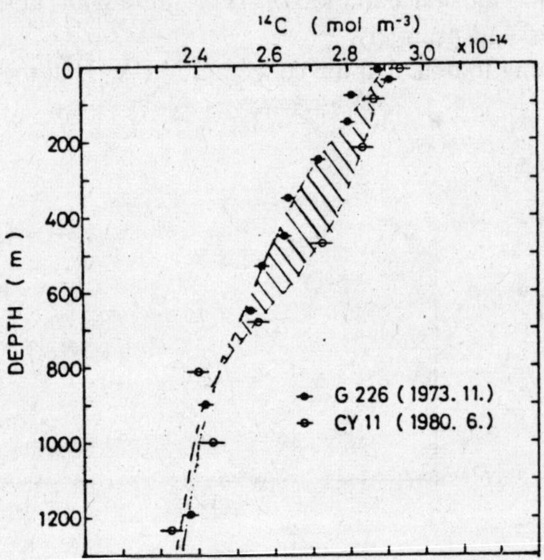


Fig. 3. $^{14}$C content at 30°N, 170°E in 1972 and 1980.

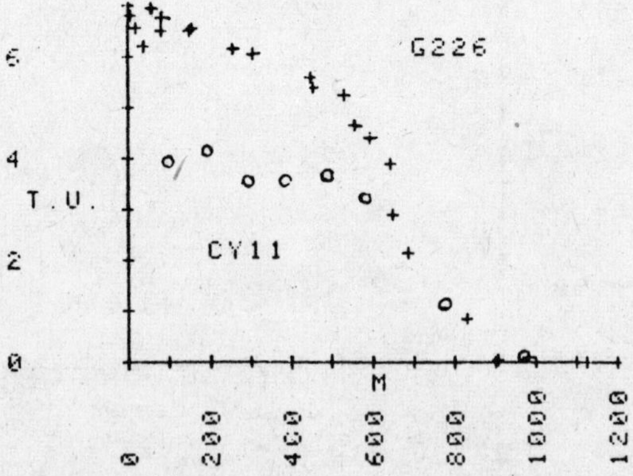


Fig. 4. Vertical distribution of Tritium at 30°N, 170°E in 1972 and 1980.

The data regarding $^{3}H$ distribution in 1972 is available for points G 227 (25°00′N, 170°05′E), G 228 (19°10′N, 169°21′E) and G 229 (12°53′N, 173°28′E) which are farther southward than CY 11. The data regarding $^{14}C$ is also available at G 227, G 229. The *Hakucho-maru* plans to sample waters at these locations during its travel in January–March 1982. Until then let's look forward to the results.

## REPORTS

### Minutes of the Fourth Executive Committee Meeting held on September 25, 1981 at the Institute of Oceanography, Tokyo University

Mombusho (Ministry of Education) has been informed to allocate the same amount of funds during this fiscal year as in the previous year.

The general plans for next year should be made on the same lines as this year, the first year of execution, and the budget should be prepared accordingly.

The committee for using sophisticated equipment (mass spectrometry) has proposed the standards for equipment utilization. They also studied the use of large equipment in general.

Mr. Kajiura, a researcher, had discussions with Mr. Marumo of the Marine Biology Study Group regarding future plans of the study. It was decided to continue such discussions in future.

The tripod cover was proposed to be imported from the United States but as the company has stopped manufacture, the administrative office shall procure it indigenously.

The members studied the proposals to be submitted at the 2nd meeting of the Managing Committee which is centered round the above points.

### Minutes of the Second Managing Committee Meeting held on October 13, 1981 at Minato-ku, Tokyo

As proposed in the first meeting, two subgroups, viz., a group for using the sophisticated equipment (concerned with hardware) and a working group (concerned with software), were formed.

The working group was also given the responsibility of using the sediment trap.

Each subgroup shall work under the leadership of the following members:

| | |
|---|---|
| Mass spectrometer | Yoshio Horibe |
| Mooring equipment | Toshihiko Teramoto |
| Clean laboratory | Yoshio Horibe |
| Heavy metals | Satoru Kanamori |
| Numerical experiment | Nobuo Suginohara |
| Methods of measurement | Hitoshi Mochizuki |
| Satellite applications | Toshihiko Teramoto |
| Sediment trap use | Shizuo Tsunogai |

Responsibility for the use of the sophisticated equipment procured by the administrative office and delivered to the Institute of Oceanography was entrusted to the supervisor of the same institute (Ref. Item 8). The details may be worked out by each subcommittee (Ref. Item 8).

The maintenance and transfer of equipment, during and after the specific research assignments can be decided by the Managing Committee on a later occasion.

Mr. Saito, Mr. Nasu and Mr. Koyo were asked to work on the Evaluation Committee.

While preparing the plans for next year and applying for the research grants, the same general objectives of research and execution program should be maintained as in the current year.

The name of Prof. Taira (Asst. Professor, Institute of Oceanography, Tokyo University) should be included in the list of Research Coordinators.

The content and budget for each subtopic of research, after discussions with each study exchange group should be according to the second year plan of a total three-year project.

English title of this research project shall be "The Ocean Characteristics and Their Changes".

The unutilized portion of the current year's research grant (about 4% of the total funds sanctioned) are proposed to be carried forward next year.

The meeting concluded with a vote of thanks to all the participants of this special study project and to all those others who have continued the subsequent oceanographic studies.

## Proceedings of the 5th Symposium of Young Chemical Oceanographers

**1. Objectives of the Symposium:**

The symposium aimed to advance the knowledge of junior researchers. For this purpose recent topics related to chemical oceanography and geochemistry were selected. Eminent scientists in these fields delivered lectures which were followed by discussions. In order to achieve the best results in a short time, the symposium was designed to be residential so that participants have ample time for mutual discussion.

**2. Place and time of Symposium:**

Conference room in Miho Research Laboratory of Tokyo University (No. 2 Seminar Room), 8:30 p.m. of October 12, 1981 to 3:00 p.m. of October 13.

**3. Participants:**

Thirty-nine scientists attended. The list of participants with their affiliation is appended at the end of this report.

**4. Program:**

October 12:

8:30-10:00 p.m. Session 1—Manganese Nodules.

Speaker: Prof. Harada. Asstt. Professor, Faculty of Science, Yamagata University.

Discussions:

10:00-11:00 p.m. Free discussions in small groups.

October 13:

9:00-10:30 a.m. Session 2—History of Environmental Changes as indicated by deposits.

Speaker: Prof. Matsumoto. Central Laboratory of Geological Survey.

Discussions:

Moderator— Ueda Masashi.

10:30-12:00 a.m. Session 3—Modern trends and future developments in chemical oceanographic studies. Based on the Gordon Research Conference.

Speaker: Prof. Nozaki Yoshiyuki. Asstt. Professor, Oceanographic Laboratory, Tokyo University.

Discussions:

Moderator—Matsueda Hidekazu.

Lunch Break

1:00-3:00 p.m. Concluding remarks. Summary of Symposium and future plans.

Moderator: Prof. Kato Yoshihisa.

**5. Summary of the Lectures and Subsequent Discussions**

**Session 1:**

The scope of earth science studies includes nothing but "field science." As such the care to be taken while recording and classifying the samples used during studies is clearly illustrated in the studies related to conventional manganese nodules. Correct reporting helps identify the historical and geological properties of the sample. That is to say, it is possible to know where over the earth's history of 45 billion years the given sample can be placed and under what environmental conditions. This type of information is essential to carry the studies further. However, all the studies related to manganese nodules so far were rather vague in recording. According to the speaker, there are more than 5000 reports available from around the world regarding manganese nodules, but the basic study about the formation of manganese nodules is said to lag very much behind.

After placing this background before the audience, the speaker discussed the latest information about manganese nodules using photographs of the seabed or slides obtained with electron microscopes. Manganese nodules can be classified according to their outer shape and internal structure. The age (of the nodule) as determined by radioactive isotopes and from patterns of micro-fossil accumulation are not in agreement with each other. The speaker also discussed the structure of the seabed interface under which manganese nodules could grow.

**Session 2:**

The substances which deposit continuously under a given environment, such as deposits at the seabed or water bed in polar region or coral reef, etc., are formed while being affected by environmental changes in one way or another. Periodic observation (of these substances) provides a time dimension to these variations. The period measurement using $^{210}Pb$ (half-life period of 22 years) among the deposits on the seabed can be used to analyze the phenomena in the past 100-150 years. It can be assumed that there is no environmental variation due to natural factors over this time scale and hence any variation that can be recorded can be attributed to human factors. With this background, the speaker introduced the results of his analysis of the core sample in Tokyo strait and discussed the concept further while analyzing those results.

The first problem was how to collect the core sample from seabed in a place like Tokyo strait without mixing (with other substances). For this purpose the author had developed a core sampler and its merits were discussed. Another problem is whether $^{210}Pb$ period indicates a real time period. This has been confirmed by Koide (1973) during his experiments to study the autumn effect using a nodular core (valve).

The speaker also discussed in detail the methods which take into consideration the variation in concentration of chemical elements (chemical fossils) such as heavy metals, artificial radioactive isotopes, artificial organic deposits, etc., or which take into consideration the variation in bio-phase (biological fossils) substances such as silicon or corals while classifying the index of environmental changes as fossils.

**Session 3:**

The speaker attended this year's Gordon Research Conference and he introduced the modern trends in the study of chemical oceanography based on the program of the above conference. The central theme of this conference was "material flux". For five days the conference discussed the subject in four sessions, viz., Flux between atmosphere and ocean, Flux in the sea-water column, Flux between sea water and deposits, and Flux due to rivers. The speaker explained the results of various unpublished studies which was followed by a lively discussion.

He also introduced some of the points discussed during the poster session organized during the conference. He explained the trends of future studies by citing examples of TTO, PARFLUX phase-II, the extensive research project in the 80s and by quoting the studies on the behavior of warm currents in the deep sea.

On his own research on the subject, the speaker studied the removal of materials from the behavior of $^{230}$Th and $^{231}$Pa in a sediment trap experiment. Similarly he measured Th isotopes in the accumulated sea water using the Mn fibers attached to a buoy. He discussed his observations with the unpublished data.

**Concluding session:**

The contents of the above sessions were summarized and points supporting or criticizing the expressed views were raised. There were many points regarding the Symposium proceedings and future topics for discussion, on which a general agreement was reached. In continuation, a lively discussion was held on the methods of observations or handling problems which may be faced by young researchers (particularly post-graduate students) in the field of chemical oceanography.

Many participants expressed the desire that a similar symposium be held next year as well and the general responsibility for that symposium was given to Prof. Nakayama, Asstt. Professor, Faculty of Science in Kyoto University.

**List of participants:**

| **Name** | **Affiliation** |
|---|---|
| Tanaka | Fisheries Department, Hokkaido University |
| Masatoshi Yamada | Fisheries Department, Hokkaido University |
| Yasunori Watanabe | Fisheries Department, Hokkaido University |
| Takanobu Kurata | Fisheries Department, Hokkaido University |
| Kenichi Harada | Faculty of Science, Yamagata University |
| Eiji Matsumoto | Geological Survey Department |
| Yoshimi Suzuki | Meteorological Department |
| Katsumi Hirose | Meteorological Department |
| Katsuhiko Fushimi | Meteorological Department |
| Noburu Takematsu | Scientific Laboratory |
| Junichi Oyama | College of Meteorology |
| Yoshiyuki Nozaki | Oceanographic Laboratory, Tokyo University |
| Gamo Toshitaka | Oceanographic Laboratory, Tokyo University |
| Osumi Takashi | Faculty of Science, Tokyo Institute of Technology |
| Shigara Yoshio | Faculty of Science, Tokyo Prefectural University |

| | |
|---|---|
| Toshiyuki Masuzawa | Marine Science Laboratory, Nagoya University |
| Hisashi Ikegami | Marine Science Laboratory, Nagoya University |
| Chisato Toyama | Marine Science Laboratory, Nagoya University |
| Hiroshi Sakugawa | Marine Science Laboratory, Nagoya University |
| Hidekazu Matsueda | Marine Science Laboratory, Nagoya University |
| Eiichiro Nakayama | Faculty of Science, Kyoto University |
| Hiroyuki Tokoro | Faculty of Science, Kyoto University |
| Kenji Isshiki | Faculty of Science, Kyoto University |
| Tetsuo Okada | Faculty of Science, Kyoto University |
| Masashi Ueda | Faculty of Science, Kyoto University |
| Takashi Okubo | Kobe Marcantle Marine College |
| Shigeru Kadoya | Faculty of Agriculture, Kagawa University |
| Minoru Okumura | Faculty of Science, Shimane University |
| Koji Hayase | Faculty of Science, Hiroshima University |
| Tetsuo Kodama | Faculty of Science, Hiroshima University |
| Yoshio Sato | Faculty of Marine Science, Tokai University |
| Yoshihisa Kato | Faculty of Marine Science, Tokai University |
| Kenji Yamagata | Faculty of Marine Science, Tokai University |
| Seiichi Kawano | Faculty of Marine Science, Tokai University |
| Seiichi Kim | Faculty of Marine Science, Tokai University |
| Koji Kurami | Faculty of Marine Science, Tokai University |
| Akira Kikutani | Faculty of Marine Science, Tokai University |
| Ippei Shimazaki | Faculty of Marine Science, Tokai University |
| Toshinori Tsuruoka | Faculty of Marine Science, Tokai University |

(Total 39) (Recorded by Toshitaka Gamo)

**On the use of equipment procured under Research Grants**

During this study, a request was sent by Kinjiro Kajiura, a representative researcher, to the Director of the Institute of Oceanography as mentioned below and his consent was obtained.

Noriyuki Nasu
Director, Institute of Oceanography — October 6, 1981

To
Kinjiro Kajiura
Representative Researcher
Special Study Group for "Ocean Characteristics and Their Changes"

Sub: Use of equipment procured under Research Grants

The administrative office of the special study group on "Ocean Characteristics and Their Changes" established in the Fiscal year 1981 proposes to procure the large equipment necessary to carry out these studies and make them available to concerned researchers. Actually an estimate of expenditure has been submitted to Prof. Teramoto of Institute of Oceanography, officiating researcher in the administrative office. Immediately after their procurement, the equipment shall first be registered with the Institute of Oceanography.

Scientists who wish to use the large equipment procured by the administrative office, and registered with the Oceanographic Laboratory for this special study are requested to use them jointly with the Administrative Officer of this laboratory.

Now, the solar heated water consequent to the use of equipment is requested to be returned to the Administrative Officer of Oceanographic Laboratory.

It is desirable to adopt this procedure after due discussion.

To
Seismic Research Laboratory
Prof. Kinjiro Kajiura

Todai, Institute of Oceanography,
No. 1300
October 14, 1981
(signed by) Noriyuki Nasu
Director,
Institute of Oceanography

**Use of equipment procured under Scientific Research Grants**

The above letter dated October 6, 1981 shall be considered the principal document regarding the use of equipment procured under research grants.

**On the use of the Isotope mass spectrometer**

The Isotope mass spectrometer (Finnigan Mat 250) procured this year shall be made available to all researchers participating in the special project "Ocean Characteristics and Their Changes."

Place of Equipment:

Institute of Oceanography, Tokyo University (2nd Floor, Room No. 211).

Concerned authority:

Yoshio Horibe, Institute of Oceanography, Tokyo University (Tel: 03-376-1251, Extn. 214 or 273).

Use:

For the management, operation and proper use, a special caretaker committee is formed.

Caretaker committee: Representative—Yoshio Horibe. Total members—2.

Joint use:

As a general rule, the equipment shall be used by all researchers participating in the special project. However, the actual use shall be made in consultation with the concerned authority.

**On the use of mooring equipment**

The equipment procured this year, include:

AMF Sonic Sounding meter 2 Nos.

Aanderaa Automatic Current meter 7 Nos.

Nippon Yushi (L type) Sonic Sounding meter 4 Nos.

Nippon Yushi (L type) Sonic Sounding meter usable on ship 1 No.

These shall be used by field study section 2 concerned with physical oceanography in order to conduct further studies on special research project.

Concerned authority:

Toshihiko Teramoto, Institute of Oceanography, Tokyo University.

Use:

To use the equipment, the material and design of the buoy, its working, etc., should first be studied carefully and then the equipment should be used jointly.

ANNOUNCEMENTS

## Conference on Satellite Applications in Oceanographic Studies

The idea of using a satellite as the platform for oceanographic studies is as old as the history of satellites themselves. Many methods for such use were developed and some of them are known to have been employed successfully. The methods for using a satellite have thus become more or less routine but there is much more potential in the satellite (in this regard) and a great scope in studying its effectiveness or other methods of use. Also there is a possibility that new methods may be developed in the future.

At the University, we have identified the following problems regarding the use of satellites in oceanographic studies or oceanographic studies utilizing satellites.

1) It is difficult to obtain the data by routine methods.

2) The effectiveness of data obtained by routine methods, the limits of its application or the methods to use the same have not been studied sufficiently.

3) It is not clear whether a fresh study regarding hardware or software is necessary at the University level to efficiently use for oceanographic studies satellites that are planned to be launched in future, or if necessary, how it may be carried out.

Many have expressed the desire that these points be discussed in detail. We would like to provide a platform for such discussions and if necessary prepare a research plan or propose the organizational structure.

With this background, we would like to hold a meeting around January next year, at Oceanographic Laboratory to provide an opportunity for the exchange of ideas from scientists desirous of using the data and the researchers engaged in basic methods of applications (including hardware). The details regarding date, time and program shall be announced in the next issue of the Newsletter. Any suggestions regarding the agenda of this meeting may be communicated to Prof. Teramoto, Institute of Oceanography, Tokyo University.

## Suggested Literature and Errata

Suggested literature:

Newsletter No. 1: Highlights

A long-term measurement of deep sea currents in the eastern region of Izu-Kogawara Strait by Taira and Teramoto.

## Bibliography


1. Swallow, J. 1955. A neutral-buoyancy float for measuring deep sea currents. *Deep Sea Research, 3,* 74-81.
2. Webster, F. 1969. Vertical profiles of horizontal ocean currents. *Deep Sea Research, 16,* 85-98.
3. US Polymode Organizing Committee. 1976. US Polymode Program and Plan, Office of the International Decade of Ocean Exploration of the NSF and ONR.
4. Kimura, R. 1979. Science of currents. An approach based on natural phenomena. Tokai Scientific Selection, Tokai University Publication.
5. Taira, K., S. Imawaki and T. Teramoto. 1979. Deep and bottom current measurements in an area adjacent to the proposed waste dump site. Marine Radioecology (Proceedings of the 3rd NEA seminar), 69-86.
6. Horibe, Y. (Chief Editor). 1979. Oceanographic studies as part of Environmental Science 3 (Special research project—Fundamental studies regarding the protection of ocean environment. Final report). 1.4. Measurements of currents near the processing site of low level radioactive waste materials, 30-40.
7. Nagata, Y. 1981. Physics of Ocean Currents, Blue Box, Published by Kodansha.

**Members of the Coordination Committee for the Special Research Project "The Ocean Characteristics and Their Changes"**

| Project No. 56117006 | Representative Kinjiro Kajiura | Seismic Research Laboratory,<br>1-1-1 Yayoi, Tokyo University,<br>Monkyo-ku Tokyo 113<br>Tel.: 03-812-2111, Extn. 5724 |
|---|---|---|

| Participant | Organization | Field of specialization | Organizational address | Participates as |
|---|---|---|---|---|
| Kajiura, Kinjiro | Professor, Seismic Research Laboratory, Tokyo University | Physical Oceanography | Monkyo-ku<br>Tokyo 113,<br>Tel.: 03-812-2111, Ext. 5724 | Coordination and Executive Committee Member |
| Aota, Masaaki | Asst. Professor, Low-Temperature Laboratory, Hokkaido University | Physical Oceanography | 6, Minamigaoka,<br>Mombetsu Shi,<br>Hokkaido 094<br>Tel.: 01582-3-3722, Ext. 8 | |
| Asada, Toshi | Professor, Development Laboratory, Tokai University | Seismology | 1117, Kita Kanemoku<br>Hiratsuka-shi 259-12<br>Tel.: 0463-58-1211, Ext. 2299 | |
| Kagami, Hideo | Asst. Professor, Marine Research Laboratory, Tokyo University | Marine Geology | 1-15-1 Minami-dai<br>Nakano-ku Tokyo 164<br>Tel.: 03-376-1251, Ext. 255 | |
| Kanamori, Satoru | Asstt. Professor, Oceanographic Laboratory, Nagoya University | Chemical Oceanography | Huromachi, Chigusa-ku,<br>Nagoya 464<br>Tel.: 052-781-5111,<br>Ext. 5716 | |
| Kitano, Yasushi | Professor, Oceanographic Laboratory, Nagoya University | Geochemistry | Huromachi, Chigusa-ku,<br>Nagoya 464 Tel.: 051-781-5111, Ext. 5715 | Executive Committee Member |

| | | | | |
|---|---|---|---|---|
| Kunishi, Hideaki | Professor, Faculty of Science, Kyoto University | Physical Oceanography | Oiwake-cho, Kita-Shirogawa, Sakyo-ku, Kyoto 606 Tel.: 075-751-2111, Ext. 3910 | Executive Committee Member |
| Kobayashi, Kazuo | Professor, Oceanographic Laboratory, Tokyo University | Ocean-bed Physics | 1-15-1 Minami-dai, Nakano-ku, Tokyo 164 Tel.: 03-376-1251, Ext. 365 | |
| Takayanagi, Yokichi | Professor, Faculty of Science, Tohoku University | Paleontology | Aoha, Aramakiaza, Sendai 980 Tel.: 0222-22-1800, Ext. 3419 | Executive Committee Member |
| Tsunogai, Shizuo | Professor, Department of Fisheries, Hokkaido University | Chemical Oceanography | 3-1-1 Minato-machi, Hakodate-shi 041 Tel.: 0138-41-0131, Ext. 309 | |
| Teramoto, Toshihiko | Professor, Oceanographic Laboratory, Tokyo University | Physical Oceanography | 1-15-1 Minami-dai, Nakano-ku, Tokyo 164 Tel: 03-376-1251, Ext. 251 | Executive Committee Member (Administrative Office) |
| Toba, Yoshiaki | Professor, Faculty of Science, Tohoku University | Physical Oceanography | Aoha, Aramakiaza, Sendai 980 Tel.: 0222-22-1800, Ext. 3342 | |
| Nasu, Noriyuki | Professor, Oceanographic Laboratory, Tokyo University | Marine Geology | 1-15-1 Minami-dai, Nakano-ku, Tokyo 164 Tel.: 03-376-1251, Ext. 256 | Executive Committee Member |
| Handa, Nobuhiko | Professor, Oceanographic Laboratory, Nagoya University | Marine Biochemistry | Huro-machi, Senshu-ku, Nagoya 464 Tel.: 052-781-5111, Ext. 5720 | |

(continued)

| Participant | Organization | Field of specialization | Organizational address | Participates as |
|---|---|---|---|---|
| Fukuoka, Jiro | Professor, Department of Fisheries, Hokkaido University | Physical Oceanography | 3-1-1 Minato-Machi, Hakodate-shi 041, Tel.: 0138-41-0131, Ext. 373 | |
| Fujita, Yoshihiko | Professor, Oceanographic Laboratory, Tokyo University | Physical Oceanography | 1-15-1 Minami-dai, Nakano-ku, Tokyo 164 Tel.: 03-376-1251, Ext. 286 | |
| Fujiwara, Shizuo | Professor, Faculty of Science, Chiba University | Analytical Chemistry | 1-33 Yayoi-cho, Chiba-city 260 Tel.: 0472-51-1111, Ext. 2618 | |
| Horibe, Yoshio | Professor, Oceanographic Laboratory, Tokyo University | Chemical Oceanography | 1-15-1 Minami-dai, Nakano-ku, Tokyo 164 Tel.: 03-376-1251, Ext. 273 | Executive Committee Member |
| Marumo, Ryuzo | Professor, Oceanographic Laboratory, Tokyo University | Marine Biology | 1-15-1 Minami-dai, Nakano-ku, Tokyo 164 Tel.: 03-376-1251, Ext. 234 | |
| Mochizuki, Hitoshi | Professor, Dentsu University | Communication Engineering | 1-5-1 Chofugaoka Chofu-shi Tel.: 0424-83-2161, Ext. 406 | Executive Committee Member |
| Yamazaki, Hiroo | Professor, Faculty of Engineering Tokyo University | Engineering Measurements | 7-3-1 Hongo Bunkyo-ku, Tokyo 113 Tel.: 03-812-2111, Ext. 6915 | |

## Guidelines for Authors

Newsletter welcomes any enterprising article.

The contents are generally classified as:

| | |
|---|---|
| Highlights: | Quick results of this Special Research Project |
| Bulletin: | Domestic or international news reports closely related to this Special Research Project |
| Reports: | Reports regarding the activities of each section or working group |
| Announcements: | Announcements regarding the proposed meetings or conferences of each group |
| Bibliography: | Listing of papers published by the members on this Special Research Project or other reports. |

Manuscripts should be written on standard 400 character size paper (handwritten material also acceptable). Figures should be drawn in final form ready to be sent to press.

Manuscripts may be sent to this office through the respective group leader.

Manuscripts should reach this office a minimum of 20 days before the scheduled date of publication.

The future publication schedule is December 20, 1981, January 30, 1982 and March 10, 1982.

**Note:**

When reproducing or citing material published in this Newsletter, due acknowledgment should be given. Authors' names, wherever mentioned, should also be quoted. A copy of the publication containing such a reproduction may be sent to the Editorial Section of the Administrative Office.

Published by: Administrative Office,
Special Research Project "The Ocean Characteristics and Their Changes."

Published at: Administrative Office,
Physical Oceanography Department,
Institute of Oceanography,
Tokyo University,
1-15-1 Minami-dai, Nakano-ku,
Tokyo 164
Tel.: 03-376-1251, Ext. 251.

Special Research Project

# The Ocean Characteristics and Their Changes

NEWSLETTER NO. 3, DECEMBER, 1981

# Contents



Researchers engaged in this Special Research Project, who wish to submit their papers in English and acknowledge receipt of a research grant for this special study, should indicate the above title of the project.

**HIGHLIGHTS**

# Analysis of Waters below 1,500 m

*Ikuo Kaneko and Toshihiko Teramoto*

(Institute of Oceanography, Tokyo University)

(Received: November 2, 1981)

Since 1977, we have often studied the ocean currents at intermediate and deep layers around Izu. A method based on material tracers developed by our Institute was employed during these studies. Basically this method consists of a quantitative evaluation of the dissolved oxygen ($DO_2$) and dissolved silica (Si) in sea water and an analysis of iso-$\sigma_\theta$ plane using CTD Rozet waters.

The deep sea layers (below 1,500 m) in the Pacific Ocean have a much more uniform material distribution in the horizontal plane than that in the Atlantic Ocean, but with the measurement accuracies presently available, it is difficult to obtain an ordinary T-S analysis. (There are many facets of CTD that can be improved but for the proper analysis of ocean waters around Japan, it is necessary to obtain S with an absolute accuracy of 0.001‰). If we look at the models proposed by Stommel-Arons (1960) and Fiadeiro-Craig (1978) we notice that they point out the same source responsible for deep sea currents in the Pacific because of the technological limitations. As such the very meaningfulness of conducting T-S analyses in the horizontal plane is being questioned now. The moderate changes in T-S values which are observed as we move from the southwest to the northeast region of the Pacific are a result of the vertical dispersion of both parameters following sea-water ageing. This cannot be considered as a transition to another water mass of a different source. If the T-S relation is to be used to estimate water movements in the horizontal plane, then $\sigma_\theta$ can be used as a reference level and then the value of S (or $\theta$ ) in that plane can be compared for each measuring interval. However, if the distribution on $\sigma_\theta$ plane over a certain region of sea appears as if a quantum is subtracted, it may be mostly because $\sigma_\theta$ does not exist over that region as a result of the vertical distribution of T-S. If we consider a time scale over which $\sigma_\theta$ does exist in the same region then naturally we cannot obtain any distribution.

Now, if we limit our points of observation a slightly different situation is encountered. In the deep sea layers of the Pacific Ocean, a variation in T-S, dissolved oxygen in water, or other nutritive salts do not vary uniformly as we go from the southwest region to the northeast region but there are sudden changes in concentration at points. This is probably because the Pacific has been divided into a number of basins and the path of deep-sea waters entering each basin is different. Thus a difference in material concentration can be attributed to the difference in age as well as to the material distribution front formed at the boundaries between two basins.

Another factor for this is a non-uniform material distribution. In such seas, a minor difference in the properties of water masses of each basin can be considered the basis and if the phenomena (to be observed) are restricted to a short period and small size over which there are no changes in properties, the analysis of water masses of the deep Pacific layer may also be meaningful.

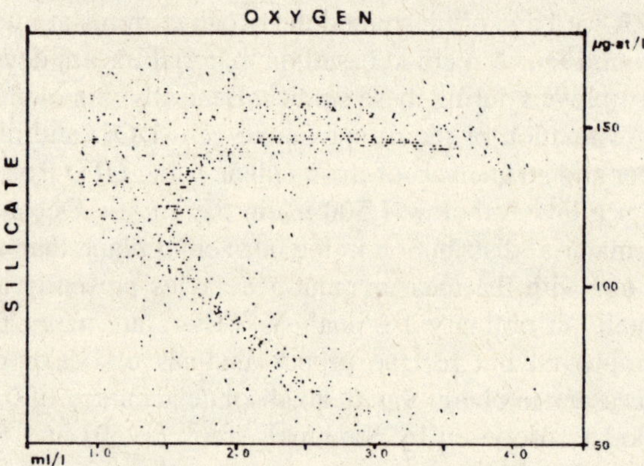


Fig. 1. Dissolved oxygen ($DO_2$) - Silicate (Si) diagram. Raw data plot.

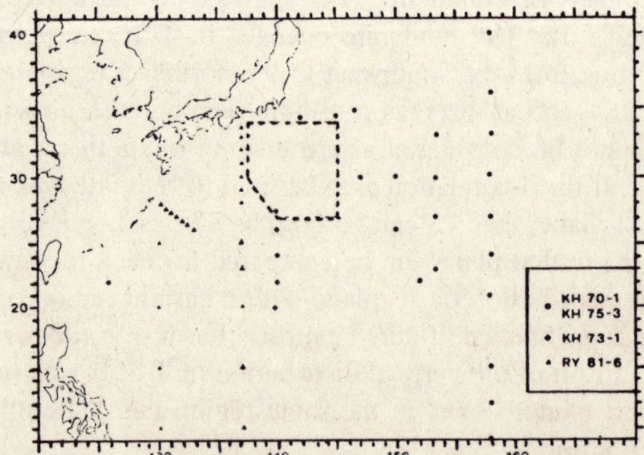


Fig. 2. Locations of observation points. RY 81-06 corresponds to a cruise by the Meteorological Department. KH 70-1·75-3 is a cruise by the Inorganic Chemistry Department of Institute of Oceanography, Tokyo University and KH 73-1 is by the Biology Department of the same institute. The extended diagram of the area is shown by the dotted line in Fig. 3.

Now, if we use few parameters which are independent of each other, it may not be necessary to persist in the T-S analysis which is difficult from accuracy considerations. As far as Izu Ridge and the surroundings, which the authors observed, are concerned, we have the northern tips of Shikoku and the Philippines basin lying on the western side of the ridge extending further south and these water masses have different properties than those on the eastern side. The difference in the T-S plane may not be much but is considerable for dissolved oxygen and silica.

Figure 1 shows the relation between $DO_2$–Si prepared from the data of Figs. 2 and 3. Si increases with depth and attains a small peak around 2,000 m. The $DO_2$–Si curve around Japan shows that Si disperses above a level of 100 μg·at/l and distributes with a certain width at the deepest parts. As this point becomes deeper than 1,500 m then the T-S curve becomes symmetric. When the data of all observational points is extrapolated for different values of $\sigma_\theta$, we can plot $DO_2$-Si for each $\sigma_\theta$ corresponding to depths of 1,500 m, 2,000 m and 3,000 m. This is shown in Fig. 4. (Since the data of KH 81-2 is obtained by direct samplings of the water at given $\sigma_\theta$ planes using a real time $\sigma_\theta$ monitoring method based on N.B.CTD-HP 45, it is not necessary to extrapolate any data.)

The area around Izu Ridge, shown by a dotted line in Fig. 2, has a considerable variation in water types and needs particular attention. The analysis of data has just begun and we shall conclude with some comments on typical distributions.

$\sigma_\theta$ = 27.55 (1500 m layer)

The area around Izu Ridge shows a linear distribution (point $A_1$). This

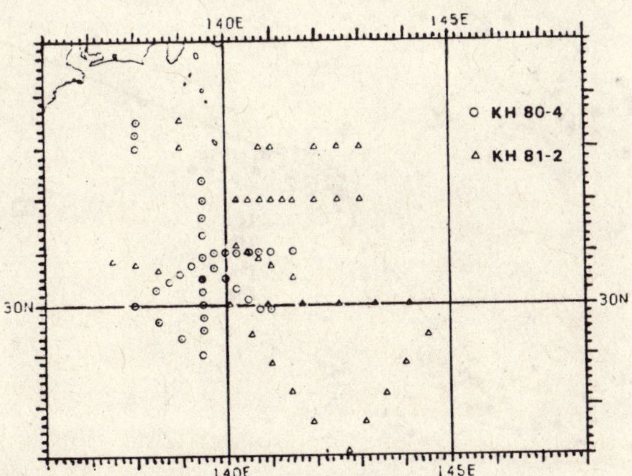


Fig. 3. Sampling points for CTD RMS sampling used by Physical Oceanography Department, Institute of Oceanography, Tokyo University.

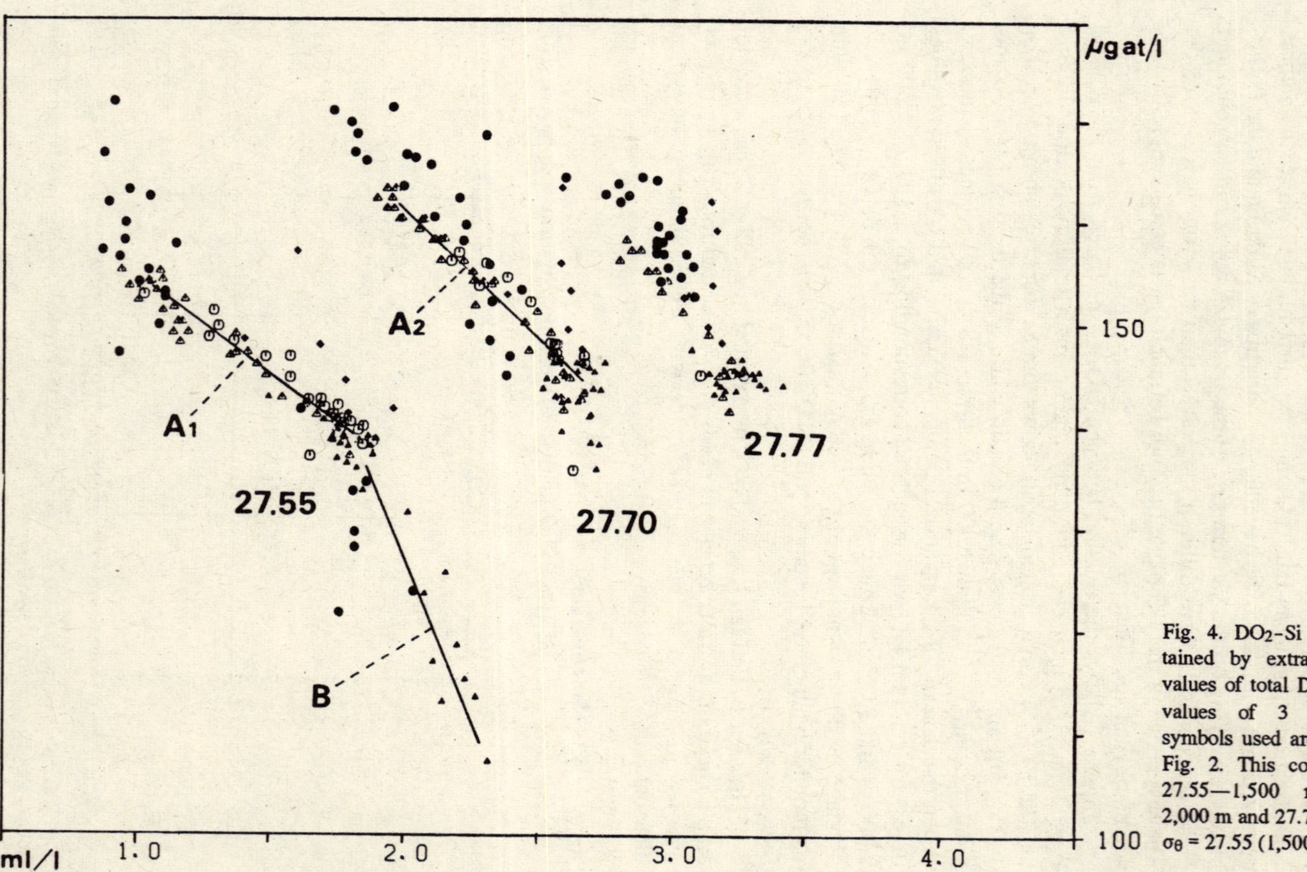


Fig. 4. $DO_2$-Si diagram obtained by extrapolating the values of total $DO_2$, Si to $\sigma_\theta$ values of 3 layers. The symbols used are same as in Fig. 2. This corresponds to 27.55—1,500 m, 27.70—2,000 m and 27.77—3,000 m. $\sigma_\theta$ = 27.55 (1,500 m layer)

shows that in this ocean area, the changes in water types are mainly caused by dispersion along the $\sigma_\theta$ plane.

The water type on the right edge of line $A_1$ is restricted to the Philippine Sea beyond 25°N. The water from the lower layers of Kuroshio on the west side of Izu Ridge also collects in this area. The waters south of 20°N cannot be grouped as an extended $A_1$ line (as it distributes in the direction of B) and the fluctuations in these water types show a different phenomena than that observed in ridge area.

$\sigma_\theta$ = 27.70 (2,000 m layer).

The linearity around Izu ridge ($A_2$) is particularly good.

There is a tendency to cluster at the right edge of $A_2$ line in the Philippine Sea. As we can see also from $\sigma_\theta$ = 27.77, the horizontal distribution of $DO_2$, Si in the Philippine Sea below this layer is extremely uniform.

As against this, the distribution along the $A_2$ line is quite irregular ranging from low Si, high $DO_2$ to high Si, low $DO_2$.

If we plot this distribution of $DO_2$–Si after proper classification, then we come across various other interesting properties. It is quite interesting to note that such a distribution is developed as a result of the flow conditions. We would like to analyze the data from various aspects (to form a clearer picture).

We have used the data obtained by the Meteorological Department's research vessel during its 81st cruise of West Pacific during summer, by the *Hakucho-maru*—the vessel of the Institute of Oceanography, Tokyo University during the cruise organized by the Inorganic Chemistry Department KH 70-1, 75-3, the cruise organized by the Marine Biochemistry Department of the same institute KH 73-1, as the data source. We would like to express our sincere thanks to the concerned persons. Thanks are also due to Mr. Yukio Kodama of the Ocean Inorganic Chemistry Department for his continuous guidance during the precision analysis of $DO_2$·Si.

## Bibliography

Broecker, W.D. Geochemical Tracers and Ocean Circulation. Evolution of Physical Oceanography, pp. 434–460.

Fiadeiro, M.E. and H. Craig. 1978. Three-dimensional modeling of tracers in the deep Pacific Ocean—I. *Journ. Mar. Res., 36,* pp. 323–355.

Reid, J.L. On the mid-depth circulation of the world ocean. Evolution of Physical Oceanography, pp. 70–111.

Stommel, H. and A.B. Arons. 1960b. On the abyssal circulation of the World Ocean—II. *Deep-Sea Res., 6,* pp. 217–233.

BULLETIN

# Data Transmission with Satellite in Oceanographic Studies

*Jiro Segawa*

(Institute of Oceanography, Tokyo University)
(Received: November 30, 1981)

It was quite some time before the man-made satellite was first used for oceanographic studies. However, that was in America or Europe. These days a considerable information network has developed and discussions in other countries are available every day like the tides. We have thus come to an age when the studies of other people feel like our own. As regards the rationalization of oceanographic observations using satellites, the situation in Japan is not yet satisfactory.

These days "cost/performance" has become the key word and the studies or plans which cannot cross this barrier are immediately discarded. As the use of satellites is very expensive, if the performance is not properly evaluated it would not be possible to even commence any studies. These days, the absolutely essential equipment for oceanographic studies is the vessel and no one says the studies would not be possible without a satellite. However, there are various opinions expressed about satellite use, ranging from "it would be better for progress in research methods" to "it may be difficult to carry out studies without satellites." Looking back at a point when it became difficult to carry out studies without satellites, we notice that the satellite has contributed in a variety of ways to progress. The use of computers, which is so frequent today, or the tray equipment for the seabed, etc. have developed following a similar course.

It is almost five years now since a number of scientists first felt the use of satellite data communication in their studies. It was around the time when the data collection system (DCS) faculty built in TRIOS and NOAA satellites—meteorological satellite transmitting pictures—was first conceived. This system was entirely unsatisfactory as far as data communication faculty was concerned. However, the idea that we could communicate to the satellite at will, without spending much money, was quite catching. Here the cost/performance concept is temporarily shelved from our minds.

A considerable economy is possible by using the satellites launched by NASA. On the other hand, a lot of difficulties are faced when using satellite launched by other agencies, particularly when the structural details are not known to the users. Over the last few years however, image processing was studied in further detail and as a result it was possible to learn the DCS thoroughly. In addition, the transmitting equipment was manufactured locally and there were even signs of local production of receiving equipment. Thus, immediately after this special study project was proposed, we started thinking of using a satellite this time, with great expectations.

There are two merits of data communication through satellite. One is compared with the normal communication system it is possible to send the signals to a longer distance with less electrical power and the other is that it is possible to conduct long-term monitoring which is essential in oceanographic studies. For the effective utilization of (satellite) data transmission, it is necessary to send the data clearly and receive the same at a familiar place. The theme of our special research project is the development of a receiving station (Local Users Terminal or LUT). Recently we have installed an antenna for satellites on the roof of the Institute of Oceanography, in Tokyo University. The satellite signal contains a complicated code such as a cipher and it is

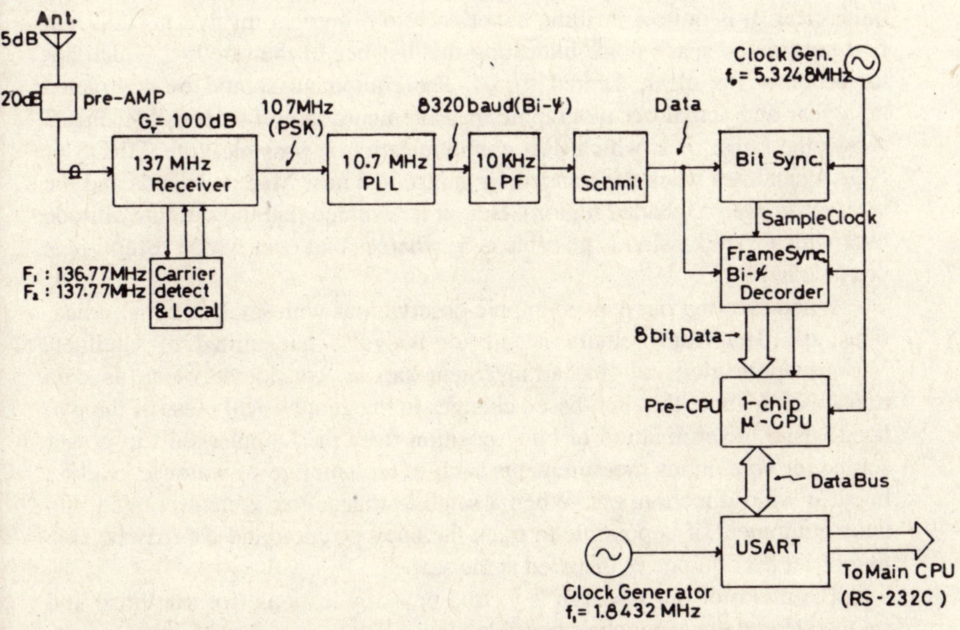


Fig. 1. The circuit layout of the VHF signal section and signal recovery section of LUT (Local Users Terminal).

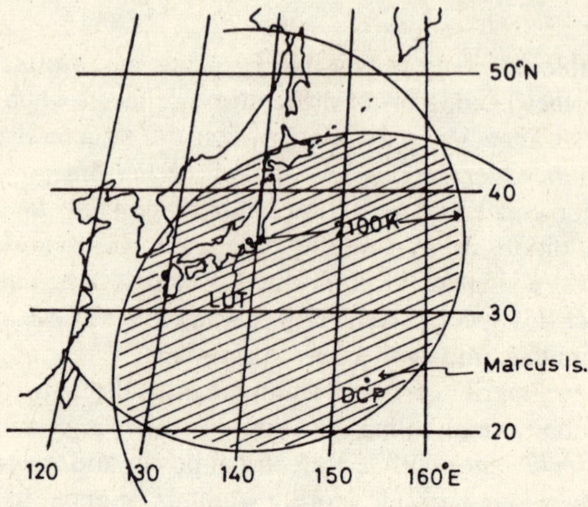


Fig. 2. The range over which satellite communication is possible when the satellite transmitting equipment (also called Data Collection Platform or DCP) is installed on Marcus Islands and the LUT in Tokyo (shaded region). It is assumed that transmission is possible within 5° of the satellite altitude and reception is possible within 10°.

quite difficult to recover our data from that code. Currently we are receiving the satellite data using the circuit layout shown in Fig. 1. However, in a place like Shinjuku, the interference from the daytime TV signal is considerable and it is extremely difficult to filter that noise. The midnight records are, however, quite clear. It is quite a thrilling experience to observe a mysterious noise, at first considered space noise indicating the distance of the satellite, which latter becomes the clear, desired signal. The equipment should be completed next year and actual oceanographic measurements should be possible. Fig. 2 shows the range over which data communication is possible with TIROS or NOAA satellites when the transmitter is installed near Marcus Islands and the receiver in Tokyo (shaded region). Here it is assumed that the satellite altitude over which transmission is possible is 5° whereas that over which reception is possible is 10°.

When carrying out oceanographic observations with satellite communications, the data from solitary islands or buoys is transmitted to satellites. Various applications of this arrangement can be considered, such as data transmission about the time-based changes in the geophysical mass of the isolated island, determination of buoy position from the Doppler-shift in power source, or continuous measurements such as temperature of water as read by buoy or wind direction, etc. When a satellite transmitter is incorporated into buoy equipment, it is possible to track the buoy position and we may be able to recover the equipment installed at the seabed.

The motto of our research is to find new applications (for satellites) and try to extend them as much as possible.

## REPORTS

### Fifth Executive Committee meeting was held on November 16, 1981 (Monday) at the Oceanographic Laboratory of Tokyo University

Profs. Saito, Nasu and Mitsuyasu were formally requested to accept membership in the Evaluation Committee.

As there was some unspent (920,000 yens) research grant this year, the financial assistance needed for the research project for the fiscal year 1981 was revised and submitted.

The research plan for fiscal year 1982 proposed. Prof. Taira (Asst. Professor, Institute of Oceanography, Tokyo University) was included in the research team.

It was decided to hold the third meeting of the managing committee on February 12, 1982 in which the leader of each study group in this project shall report the progress of their studies and the future course of action for the overall coordination of the project will be discussed.

### Symposium on the Paleoenvironments of the Ocean Floor and an Accurate Determination of the Center of Gravity

The participants in this symposium were those mainly engaged in studies of "fourth century environments at the seabed" and "accurate determination of the center of gravity of the ocean bed", as well as those concerned with similar projects from the government and other organizations.

Place and date of symposium: Oceanographic Laboratory, Tokyo University, November 13 and 14, 1981 (10:00–17:00 hrs).

Total No. of participants: 89.

Program:

Welcome: Takayanagi (Physics Department, Tohoku University)

1. Seismological behavior of the Japan seabed: Minamigumo, Kasahara, Tsutsumizawa, Nishi (Earthquake Laboratory, Tokyo University).

2. An accurate distribution of epicenters (of earthquakes) in and around the Japan Sea region: Hirata (Science Faculty, Tokyo University), Shimamura (Science Faculty, Hokkaido University), Yamada (Science Faculty, Tokyo University), Suchiro (Science Faculty, Tohoku University).

3. Warm currents in the northern Philippine Sea and South Sea trough:

Ueda, Yamano, Fujizawa, Honda (Earthquake Laboratory, Tokyo University).

4. Comparison of sedimentation phenomena in the Japan Sea and South Sea trough: Kagami, Tokuyama, Nyu, Igarashi, Nasu (Institute of Oceanography, Tokyo University).

5. Seabed profile around the South Taito Islands and abnormalities in the earth's magnetism: Matsuda, Yasukawa, Isezaki, Matsubara, Yamaguchi (Faculty of Science, Kobe University).

6. Abnormalities in the gravitational force of seas around Japan: Tomoda (Institute of Oceanography, Tokyo University).

7. Survey regarding the seabed structure around the Suruga-South Sea trough (Nankai trough): Kinoshita (Faculty of Science, Chiba University).

8. Paleoenvironments as indicated by deposits in the Hidakagawa area, particularly in the Miyama region: Kimifumi, Nakazawa (Faculty of Science, Kyoto University).

9. Problems regarding the history of the formation of the Yonjuman belt-exterior belt of southwest Japan: Kanmai (Faculty of Science, Kyushu University).

10. Basic problems in the determination of historic environments at the seabed from the deposits: Okada (Faculty of Science, Shimoka University).

11. Present condition of the 4th group isotopes stratigraphy and the problems: Saito (Faculty of Science, Yamagata University).

12. The upper 4th group isotopes in seas around Japan: Oba (Kanezawa University).

13. Fluctuations of ocean conditions in the late glacial period off Boso peninsula based on the piston core: Shizuishi (Faculty of Science, Tokyo University), later glacier paleotemperature group.

14. Distribution of planktonic foraminifera in surface layer sediments on the seas east of Honshu. Basic studies to analyze the paleoenvironments of the ocean: Takayanagi, Oda (Faculty of Science, Tohoku University).

15. Distribution of benthonic foraminifera in the northwest Pacific—an approach to find out the environment of the 1st period: Tekiba (Department of Mining, Akita University).

16. Excavation of a shallow coral reef: Konishi (Faculty of Science, Kanezawa University).

*Summary of the Proceedings of the Symposium*

There were eight presentations on the first day, each followed by detailed discussions.

The theme of the first day's discussion was an accurate survey of important locations in the seabed. The Sea of Japan-South Sea trough, including the Yommanju belt and Taito Sea, were considered important locations and various geological and geophysical observations were carried out at these locations for three years. The original objective was to study high density or high disintegration phenomena. But, during those first three years it was proposed to gather seismological data and determine the structure of sediments, structure of the earth's crust or the distribution of earthquake epicenters, etc. in this region.

The discussions are summarized below by broad classification into four groups.

1. Seismological activities in the Japan gulf region

Buoy type earthquake measuring equipment was installed in the submerged zone, including the gulf area, and (later) recovered safely. As a result, it was possible to determine the distribution of earthquake epicenters accurately within 2-3 km. The interesting finding about this distribution is that earthquake is rare toward the east side of the outer swell center as well as toward the west of the gulf axis. The depth at which earthquake occurs is less than 10 km in the gulf region and it increases as we move westward (Hirata and Shimamura). In previous reports it was indicated that the earthquake occurs at a depth of 50-100 km around the gulf region. This distribution therefore shows an entirely different situation. This phenomenon may be a result of resosphere bending. The marine deposits present on the periphery of the gulf indicate $V_s$ = 0.34 km/sec and thus indicating clearly that (their) Poisson's ratio is extremely small (Minamikumo). One may conclude that the properties of rocks and acoustic behavior may not agree well.

2. Warm currents in the South Sea trough, gravitational force and structure of deposits

While studying the warm currents twin phenomena were observed. Near the South Sea trough axis there was a high zone of 3.5 HF which went down toward the land side and another high zone reaching a level of 2.5 HF was seen near the land that connects a cape. It is particularly interesting to note that around the trough an eyeball type distribution was observed off the cape. A multicomponent reflection seismic survey made it clear that a water-soluble gas layer surrounds the continental slope on the west side of Shio Cape. A layer of (water soluble gases) that thick was generally considered to increase with the increase in water depth but in this region, exceptionally, the layer becomes thin at the bottom of a deep trough. The peculiar thermal structure of the earth's crust at that point was found to be responsible for this behavior (Ueda and Yamano).

As regards the gravitational forces, an outer gravity high is clearly seen (20 mgal <) in the Sea of Japan but in the South Sea trough it was only in fragments. Also as we move eastward, it becomes more uncertain and clearly does not continue to Izu Ridge (Tomota).

The structure of deposits at the Sea of Japan and at the South Sea trough are quite typical. The horst or trough that develops in an oceanic crust propagates in a model way in the Sea of Japan. The development of a sliding surface with minimum angle, as seen within the appendages, is accompanied by tectonic erosion and as a result the width of the appendages in the Japan trench is about 23 km, which is just half the width seen in the South Sea trough (Nankai trough). In the Nankai trough, a thrust develops in appendages at an angle of 60°~30° after which the sliding surface with a minimum angle is developed. In this case it became clear that the upper layer of the ocean bed plays the role of the sliding surface and the pelagic sedimentation layer which is formed above, starts the deformation of appendage prisms by micro-accretion (Kagami and Nasu).

3. Abnormalities in the earth's magnetism in the Taito Sea region

In this region, we measured the total magnetic force of the island and in space. As a result, it was possible to prepare maps showing the distribution of the total magnetic force around the southern part of Taito Island; and subsequently determine the shape of volcanic bodies. Using this data, model calculations for a certain magnetic volcanic mountain lying above a depth of 1,700 m were carried out (Matsuda, Yasukawa).

4. Models of Yomanju belt including the South Sea trough

The structure of the stratigraphic layers in the Yomanju belt has become clear with the progress of radiolarian stratigraphy. During the studies of Kii Peninsula and Yomanju belt, the formation period of the olistostrome including marine basalts (a large sliding layer around the crust) was determined to be in the early part of the Cretaceous period (Official document: Nakazawa).

These sunken tile structures show a northern layer in one of the inverse layer blocks. Broadly speaking, the period becomes modern as we go southward and as a continuation of this process, the Nankai trough must have been formed. In this case, one can also point out the advance of the trench axis with time (Sadamero).

On the second day, there were many presentations about the paleoenvironments of the ocean floor. Among them, Okada compared the environments as determined by two approaches, viz., a study of seabed sediments and of photographs. He mentioned the present state and future direction. Ando considered the history and current status of the development of the stratigraphy of ocean floor sediments using the oxygen isotope ratio and the theoretical meaning of the periodic fluctuations of the isotope ratio with reference to paleoenvironments. Oba explained the fluctuations of the oxygen isotope ratio after the last glacier period, as seen in sediment cores from the Sea of Japan and Pacific Ocean, and discussed the results in the context of paleoenvironments. Shizunishi discussed the variation in micropaleontological substances from the last glacier period to date in sediment cores off the Boso Peninsula and the fluctuation of the oxygen isotope variation. In order to clarify the paleo-oceanic and paleo-climatic changes quantitatively, using microfossils of plankton, Takayanagi and Oda investigated the spatial distribution of planktonic foraminifera in surface layer sediments, i.e., both horizontally and vertically. Tekiba discussed the paleoenvironments in the Japan seabed based on the distribution of protozonic benthonic foraminifera. Onishi presented information on the marine thermometer use of coral reefs to measure the fluctuation of sea level and the present condition of studies of boring in shallow layers. Thus various subjects were discussed and the results were presented in detail. These were followed by lively discussions and, as far as this special project research is concerned, the results of the first year can be said to be much above expectations.

The various papers presented on the second day of the seminar are proposed to be published separately.

(Coordinator: Takayanagi and Kagami)

## ANNOUNCEMENTS

### 1. First Meeting of Numerical Experiment Group

The coordination committee of the Special Research Project "The Ocean Characteristics and Their Changes" has formed a numerical experiment group in April which has so far conducted numerical models to describe the mechanism of serpentine waves at Kuroshio. During this first year of study, it aimed to prepare a simple model and conduct preliminary studies which would be useful for subsequent years. Using the reduced gravity model, it was possible to obtain a serpentine pattern very similar to the serpentine waves at Kuroshio with quite an interesting result.

During this meeting, we would like to critically examine the results available so as to proceed further in presenting an accurate numerical model to express serpentine motion.

Date: January 22, 1982.

Place: Conference Hall, Oceanographic Laboratory, Tokyo University.

Program:

1) Report on the development of a numerical model for Kuroshio (13:00–15:30 hrs): Ii, Suginohara (Faculty of Science, Tokyo University).

2) Future plan for developing a numerical model for Kuroshio (15:40–17:00 hrs): Imawaki (Faculty of Science, Kyoto University), Kubota (Faculty of Ocean Science, Tokai University), Sekine (Faculty of Science, Tohoku University), Fukazawa (Oceanographic Laboratory, Tokyo University), Masuda (Application Studies, Kyushu University), Suginohara (Faculty of Science, Tokyo University, Ii (Faculty of Science, Tokyo University).

For further information regarding the meeting contact Jang-Huan-Yoon (812-2111, Ext. 4285).

### 2. Conference on Satellite Use in Oceanographic Studies

As informed in Newsletter No. 2, this conference is proposed to be conducted as given below. Interested persons are requested to attend.

As mentioned earlier it is felt that a large number of researchers would be interested in this problem, including those connected with university and other organizations that are using satellite data for oceanographic studies or those who would like to use it, those people who are connected with satellite use itself (both software and hardware aspects), and others who would like to pursue this problem.

However, earlier there was not much activity regarding this problem. This conference, therefore, should provide an opportunity to exchange ideas in this respect and it may be the first step toward future studies if a need exists and a desire is expressed.

Date and time: January 27, 1982 (Wednesday) (10:00–17:00 hrs).

Place: New Conference Hall, Institute of Oceanography, Tokyo University.

Program (since this includes all points currently being considered, there may be some changes in the actual agenda):

1) State of the art use of satellites in oceanographic studies at universities (10:00–11:30 hrs): Taira (Institute of Oceanography, Tokyo University), Segawa (Institute of Oceanography, Tokyo University), Sugimori (Tokai University) and others.

2) Methods of measurement by satellite and studies on information processing at universities (12:30–14:00 hrs). Takagi (Biological Research Laboratory, Tokyo University, Suzuki [Dentsu (Electrical Communication) University], Hirozawa (Space Studies Center) and others.

3) Future plans for using satellites in oceanographic studies at universities (14:00–15:00 hrs): Toba (Tohoku University), Teramoto (Institute of Oceanography, Tokyo University) and others.

4) General discussion (15:30–17:00 hrs).

*Note:* For additional information regarding this conference please contact Prof. Taira at the Institute of Oceanography, Tokyo University (Tel. 03-376-1251, Ext. 254).

### 3. Symposium on "Future Geoscience"

This symposium shall be conducted as stated below.

The objective of this symposium is to exchange views on what should be the (aspect of) geoscience beyond oceanographic studies during the 21st century, with no restriction to a particular field or grade and with a special reference to the execution of the study project "Environments at the seabed during the 4th period" (Project Supervisor: Prof. Takayanagi) being conducted as a part of the Special Research Project "The Ocean Characteristics and Their Changes" (Project Supervisor: Prof. Kajiura).

During the symposium, lectures and discussions will be divided in four groups as stated below. Persons interested in attending the symposium (with an indication of study group) should contact the coordinator. All inquiries regarding the symposium should also be directed to him.

Date and time: From 17:30 hrs on January 24, 1982 (Sunday) to 13:00 hrs on January 25, 1982 (Monday).

Place: Seminar Hall of Kanto Regional University (15 minutes by bus from JNR Hachioji Station on the Central Route).

Fees: About ¥ 6000 (including lodging and transport).

Topics for discussion:

(Proposed)—Paleontology and Marine Geology: Koizumi (Department of Education, Osaka University), Ryokaku (Osaka Municipal Museum of Natural History), Akiba (Technical Development Department, Petroleum Resource Development Corporation Ltd.).

Deposits and Marine Geology: Fujioka (Seabed Deposits Section, Institute of Oceanography, Tokyo University), Nishimura (Ocean Geological Section, Department of Geological Survey).

Chemical Considerations of Deposits at Seabed: Kato (Ocean Chemistry Department, Tokai University).

Earth's Structure at Seabed: Yamano and Kodama (Earthquake Research Laboratory, Tokyo University), Matsumoto (Institute of Oceanography, Tokyo University).

Coordinators:

Chief coordinator: Takayanagi (Faculty of Science, Geopaleontology Department, Tohoku University). Aoha, Aramaki-aza, Sendai-city, Miyagi Prefecture 〒980. Tel.: 0222-22-1800, Ext. 3419.

Oda (Faculty of Science, Geopaleontology Department, Tohoku University). Aoha, Aramaki-aza, Sendai-city, Miyagi Prefecture 〒 980. Tel.: 0222-22-1800, Ext. 3421.

Harada (Faculty of Science, Geoscience Department, Yamagata University). 1-4-12 Koshirokawa, Yamagata-city, Yamagata Prefecture 〒 990. Tel.: 0236-31-1421, Ext. 2592.

Kitazato (Faculty of Science, Geoscience Department, Shizuoka University). 836, Otani, Shizuoka-city, Shizuoka Prefecture 〒 422. Tel.: 0542-37-111, Ext. 596.

## Bibliography

Keisuke Taira and Toshihiko Teramoto. 1981. Velocity fluctuations of the Kurushio near the Izu Ridge and their relationship to its current path. *Deep-Sea Research,* Vol. 28A, No. 10, pp. 1187–1197.

## Guidelines for Authors

Newsletter welcomes any enterprising article.

The contents are generally classified as:

| | |
|---|---|
| Highlights: | Quick (intermediate) results of this Special Research Project |
| Bulletin: | Domestic or international news reports closely related to this Special Research Project |
| Reports: | Reports regarding the activities of each section or working group |
| Announcements: | Announcements regarding the proposed meetings or conferences of each group |
| Bibliography: | Listing of papers published by the members on this Special Research Project or other reports. |

Manuscripts should be written on standard 400 character size paper (handwritten material also accepted). Figures should be drawn in such a way that they can be reproduced as they are.

Manuscripts may be sent to this office through the respective group leader.

Manuscripts should reach this office at least 20 days prior to the scheduled date of publication.

The future publication schedule is January 30, 1982 and March 10, 1982.

**Note:**

When reproducing or referring to material published in this Newsletter, due acknowledgment should be given. Authors' names, wherever mentioned, should also be quoted. A copy of the publication containing such a reproduction may be sent to the Editorial Section of the Administrative Office.

| | |
|---|---|
| Published by: | Administrative Office, Special Research Project "The Ocean Characteristics and Their Changes" |
| Published at: | Administrative Office,<br>Physical Oceanography Department,<br>Institute of Oceanography,<br>Tokyo University,<br>1-15-1 Minami-dai, Nakano-ku,<br>Tokyo 164<br>Tel.: 03-376-1251, Ext. 251. |

Special Research Project

# The Ocean Characteristics and Their Changes

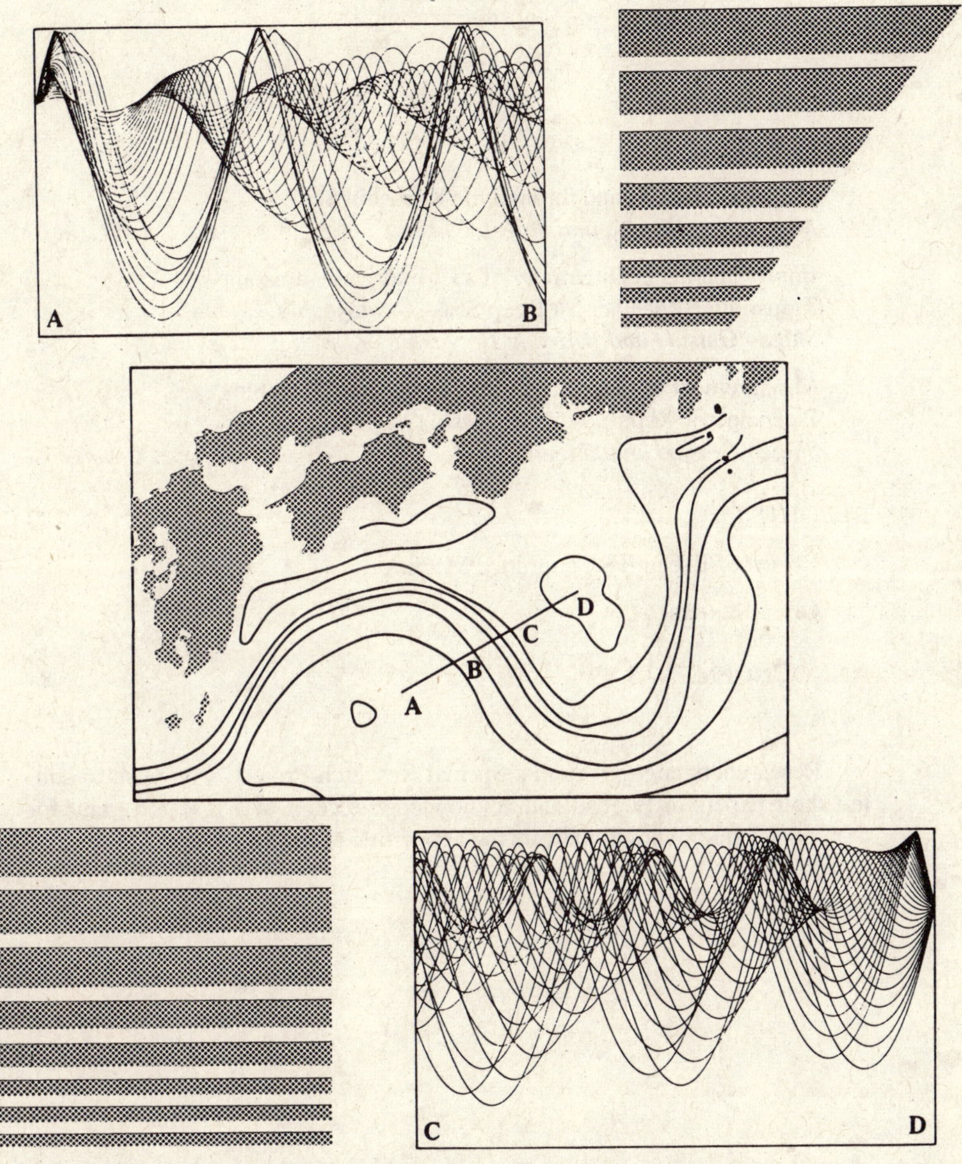


NEWSLETTER NO. 4, JANUARY, 1982

# Contents



Researchers engaged in this Special Research Project, who wish to submit their papers in English and acknowledge receipt of a research grant for this special study, should indicate the above title of the project.

HIGHLIGHTS

# Propagation of Sound through the Kuroshio Sea

*Motoyoshi Okujima and Shigeo Ohtsuki*

(Precision Engineering Research Laboratory, Tokyo Institute of Technology)
(Received: December 2, 1981)

A method of oceanographic measurements has been proposed in the USA whereby a sound wave is allowed to propagate over a long distance in the horizontal and the time required for such propagation is measured. This, with the CT (Computer Tomography) technique, is used to estimate the distribution of the sound velocity and water temperature over a wide region. The authors planned to use the same technique to measure variations in the Kuroshio seas around Japan. To begin with, we calculated the path traversed by sound and the propagation time when it passes through Kuroshio to know the condition of sound waves.

Figure 1 shows the route followed by the *Hakucho-maru* observation vessel (September–October, 1976), to which we also refer. The computations were carried out assuming that the sound wave is made to propagate between CTD observation point C 33 (warm water center) and C 88 (cold water center). Using the measured values of CTD at each observation point between C 33 and C 88 (385 km) and using the equation

$$c = 1448.6 + 4.618T - 0.0523T^2 + 0.00023T^3 + 0.018D$$

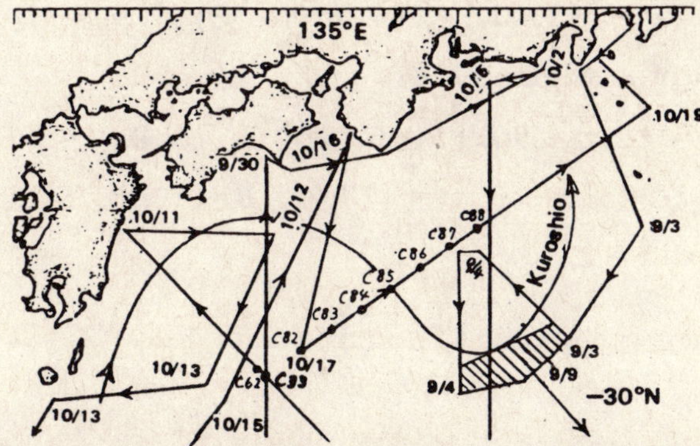


Fig. 1. Path traversed by *Hakucho-maru* KH-76-4 (September 1–October 20, 1976).

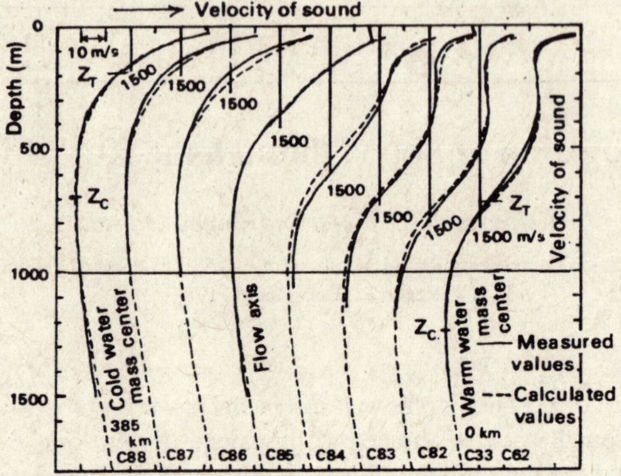


Fig. 2. Vertical distribution of sound velocity at each observation point (measured values and calculated values obtained by approximate equation).

the sound velocity at each point was calculated. The vertical distribution is shown by solid lines in Fig. 2. Here $c$ is the velocity of sound (m/s), $T$ is temperature of water (°C) and $D$ is depth (m). Since it is easy to calculate the sound propagation in the horizontal, we arrived at the following approximate equation to estimate the sound velocity distribution. The values calculated from this equation are shown by the dotted line in Fig. 2.

Horizontal distance:

$$x \text{ (m)}, \Theta = x\pi / 385{,}000.$$

The sound velocity in the sound channel:

$$c_C = 1{,}482.3 + 0.94\,(6 - \cos^2 \Theta) \cos \Theta.$$

The depth of sound channel:

$$z_C = 960 + 54\,(6 - \cos^2 \Theta) \cos \Theta.$$

$$z_T = z_C - 500, \quad A = 17.7033, \quad B = -0.100767.$$

i) For depth $z > z_T$:

$$c = c_C + 4.77\left(\frac{z - z_C}{265} + e^{-\frac{z - z_C}{265}} - 1\right).$$

ii) For $z_T > z > 50$:

$$c = c_C + A + 2\left(\frac{z_T}{z} + \frac{z}{z_T} - 2\right) + B(z - z_T) + 0.00007\{(1 - \cos\Theta)^3 - 2\}\ (z - z_T)^2$$

iii) $50 > z$:

$$c = c_C + A + 2\left(\frac{z_T}{50} + \frac{50}{z_T} - 2\right) + B(50 - z_T)$$

$$+ 0.00007\{(1 - \cos\Theta)^3 - 2\}(50 - z_T)^2$$

$$+ (z - 50)\left[\frac{2}{50}\left(\frac{50}{z_T} - \frac{z_T}{50}\right) + B + 0.00014\{(1 - \cos\Theta)^3 - 2\}(50 - z_T)\right].$$

Figure 3 shows the sound velocity distribution between points C 33 (0 km) and C 88 (385 km) as obtained from the above equations. It can be seen from this figure that the change in the vertical distribution of the sound velocity is comparatively less at depths greater than 1,500 m whereas in shallower regions, velocity changes considerably depending on the places.

A (sound source) transmitter was placed at the sound channel of C 33 (depth 1,230 m) and a receiver at the sound channel of C 88 (depth 690 m). The path followed by sound waves under this condition is shown in Fig. 4a, b. The sound code in this Fig., e.g., a code +10 (−10.5) indicates a sound wave which was propagated from C 33 in the upward (downward) direction and which reached C 88 by dividing the sound channel into 10 cycles (10.5 cycles).

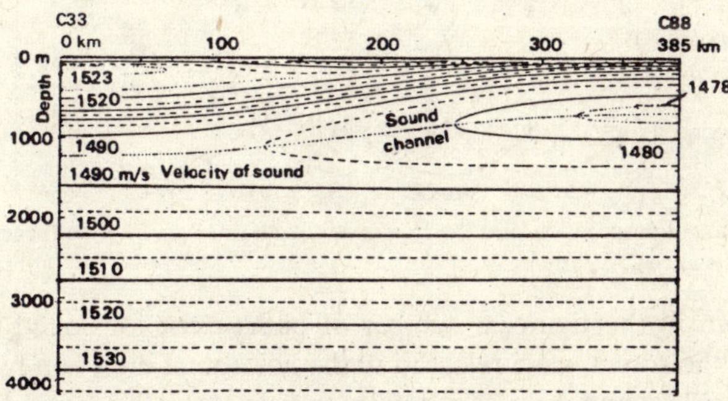


Fig. 3. Sound velocity distribution between C 33–C 88 (empirical equation).

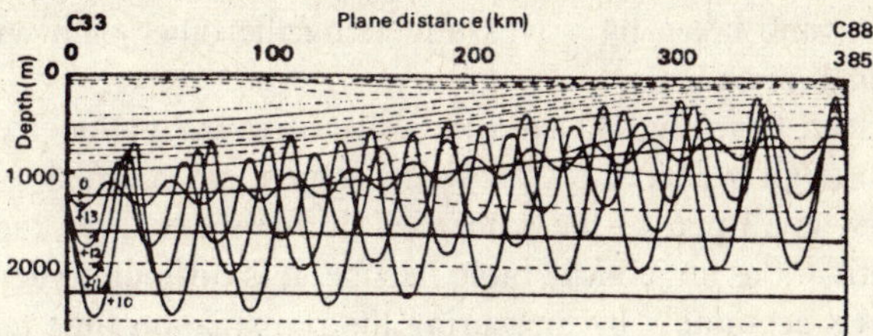


Fig. 4a. Path traversed by sound waves propagating from C 33 (1,230 m) to C 88 (690 m) (sound code 0, +13, +12, +11, +10).

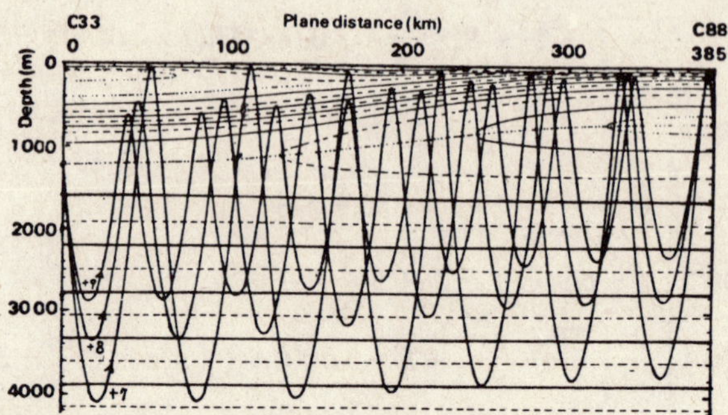


Fig. 4b. Path traversed by sound waves propagating from C 33 (1,230 m) to C 88 (690 m) (sound code +9, +8, +7).

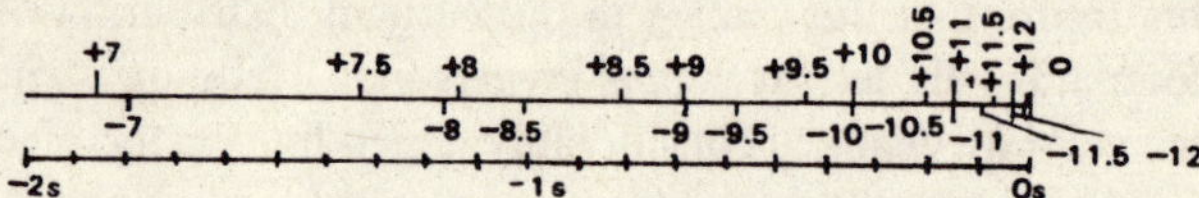


Fig. 5. Propagation time difference between different paths (sound code 0 is assumed the base).

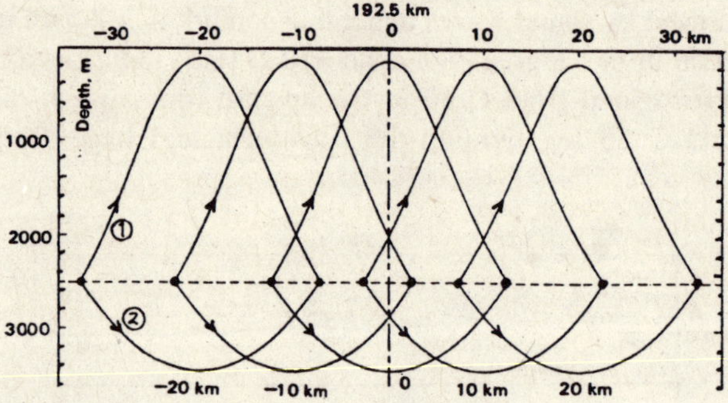


Fig. 6. Path of sound propagation near the flow axis (depth of transmitter and receiver 2,500 m, and horizontal distance 25.2 km).

As shown in the figure, a number of paths exist for sound waves, but among them the sound wave released in the horizontal direction by the sound channel of C 33 (sound code 0) generally travels along the sound channel and reaches the sound channel of C 88 last. Fig. 5 shows the difference between the propagation time taken by waves traversing the other path with the one following the path of code 0.

It can be seen from Fig. 4 that a majority of sound paths of each sound wave passes through a deeper region where acoustical scintillations are less and only a very small part of the path goes through a shallow region where acoustical scintillations are considerable. Hence, it is difficult to determine the shift in flow axis accurately by measuring the propagation time taken by the sound wave to cover a long distance.

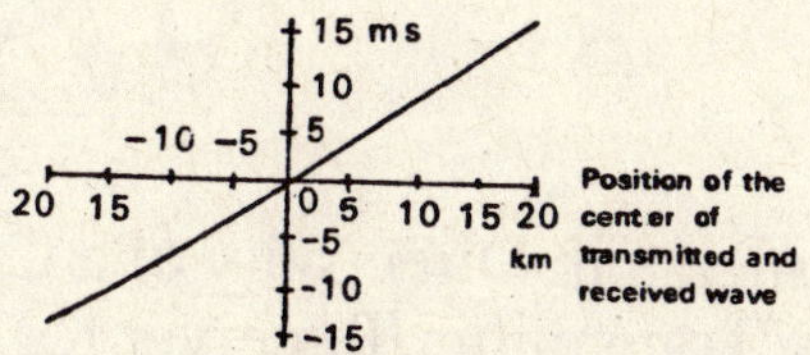


Fig. 7. Variation in propagation time of sound wave (1) of Fig. 6 (for line (2), variation is less than 1 mS).

We therefore measured the sound wave propagation time between two points of equal depth separated by a shorter distance near the flow axis. Fig. 6 shows the sound ray between these two points which are 2,500 m deep and 25.2 km apart. The sound waves which were released upward traveled up to 150 m. As the center between source and receiver shifts from flow axis, the propagation time of wave 2 varies as shown in Fig. 7.

Thus, if we could measure the propagation time with an accuracy of a few mS, then it may be possible to measure the shift in flow to a few km. Now, the propagation time of a sound wave which is released downward varies only by 1 mS as the position of transmitter and receiver is varied. Thus the measurement of the difference in the propagation time along both paths could be a very practical method to determine the shift in flow axis.

A similar result was obtained when the depth of the transmitter and receiver was varied between 1,500–3,500 m and the distance between 15.3–41.6 km.

Subsequently, we would like to determine the propagation path and time for different depths of transmitter and receiver and for different distances between them to propose a detailed method to estimate the shift in flow axis or variation in water temperature at shallow regions, etc.

# Improvements in Directivity of a Cylindrical Ultrasonic Transmitter/Receiver for Deep Seas

*Motoyoshi Okujima, Shigeo Ohtsuki and Hiroyuki Hachiya*

(Precision Engineering Research Laboratory, Tokyo Institute of Technology)
(Received: December 2, 1981)

## 1. Introduction

The ocean currents considerably affect the fisheries industry and weather. For a long-term monitoring of these oceanic currents, monitoring equipment is installed at various depths and continuous recordings are taken. To recover this equipment, sonic cutters are very useful as they can detach the weights at the seabed using sonic directivity principles (Fig. 1).

The directivity of the cylindrical transmitter/receiver used in this cutter instrument, at present, has a low sensitivity in one direction—3 lobes (Fig. 2). As a result, it is not possible to obtain response for a sound signal from the ship in the low sensitivity direction and the position (of monitoring equipment) cannot be determined. It is desirable to change this directivity into non-directivity (uniform response in all directions). In order to achieve this, a material called "synthetic foam" is coated on the inside of the oscillator and by controlling the phase of sound waves incident on the inner side, the directivity can be improved. This technique is explained below.

## 2. Experimental Details and Results

The inner surface is coated with styrol foam and the arrangement is such that the inner surface is not exposed to sound waves but only the outer surface of the oscillator. The directivity under these conditions is shown in Fig. 3. Similarly the directivity, when the outer surface is coated and only the inner surface receives sound waves, is shown in Fig. 4. Both indicate a good directivity but the response of the outer surface of the oscillator is good over a wider region. When both inner and outer surfaces are allowed to receive the sound waves, the waves incident on these surfaces interfere with each other and create a tri-lobular response characteristic, as shown in Fig. 2. When

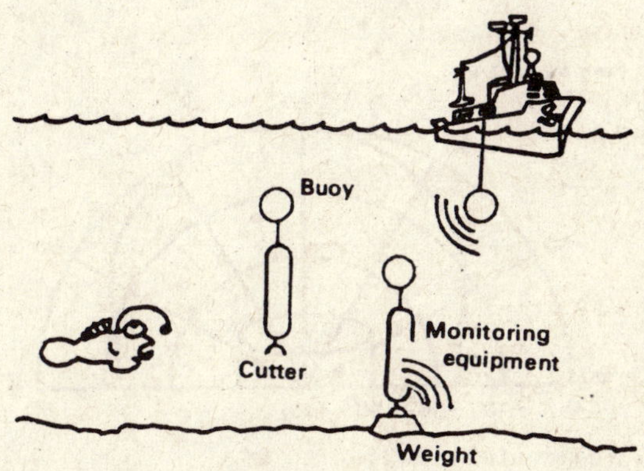


Fig. 1. Basic principles of sonic cutter equipment.

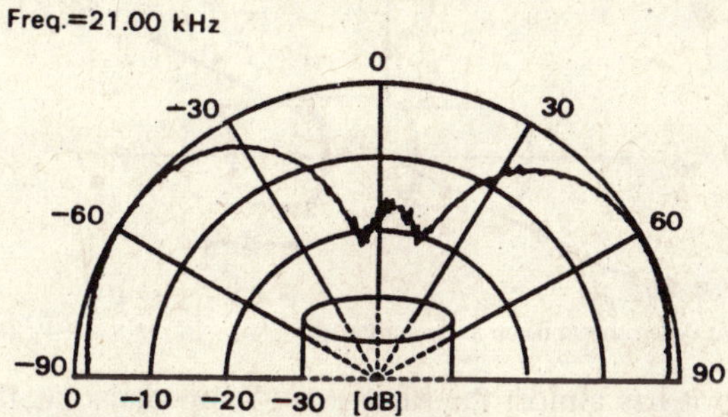


Fig. 2. Directivity of cylindrical transmitter/receivers.

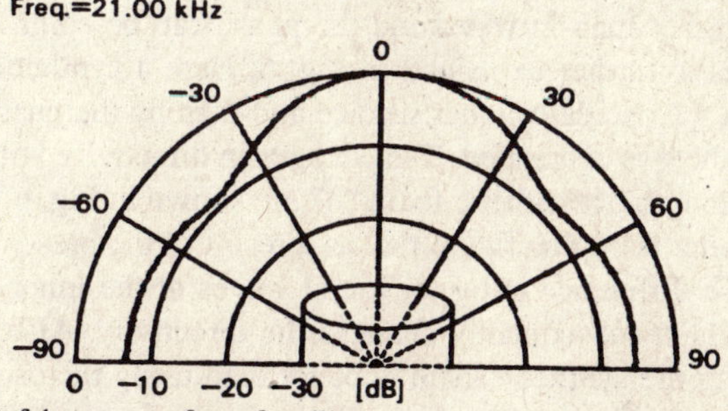


Fig. 3. Directivity of the outer surface of oscillator.

styrol foam is coated on the inner or outer surface, we do not get a better sensitivity over a wider region as shown in Figs. 3 and 4 and this styrol foam cannot withstand the high pressures at deep sea. We therefore thought of using a material known as "synthetic foam" on the inner surface of the oscillator to control the phase. This synthetic foam consists of numerous microglass balloons suspended in epoxy resin which can withstand the deep sea

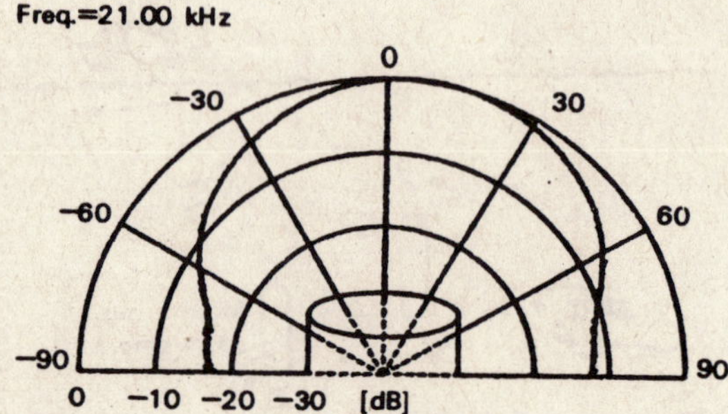


Fig. 4. Directivity of inner surface of oscillator.

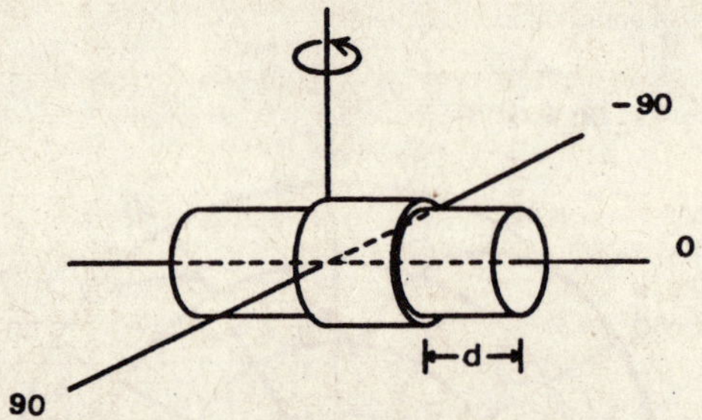


Fig. 5. Arrangement of synthetic foam and oscillator.

pressures. Also it has almost the same sound impedance as that of water so that sound waves are not reflected at the interface of the water and synthetic foam but are easily transmitted through. The velocity of sound in this foam is 2,500 m/sec, higher than in water and the phase can be controlled using this difference. This is further explained in Fig. 5. Here a synthetic foam tube is passed through the oscillator inner surface and thereby the phase of the sound wave arriving there is controlled. The changes in directivity with the variation in a protruded length of synthetic foam "d" are shown in Fig. 6.

It can be seen from this figure that as a result of the presence of synthetic foam, the phase difference between sound waves at the inner and outer surfaces varies, which subsequently changes the directivity. Also the sensitivity of the unit using the synthetic foam is better than using the oscillator alone if we consider the absolute value of sensitivity in the forward direction. Fig. 7 shows the relationship between the distance "d" and the difference between maximum and minimum sensitivity depending on the direction of incident sound wave. The smaller the difference in sensitivity the closer we are in achieving nondirectivity. It can be seen from the figure that d = 4 cm, the equipment responds most uniformly in all directions. In this case, the difference between maximum and minimum sensitivity was 8 dB. The direc-

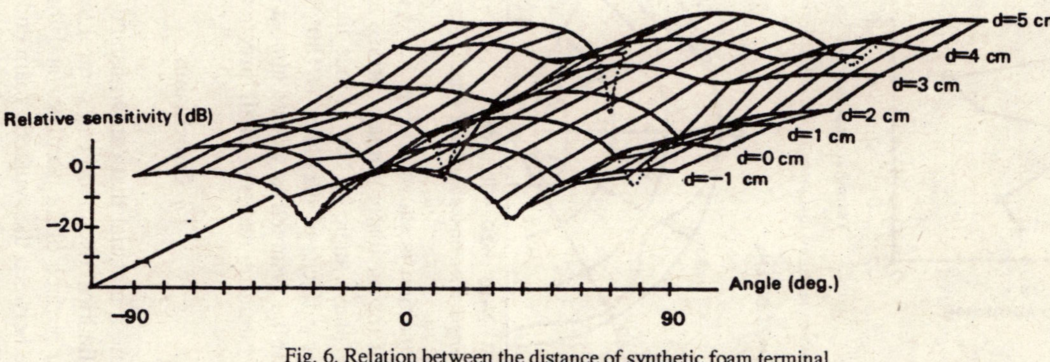


Fig. 6. Relation between the distance of synthetic foam terminal and oscillator terminal "d" and directivity.

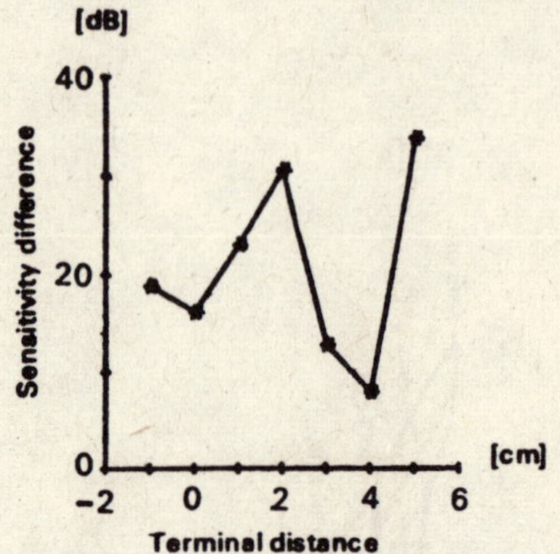


Fig. 7. Relationship between terminal distance "d" and the difference in sensitivities.

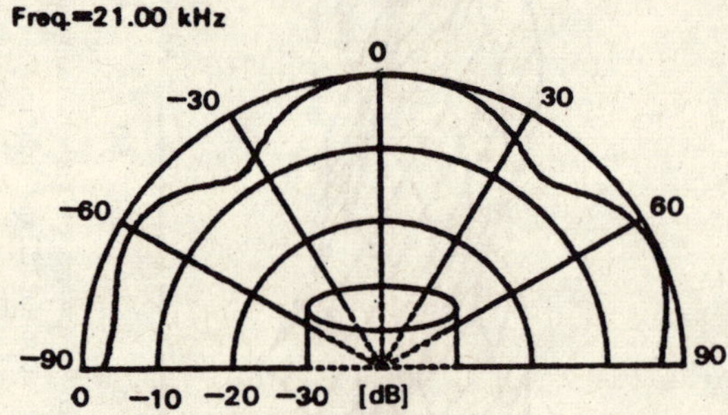


Fig. 8. Directivity property at a terminal distance of 4 cm.

tivity under this condition is shown in Fig. 8. At d = 4 cm, the distance between the center of the oscillator's inner surface and the synthetic foam end corresponds to the phase angle of 166°. Since the phase difference between the inner and outer surfaces of oscillator is 180°, the inner surface and synthetic foam end can be assumed to vibrate almost in the same phase. That is why the unit shows uniform response in all directions.

## 3. Conclusion

Synthetic foam, a material that can withstand deep sea pressures, on the inner side of a cylindrical oscillator, can control the phase of sound wave arriving at the inner surface of an oscillator and thus improve the directivity. When the distance between the synthetic foam end and the oscillator cylinder end is 4 cm, the difference in maximum and minimum sensitivity becomes 8 dB which is quite close to the uniform directivity response. Currently, studies on the shape of synthetic foam or its arrangement to achieve further non-directivity are in progress simultaneously with the theoretical analysis of the behavior.

# Measurement of the Piezoelectric Constant of Complex Piezoelectric Materials

*Motoyoshi Okujima and Shigeo Ohtsuki*

(Precision Engineering Research Laboratory, Tokyo Institute of Technology)

*Shigeru Okada*

(Shibaura Institute of Technology)

(Received: December 2, 1981)

## 1. Introduction

Ultrasonic waves are widely used for remote sensing of the ocean and to transmit signals for remote control. At present lead-zirconium titanate (PZT) is widely used as a piezoelectric porcelain material in the transmitter/receiver for the transmission/reception of ultrasonic waves.

Among the new piezoelectric materials, we feel that a composite material made from piezoelectric porcelain powder and rubber (piezoelectric rubber) has better properties to receive the ultrasonic signals over a wide frequency and hence the studies are being made to make it a practical material. Of the various properties of piezoelectrical materials, we measured the voltage output coefficient which is related to the sensitivity of a receiver. We also studied the sensitivity of a cable-shaped receiver made with the properties of soft materials. An outline of our studies follows.

## 2. Voltage Output Coefficient of a Piezoelectric Sheet

The piezoelectric sheet is a rubber sheet coated with electrodes (conductive rubber) on both sides. The system to measure the voltage output coefficient of such a material is shown in Fig. 1. Here the sample is fixed to the phone end of a ferrite vibrator which vibrates at a resonance frequency of 28 kHz. Under these conditions, the voltage output coefficient $g_{33}$ can be calculated with the following equation, using the acceleration $a$ acting on the piezoelectric rubber and the voltage $E$ that is developed between the electrodes:

$$g_{33} = \frac{E}{\left(M + \frac{m}{2}\right)\frac{t}{A}a} \tag{1}$$

where $m$ is the mass of piezoelectric rubber;
$M$ is the mass of one electrode;
$t$ is the thickness of piezoelectric rubber sheet;
$A$ is the area.

The dimensions of the piezoelectric sample used in the above measurement are shown in Fig. 2. Masses of piezoelectric rubber and electrodes were respectively $m$ = 118 mg and $M$ = 40 mg. The specific permitivity of piezoelectric rubber was 26.

The deflection of both surfaces of the piezoelectric sensor was measured by optical methods and the acceleration was determined therefrom. The results of measuring the acceleration and the corresponding voltage developed across the electrodes are shown in Fig. 3. The mean acceleration of both surface was assumed to be the acceleration acting on piezoelectric rubber. The voltage output coefficient $g_{33}$ as determined from equation (1) was $54\times10^{-3}$ Vm/N ($g_{33}$ for PZT is 12 ~ $15\times10^{-3}$ Vm/N.

### 3. Voltage Output Coefficient of Cable-shaped Piezoelectric Material

In addition to a sheet, piezoelectric rubber material was also formed to act as the insulating material between the inner and outer conductors of a coaxial cable and thus cable-shaped piezoelectric material was obtained, as shown in Fig. 4. In this case, the voltage output coefficient is determined from

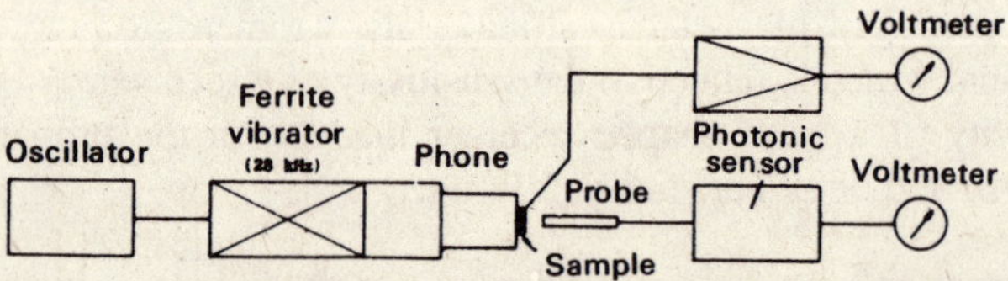


Fig. 1. System set up to measure the voltage output coefficient of sheet-type piezoelectric material.

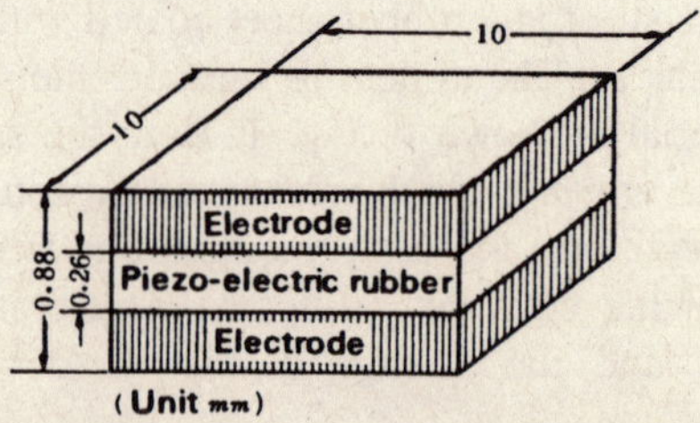


Fig. 2. Shape of the sheet-type piezoelectric material used in the measurement.

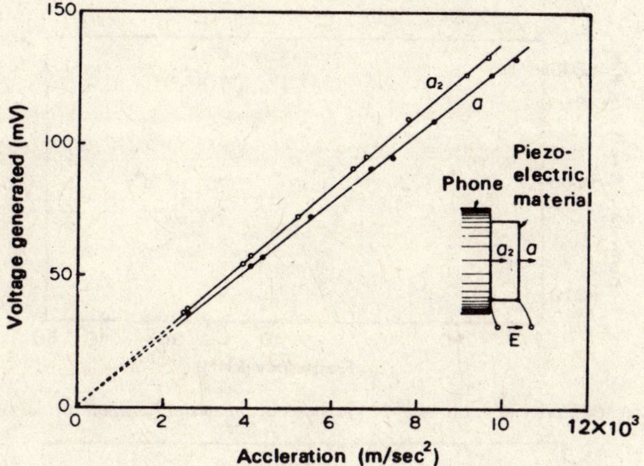


Fig. 3. Relationship between acceleration and voltage generated (temp. 24.5°C).

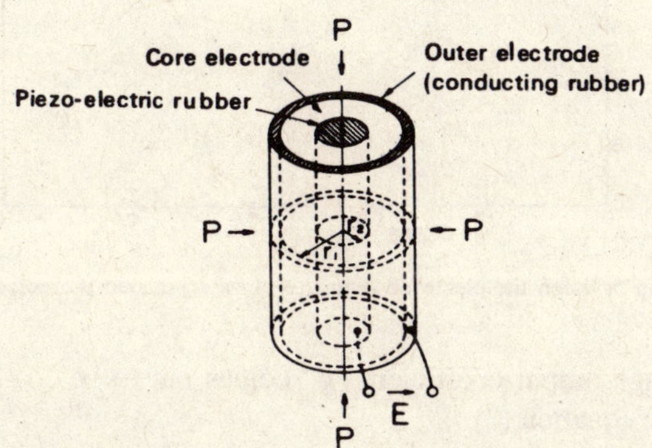


Fig. 4. Cable-shaped piezoelectric material.

the reception sensitivity. When uniform pressure $P$ is applied on the outer surface of the piezoelectric material, then the voltage output coefficient ($g_{33}$ + $2g_{31}$) $g$ is given by

$$g = \frac{1}{r_1 - r_2} \frac{E}{P} . \tag{2}$$

Here $r_1$, $r_2$ are the external and internal radii of the piezoelectric rubber, $E/P$ is the voltage sensitivity.

The piezoelectric material used for the measurement had $r_1$, $r_2$ of 2.4 mm and 0.8 mm, respectively. Here the receiver was made by cutting the cable into a length of 15 cm and the reception sensitivity was measured in the water tank. The sensitivity was found to be –205 dB (0 dB: 1V/μPa) and is fairly constant over the frequency range of 10–40 kHz, as shown in Fig. 5. Accord-

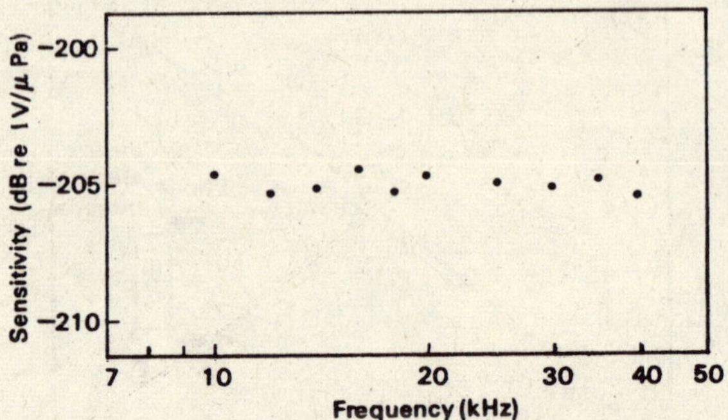


Fig. 5. Reception sensitivity of cable-shaped piezoelectric material (temperature of water 19.5°C).

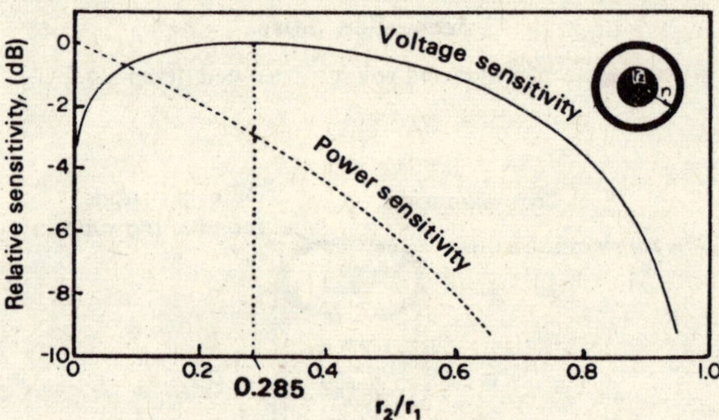


Fig. 6. Relationship between the electrode diameter of cable-shaped piezoelectric material and sensitivity.

ingly, the voltage output coefficient "*g*" comes out to be $35{\times}10^{-3}$ Vm/N by using the above equation (2).

## 4. The Optimum Diameter of a Core Conductor in a Cable-shaped Piezoelectric Material

In the cable-shaped piezoelectric material, the sensitivity changes as the diameter of the core electrode is varied. Let us, therefore, find the diameter of core electrode at which the electrical sensitivity is maximum (the sensitivity whereby maximum electrical power is obtained from a receiver). The electrical power sensitivity $M_p$ of a cable-shaped piezoelectric material of length $L$ is given as

$$M_p = 2\pi g r_1 (\varepsilon f L)^{1/2} \frac{1 - r_2/r_1}{[1\, n(r_1/r_2)]^{1/2}},$$

where ε is the dielectric constant of piezoelectric rubber,

$f$ is the frequency.

The relative value of $M_p$ varies with the ratio of the radii of the core and outer conductors $r_2/r_1$ as shown in Fig. 6 and attains a maximum value for $r_2/r_1 = 0.285$ (the voltage sensitivity at this ratio drops 3 dB below the voltage sensitivity at $r_2/r_1 = 0$). Accordingly, if the ratio of the core conductor and the outer conductor radius is around 0.2~3.3, the drop in voltage sensitivity is not much and at the same time a higher power sensitivity can be achieved. For a material used to measure the voltage output coefficient, $r_2/r_1$ was 0.32.

## 5. Addendum

As discussed above, we measured the voltage output coefficient that is necessary to estimate the sensitivity of a composite piezoelectric material for use as underwater receiver. Based on the results of these measurements, a prototype receiver is being made which can be used for directional measurements in the sea.

We would like to express our gratitude to Nippon Takushutsu K.K. for supplying the sample materials.

## REPORTS

### The Sixth Executive Committee meeting was held on December 18, 1981 at the Institute of Oceanography, Tokyo University

Preparation of a report on the study made during this year was discussed.

The administrative department informed the meeting that the additional funds for the year are being utilized as planned.

The future plans of action for the Special Research Project "Biological Processes in Ocean" were discussed. (Later, according to the spokesman, it was decided to conduct a conference on this subject at the Institute of Oceanography, Tokyo University on January 12, 1982 which is hoped to be attended by all the coordinators and members of the executive committee.)

### The Seventh Executive Committee meeting was held on January 11, 1982 at the Institute of Oceanography, Tokyo University

The points raised at the third meeting of the management committee were discussed.

Each group leader was asked to give views on the future course of studies.

The future plan of studies on the Special Research Project "Biological Processes in Ocean" were discussed with the concerned members.

Papers concerning the report of studies during this year with an abstract in English should be sent to the administrative office by February 20, 1982 and will be published subsequently. (Request letters to each group leader of the plan research project and non-plan research project were sent along with the blank manuscript papers.)

Following are the members of the committee on "Application of a Sediment Trap."

Group Leader: Shizuo Tsunogai (Fisheries Department, Hokkaido University)

Members: Nobuhiko Handa (Institute of Oceanography, Nagoya University)

Satoru Kanamori (Institute of Oceanography, Nagoya University)

Yoshiyuki Nozaki (Institute of Oceanography, Tokyo University)
Yokichi Takayanagi (Faculty of Science, Tohoku University)
Akira Azumo (Faculty of Eng., Tokyo University).

Following are the members of the committee on "Heavy Metals."

Group Leader: Satoru Kanamori (Institute of Oceanography, Nagoya University)
Members: Hiroyuki Tsubota (Department of General Science, Hiroshima University)
Magayo Murozumi (Muroran Institute of Technology)
Yukio Kodama (Institute of Oceanography, Tokyo University).

## ANNOUNCEMENTS

### Conference on Oceanographic Measurements

This conference is a part of the Special Research Project "Ocean Characteristics and Their Changes" (Chief Coordinator Prof. Kajiura Kinjiro) and is primarily meant for researchers concerned with oceanographic measurements. However, other researchers are also welcome to express their views on the future plans of development in the oceanographic measurements and related fields.

Interested persons are most welcome to attend this conference.

Date: April 13, 1982.

Place: First Conference Room in Electrical Communication Institute (Denki Tsushin Daigaku).

Program (Tentative)

Chairman: Prof. Tomoda (Institute of Oceanography, Tokyo University)

9:30: Measurement of velocity and distance using image—Mihashi, Mochizuki (Denki Daigaku)

9:50: Measurement of temperature and salt concentration (salinity) of ocean waters using microwaves—Suzuki, Arai, Watanabe and Ishikawa (Denki Daigaku)

10:10: Movement of Buoys—Azuma and Nakamura (Faculty of Eng., Tokyo University)

10:30: Response of ion sensor in sea water—Fujihara (Faculty of Science, Chiba University)

Chairman: Prof. Teramoto (Institute of Oceanography, Tokyo University)

10:50: Passive telemetry using a radar reflector with variable reflection ratios—Ogami (Biological Research Institute, Tokyo University) and Hasebe (Faculty of Eng. Science, Nichidai)

11:10: Prototype of local user's terminal and results of experiments—Segawa (Institute of Oceanography, Tokyo University)

11:30: Dropping and revolution properties of the electromagnetic flow velocity profiler—Kanenari (Faculty of Science, Hokkaido University)

11:50: Measuring apparatus for potential differences at the seabed—Tomoda (Institute of Oceanography, Tokyo University); Ogami Asada (Tokai University)

12:10: Lunch break

Chairman: Ogami (Biological Research Laboratory, Tokyo University)

13:00: Propagation of sound waves around mixed water bodies—Okushima, Otsuki and Nario (Precision Eng. Research Laboratory, Tokyo Institute of Technology)

13:20: Measurement of velocity vector space distribution using the numerical processing of wave motion—Yamazaki and Tamura (Faculty of Eng., Tokyo University)

13:40: Measurement of ocean wave heights and their distribution by sonic methods—Ando (Faculty of Eng., Tokyo University)

14:00: Observations on oceanic changes using the vertical propagation characteristics of sound waves—Takeuchi (Denki Tsuchin University)

Chairman: Prof. Yamazaki (Faculty of Eng., Tokyo University)

14:20: Basic studies regarding the transient measurements of ocean by ultrasonic waves—Moto-oka (Chiba Technical Institute)

14:40: Emission of sound waves with limited pulse amplitude into the free space and control of wave shape—Nakamura and Nakamura (Production Eng. Research Laboratory, Osaka University)

15:00: Computer simulation of non-linear sound wave propagation—Watanabe and Tobe (Faculty of Eng., Doshisha University)

15:20: Submarine information transmission using coded sound waves—Matsuda (Faculty of Eng., University of Osaka Electrical Communication)

Chairman: Okushima (Precision Eng. Research Laboratory, Tokyo Institute of Technology)

15:40: Studies on tracing low frequency sound waves under water—Ikeya and Kamakura (Faculty of Eng., Nagoya University)

16:00: Measurement of the sound impedance of soft deposits at the seabed and an analysis of reflected waves—Tsuchiya and Sugimori (Ocean Department, University of Tokai)

16:20: Impedance characteristics of a magnetic deflection-type wave transmitter under rough waves (studies regarding the possibility of using the characteristics to hydraulic pressure)—Kumamoto (Dentsu University)

16:40: Measurement of flow velocity and waves using a parametric sound source—Endo and Tokeuchi (Dentsu University)

## Obituary

Prof. Dr. Sanshiro Kawai (Doctor of Science, Asst. Professor, Faculty of Science, Tohoku University) who was acting as the coordinator for the Special Research Project "Mechanism Behind the Oceanic Changes in Waters Around Japan," Project No. 56217004, died suddenly on November 21, 1981.

Prof. Kawai graduated in 1968 from Kyoto University, Faculty of

Science, Department of Geophysics. After securing a Master's degree he worked with Kawasaki Heavy Industries for some time. Later he worked for his Doctorate in Tohoku University and was awarded the degree for his studies on the generation of tides. He started his career as an Asst. Lecturer in Ocean Physics and had been working as an Asst. Professor since January, 1981. His achievements in analyzing the mechanism that cause tides on a steady water surface were highly appreciated all over the world. He was also largely responsible for building a Tohoku model to measure ocean tides. He was invited to present a paper at the "Symposium on Wave Dynamics and Telegraphic Depth Measurement of Sea Surfaces" held in Miami in May, 1981.

He was a coordinator of the above project, a part of this Special Research Project. He was actively involved in building a system that could use the remote sensing data of the sea surface obtained through a satellite to study ocean changes and in the working group on satellite applications. He was working on the image processing of the data obtained in collaboration with the Meteorological Satellite Data Collection Center. He was also acting as coordinator for the research group on satellite applications, consisting of researchers from five universities, as a part of this Special Research Project. With profound grief we express our deep sorrow at his sudden demise.

We would like to announce that Mr. Hanawa of the Faculty of Science of the University of Tohoku is the acting coordinator for the above research project No. 56217004, the studies of which are in progress.

The following covers of Newsletters are designed by members:

No. 1: Kitagawa (Institute of Oceanography, Tokyo University)

No. 2: Kitagawa (Institute of Oceanography, Tokyo University)

No. 3: Futakuchi (Faculty of Science, Kanezawa University)

No. 4: Okushima (Precision Engineering Research Laboratory, Tokyo Institute of Technology).

Mr. Kitagawa has been responsible for the general layout.

## Guidelines for Authors

Newsletter welcomes any enterprising article.

The contents of the issue are generally classified as:

| | |
|---|---|
| Highlights: | Quick (intermediate) results of this Special Research Project |
| Bulletin: | Domestic or international news reports closely related to this Special Research Project |
| Reports: | Reports regarding the activities of each section or working group |
| Announcements: | Announcements regarding proposed meetings or conferences of each group |
| Bibliography: | Listing of papers published by the members on this Special Research Project or other reports |

Manuscripts should be written on standard 400 character size paper (handwritten material also accepted). Figures should be drawn in such a way that they can be reproduced as they are.

Manuscripts may be sent to this office through the respective group leader.

Manuscripts should reach this office at least 20 days prior to the scheduled date of publication.

The future publication schedule is March 20, 1982 and May 10, 1982.

**Note:**

When reproducing or citing material published in this Newsletter, due acknowledgment should be given. Authors' names, wherever mentioned, should also be quoted. A copy of the publication containing such a reproduction may be sent to the Editorial Section in the Administrative Office.


| | |
|---|---|
| Published by: | Administrative Office, Special Research Project "The Ocean Characteristics and Their Changes" |
| Published at: | Administrative Office,<br>Physical Oceanography Department,<br>Institute of Oceanography,<br>Tokyo University,<br>1-15-1 Minami-dai, Nakano-ku,<br>Tokyo 164<br>Tel.: 03-376-1251, Ext. 251. |

Special Research Project

# The Ocean Characteristics and Their Changes

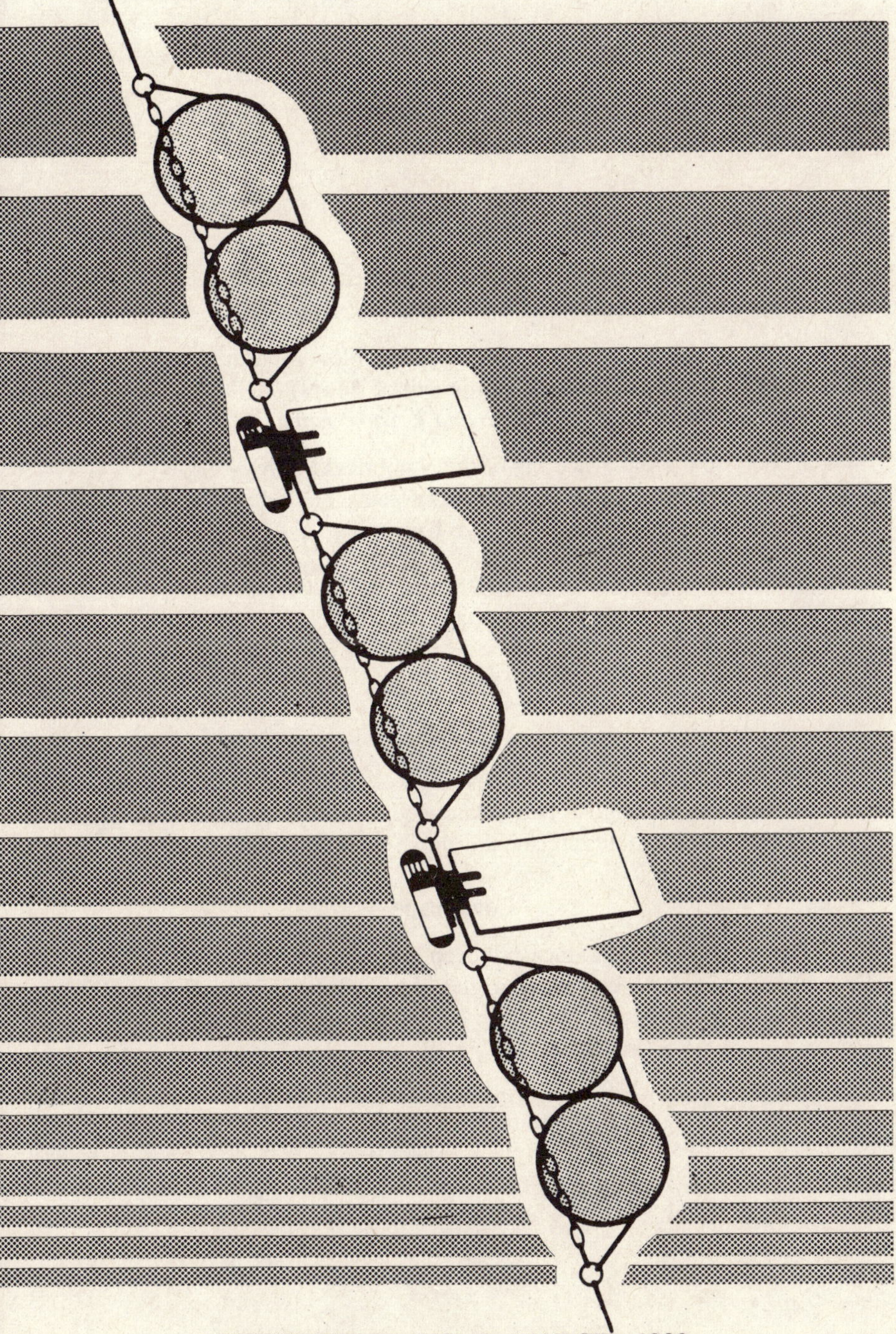

NEWSLETTER NO. 5, MARCH, 1982

# Contents



Researchers engaged in this Special Research Project, who wish to submit their papers in English and acknowledge receipt of a research grant for this special study, should indicate the above title of the project.

Cover Page

This is a buoy used to measure deep layer currents.

Floating systems consist of a glass bead chain and the flow meters are connected to the rope.

PAPERS

# Report of the Committee on "Flow Characteristics of Ocean Waters at the Continental Shelf"

*Masaaki Aota*

(Low-Temperature Research Laboratory, Hokkaido University)

(Received: January 7, 1982)

The salient structural features of water bodies in the southwest region of the Sea of Okhotsk have already been reported by Uda (1934), Kajiura (1949), Sugiura (1958), and others. According to them, the Sea of Okhotsk has three types of water bodies with different properties along the shores of Hokkaido. Thus it consists of a surface layer of low salinity which is peculiar to the Sea of Okhotsk, an intermediate cold water, and the Shyoya warm current which is actually a continuation of the Tsushima warm current and they form large tides along the shore.

This committee selected a region where these waters mix for observation, and measurements were made by the entire group on July 29-31, 1981. Prof. Kanenari of Hokkaido University (Faculty of Science) also took part in the studies under another research project (2), viz., "Development of an Electromagnetic Flow Profile Meter for Shallow Seas."

Some of the results of our observations are presented here also form the activity report of the subject committee.

## Aims and Objectives of Observations

The principal aim of these observations is to obtain the water temperature distribution in mixed water masses over the continental shelf and to explore the flow conditions. The region of observations, as shown in Fig. 3, covered about 70 miles along the coast and 30 miles off the coast centered around Monbetu.

The parameters measured were: 1) Lagrangian flow using radar buoys (4 Nos.), 2) Stationary flow measurement in a mixed region, 3) Flow measurement by GEK, 4) Detailed observations of water temperature using the improved XBT (GXBT) version developed by Prof. Kanenari, 5) Continuous measurements of surface water temperature, etc.

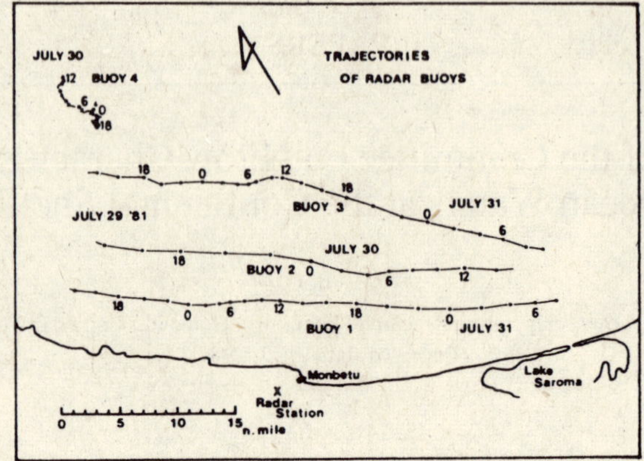


Fig. 1. Trajectories of radar buoys (July 29–31, 1981).

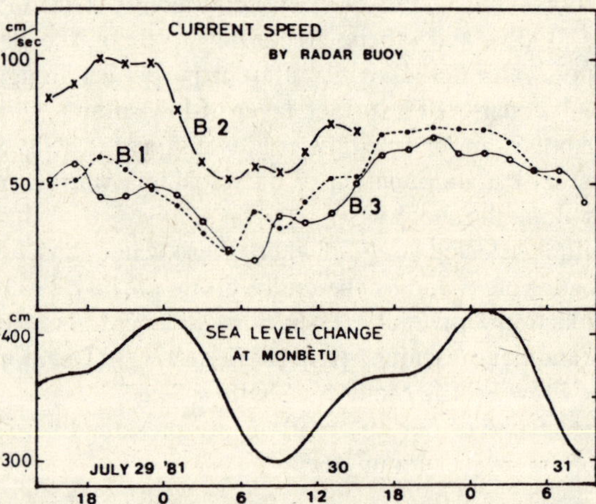


Fig. 2. Current speed calculated from the movements of radar buoys and sea level changes at Monbetu port. The buoy 20 miles off Esashi (Buoy 4 in Fig. 1) moved beyond the radar observation range after 12 noon on July 30.

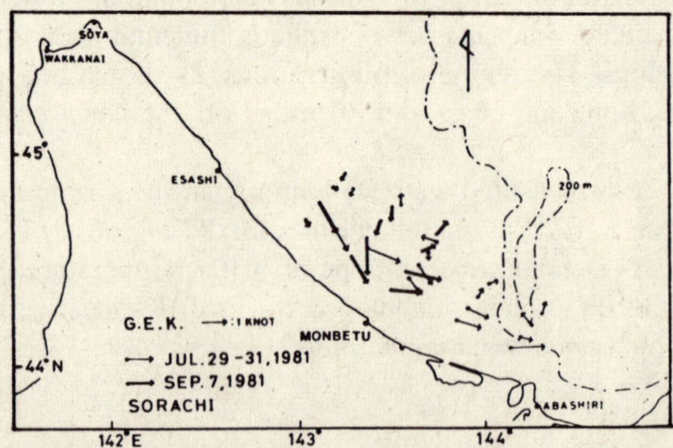


Fig. 3. Region of observations and the results according to GEK. The region of observations is the same as that used in GEK flow measurements. The figure also shows results obtained by the vessel *Sorachi* on September 7.

## Results

1. *Flow measurements using radar buoys and GEK:*

Figure 1 shows the trajectories of radar buoys (drifters) while Fig. 2 shows the Lagrangian current speeds calculated therefrom and the fluctuations of sea level at Monbetu port.

2. *Cold water bodies in the mixed region and cold water belt of the surface layer:*

A region 10-20 miles northeast of Monbetu port where flow meters were installed is a region of mixed warm and cold waters. We therefore measured the temperature of water at one mile intervals in this region using the improved version of XBT (GXBT) developed by Kanenari (Fig. 4). Similarly, the horizontal distribution of surface water temperature as obtained by GEK measurements is shown in Fig. 5.

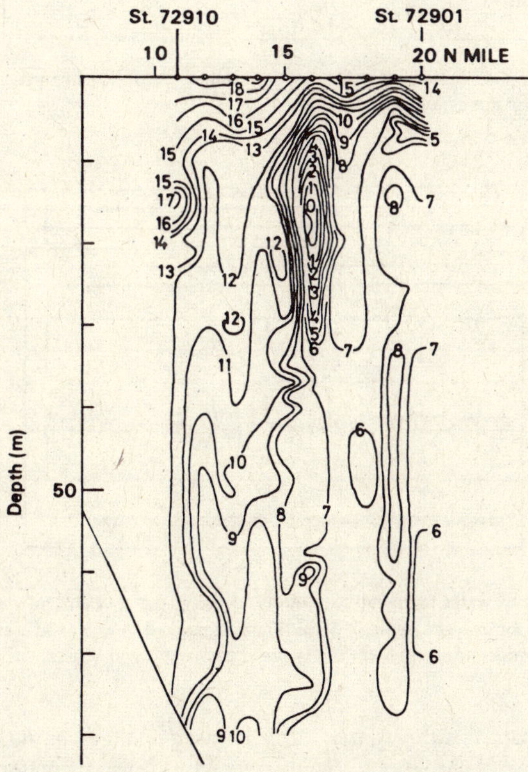


Fig. 4. Water temperature distribution in a region 10–20 miles northeast of Monbetu using the improved XBT (GXBT). The figure shows the detailed structure of temperature at one mile interval in a region 10–20 miles northeast of Monbetu. A cold water region can easily be seen at a depth of 10–30 m 16 miles northeast (abridged from original figure by Kanenari).

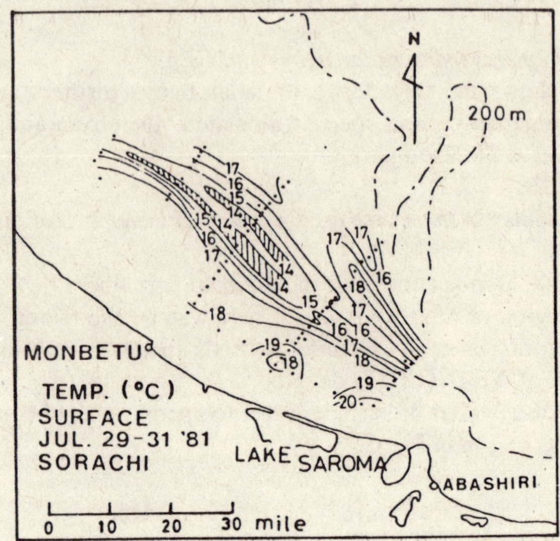


Fig. 5. Temperature distribution at surface layer obtained from measurements made during the voyage. Dark dots are the measurement points.

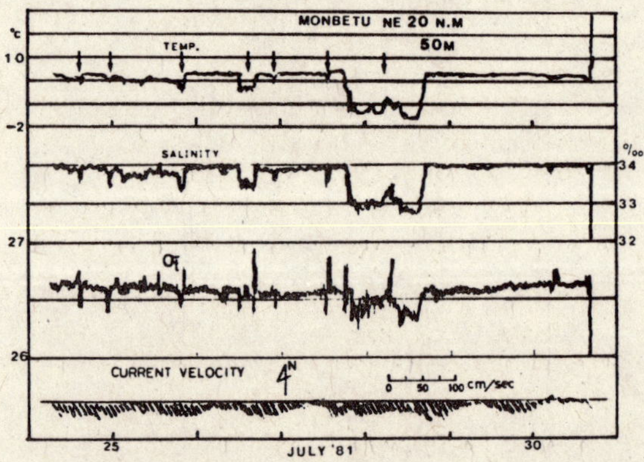


Fig. 6. Fluctuations of water temperature, salinity, density ($\sigma_T$), and current speed at a point 20 miles northeast of Monbetu at a depth of 50 m. The part marked by an arrow is supposed to be the region where low temperature and low salinity water passes through a warm current.

3. *Water temperature, salinity, and current speed changes in the mixed region:* During these observations, four Andera-type current meters were set up at 10 and 50 m depths in a mixed water region 10–20 miles northeast of Monbetu and temperature, salinity, and fluctuations in current speed were measured. Fig. 6 shows the data at a depth of 50 m at a point 20 miles away.

## Theme of Research and State of the Art

Although it was for a short period, it is still possible to observe some aspects of the complicated structure of water bodies and the corresponding fluctuations in currents in the front region of the continental reef. Of course, with such limited data, it is not possible to analyze or explain various phenomena in the continental reef part of the ocean, but at least we could see a number of peculiar phenomena which helped outline our objectives of study. We would like to plan our future observations on this basis.

1. *Generation of a cold water belt during summer (surface layer) and the cold water bodies:*

As can be seen from Fig. 5, there exists a very low temperature belt about 10-20 miles parallel to the coast during the summer. Various reports have been published about this cold belt formation. Sugiura (1958) explains this cold belt as due to the water masses of the west coast of Sakhalin that travel through the Shyoya Cape and turn southward. According to Iida (1962) this cold belt is caused as a result of balance between the Sha and Corioli of the warm currents at Shyoya and the cold offshore currents. Maeda (1968) feels the southerly winds that blow during summer are responsible for the cold belt. In their report on "Generalized studies about the Sea of Okhotsk studies about the mechanism which is responsible for the development of a cold belt in the Sea of Okhotsk," Hanzawa, et al. (1981) mention that all the above explanations may be valid at different times and different places. In addition to this cold belt, we can also see an isolated cold water mass in the mixed water region at a depth of 10-30 m as shown in Fig. 4. This type of close temperature measurement was made for the first time in this area and it is a very interesting phenomenon in the formation of the mixed water region. This author, along with Prof. Motoi of the Scientific Research Section of Hokkaido University, noticed this cold water belt and cold masses and made additional measurements of this region during mid-August to mid-September and mid-October to mid-November. A look at the three observations made during summer to autumn indicates that the isolated cold water mass is formed due to the penetration of cold waters into the warm currents at Shyoya. At present, studies are in progress to analyze the water masses, dispersion of current fluxes, convergent region, etc.

2. *Current speed in frontal region:*

The current speeds along this shore during summer are more or less clear and are shown in Figs. 1-3. Thus five miles from the shore, within the sea of the Shyoya warm currents, the speed was low—about 1 knot, whereas in the frontal region which is 10 miles off, the maximum speed was 15 knots, at 15 miles offshore it was 0.5 knots, and at 20 miles offshore the waters were steady and no current could be detected.

The relationship between the current speed in the mixed water region and the internal waves as established by the results of measurements by the GEK method is being analyzed by Prof. Nagata of Tokyo University (Faculty of Science).

Similarly the results of quantitative speed measurements (using an Andera instrument) are being studied by Prof. Inaba and Yasuda of Tokai University with particular reference to short term fluctuations.

This author, along with Prof. Motoi, noticed a sudden drop in water temperature as well as salinity (see Fig. 6) and attribute it to the passage of an isolated cold water mass as seen in observations with Kanenari's apparatus (Fig. 4) or to a shift in the frontal surface. In this context we proposed to study further the scale of the isolated cold water mass and the mechanism behind the separation or shift during the next year.

3. *Year-round measurements of sea currents:*

In addition to observations 1-3 mentioned above, the author has also measured water temperature, salinity, and current speeds at Kashibetsu near Cape Shyoya where the warm currents first enter over a period of a year, including the time when glaciers flow in. This author, along with Prof. Matsuyama of Tosui University, would like to analyze the data obtained in studying the relation between the current speed of Shyoya warm current, the outflowing water mass, and the resultant sea level at the shore.

## Reports

Discussions on the research plan on board the *Hakucho-maru* (KH-80-5) were held on December 22, 1980. The participants were Prof. Nagata of Tokyo University (Science), Prof. Matsuyama of Tosui University (Oceanography), Profs. Inaba, Yasuda, Moriya of Tokai University (Oceanography), Prof. Kanenari of Hokkaido University (Science) and Profs. Aota, Motoi of Hokkaido University (Low-Temperature Science).

The first research meeting regarding summer measurements was held on May 14, 1981 in the chamber of Prof. Nagata of Tokyo University. The meeting was attended by Profs. Nagata, Inaba, Matsuyama, Kanenari and Aota.

All members of the research group except Tabata, Hokkaido University, Low-Temperature Science (Yasuda of Tokai University and Motoi of Hokkaido University attended as observers) and Prof. Kanenari of Hokkaido University as well as Mr. Takizawa standing in for Prof. Tabata gathered on

July 26, 1981 at Monbetu. Discussions were held from July 29–31 with the Marine Security Department at Monbetu and the Patrol Vessel Section. The observational plan was discussed on board the *Sorachi.* Discussions were also held on Aug. 1 about arrangements for data collection and storage.

The results of summer observation were discussed on Oct. 14, 1981 in the chamber of Prof. Nagata at Tokyo University and at the same time, the future course of studies as well as the observational plan for next year were discussed (this was attended by Profs. Nagata, Inaba, Yasuda, Matsuyama and Aota).

Prof. Tabata of Hokkaido University (Low-Temperature Science), one of the members of this research group, passed away from a sudden illness on December 6, 1981. In his place, Prof. Ono of Hokkaido University (Low-Temperature Science) was requested to participate from next year.

We would like to conclude this report with our deep appreciation of the services of Prof. Tabata who left for his heavenly abode in the midst of this study.

# Cruise of the *Keiten-maru* for Long-term Flow Measurements of Deep Layers (Recovery Equipment for Buoys)

*Prof. Tadao Takahashi*

(Department of Fisheries, Kagoshima University)

*Prof. Akio Maeda*

(Faculty of Eng., Kagoshima University)

*Prof. Keisuke Taira*

(Institute of Oceanography, Tokyo University)

(Received: January 16, 1982)

As part of the Special Research Project "The Ocean Characteristics and Their Changes" (Chief coordinator, Prof. Kajiura), studies were undertaken from the fiscal year 1981 on the dynamics of the Kuroshio region which lies south of Japan. In this study, the deep layer currents in the Kuroshio region have been observed continuously for a long period and the structure as well as fluctuations of the currents recorded (Chief coordinator: Prof. Toshihiko Teramoto).

For the proper progress of this type of study, it is necessary to send a suitable vessel on a voyage periodically in order to recover the installed current meters. The first equipment was installed on April 8, 1981 near the Izu Ridge (IR2) by a research vessel the *Hakucho-maru*. Subsequently the same vessel installed equipment off the Kii Peninsula (CS1, CS2, CS3, CS4) on April 28 and 29.

The next retrievable equipment was installed in the sea in October, 1981 by the *Keiten-maru* (captained by Watami, 855 tons), a fishing exercise boat of the Department of Fisheries, Kagoshima University. The main research crew on this voyage consisted of Profs. Takahashi and Ichikawa from the Department of Fisheries, Kagoshima University, Profs. Maeda and Sakurai of the Faculty of Engineering of the same university; Profs. Taira, Fukazawa, Kawawata and Kitagawa of the Institute of Oceanography, Tokyo University.

In addition to them two post-graduate students and ten graduate students from Kagoshima University also participated.

On October 19, 1981, we loaded the boat with measuring equipment and fuel on the Taniyama dock-5 of Kagoshima port but at the same time typhoon 24 was moving northwest at a speed of 15 km per hour near 19°30′N, 132°50′E and therefore we left the port in the early morning of October 20, to avoid the typhoon, and sailed toward Shimabara. On the day typhoon 24 was moving with a central pressure of 950 millibar in a north-northwest direction around 21°20′N, 128°40′E but it changed its course to north-northeast thus covering the region of our studies.

While anchored at Shimabara port to avoid the typhoon fury, we prepared mooring systems by checking the buoy, adjusting the current meters or preparing the anchors, arranging the rope, etc. to minimize the installation time of the equipment.

We returned to Kagoshima port on October 25, loaded the ship with food and in the early morning on the next day began our journey to Kii Peninsula.

On October 27, the mooring systems at Kii Peninsula (CS1, CS2, CS3, CS4) were recorded and on October 28, they were again installed at three points (CS2, CS3, CS4). On the 29th, the buoy system at Izu Ridge (IR2) was recorded and reinstalled. On the following two days, i.e., October 30 and 31,

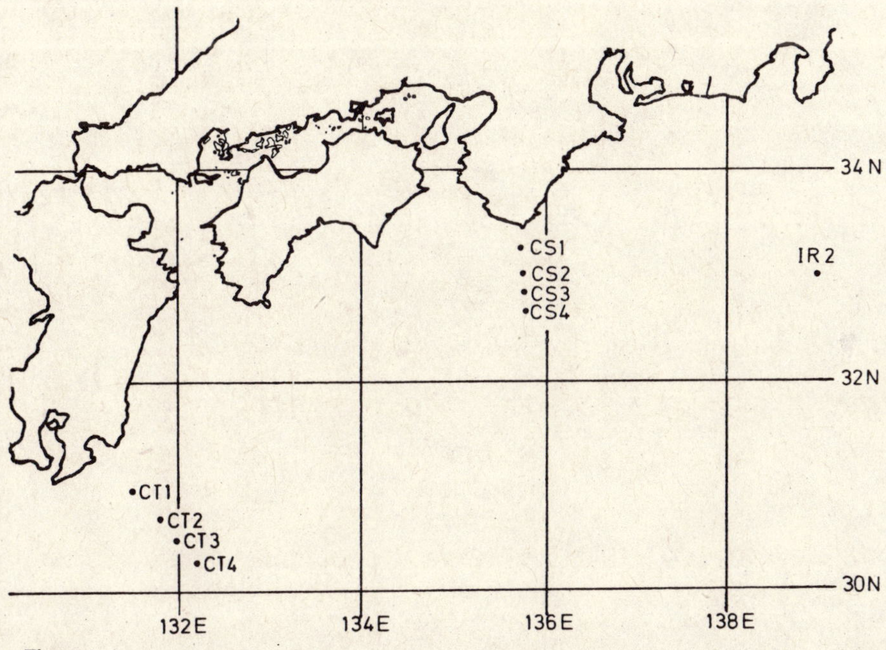


Figure 1.

new equipment was installed off Cape Toi (CT1, CT2, CT3, CT4) and CTD measurements were made on that line. The current meters were installed at depths of 1,000–5,000 m. All these points are shown in Fig. 1.

We approached Kagoshima port in the early morning on November 2, but could not berth the vessel since there was a low pressure belt in the west coast of Kagoshima and there were strong southerly winds with speeds of 30 m/sec. Therefore, we halted on the northern side of Japan Petroleum's warehouses. Our captain waited on the deck for a long time for the winds to subside. At last on November 3, we could berth at the Taniyama dock-5 of Kagoshima port.

We plan to recover the buoys installed at this time, i.e., October 1981, in August, 1982, using the same *Keiten-maru* and reinstal them at the same time. The next recovery and installation could be planned in May, 1983.

Thus we plan continuous measurements of the deep layer currents in this region, at least to the end of fiscal year 1983. The results of such long-term measurements will hopefully answer a number of questions.

## Report on the First Meeting of the Numerical Methods Group held on January 22, 1982 at the Conference Hall of the Institute of Oceanography, Tokyo University

Members attended (positions not mentioned) are:
Imawaki* (Faculty of Science, Kyoto University)
Masuda* (Department of Application Research, Kyushu University)
Sekine* (Faculty of Science, Tohoku University)
Kubota* (Faculty of Engineering, Tokai University)
Suginohara*, Yoon*, Miura and Kitamura (Faculty of Science, Tokyo University)
Taira, Sugimoto, Fukazawa*, Kawabe, Kutsuwada and Kaneko (Institute of Oceanography, Tokyo University)
Emura and Sasaki (Marine Technology Center)
Tsuchide (Department of Science and Technology)
Ishii (Marine Safety Channels)
(*Members of the Numerical Methods Group)

Preliminaries (10:00-11:30)

Only the group members met and discussed the proceedings of the meeting to be held in the afternoon.

Meeting:

1) Welcome: Nobuo Suginohara (13:00-13:10)
2) Results of the double layer model: Nobuo Suginohara (13:10-14:30)

It is possible to study the effect of the seabed profile and the inclined pressure fluctuation phenomena with this model. Since it is necessary to take into consideration the circulation between subtropical and subfrigid zones, we considered a model that would represent the west coast and the seabed profile in a broad sense extending 5,000 km east-west and 3,600 km north-south. Thus the model considers the continental slope along the west coast and the seabed profile at Izu Ridge. During the preparatory experiment the grid interval was 50 km and the phenomena were studied under different conditions, in terms of the horizontal mixing coefficient; the basic parameter indicating the phenomena. As a result we understood that it is necessary to make this coefficient lower than $3 \times 10^6$ cm$^2$/s if we want to regenerate the strong currents flowing along the south coast and the offshore warm water mass. Under steady-state conditions (including the statistical steady state) no strong currents exist in the lower layer of the Kuroshio current which flows along the slope and accordingly we can determine the course of the Kuroshio current by apparently ignoring the seabed profile effect. Subsequently the computations were repeated by reducing the grid size to 25 km (horizontal mixing coefficient of $3 \times 10^6$ cm$^2$/s. For a 50 km grid size, the computations were made with a horizontal mixing coefficient of $3 \times 10^7$ $cm^2/s$ as the initial value. As a result we obtained a pattern corresponding to shallow water waves if Kuroshio following two cold water masses regularly existing along the south Japanese coast. This cold water mass manifests a periodic separation from the cold whirls created by fluctuations in inclined pressure. Significant movements in the lower layer caused by these cold whirls were observed. During this experiment, the parameters are varied manually and hence it is not possible to establish the correlation with the above phenomena. However, one can say that this mixing which occurs for some reason is responsible for the generation of a cold water mass. In this model, the mixing following the periodic separation of cold water masses of upper currents are different from those in an isostatic model and they are linked with the cold water masses of the lower layer.

Within the Kuroshio current, the warm whirls and cold whirls were different from the wave pattern mainly due to inclined pressure fluctuations. However, we could find no trace of this separated cold whirl which extends southward to Izu Ridge. Accordingly this problem also needs to be solved in future, along with the shore separation phenomena occurring offshore in the Kuroshio region (which does not recur in an isostatic or a two-layer model).

3) Isostatic model (reduced gravity model): Yoon Jong Hwon (14:40-16:00)

In this model, the lower layer is assumed to be steady and therefore it is not possible to determine the effect of the seabed profile or inclined pressure fluctuation phenomena. We considered the north Pacific Ocean with its subtropical circulation as the model sea, selected the grid interval as 1/4 and represented the seashore profile as close to the original as possible (the depth of 200 m was considered the boundary between ocean and continent). Izu Ridge is represented as an island. The horizontal mixing coefficient was assumed to be $10^6$ $cm^2/sec$ and the wind stresses within the sea were assumed in such a way to cause a spell drop of 20, 30, 50 (Sv) and the response of the Kuroshio under those conditions was studied. At this value of horizontal mixing coefficient, any nonlinear effects are excluded and a strong current is formed along the south coast of Japan corresponding to the Kuroshio current (in the offshore region a reactive current following the warm water masses also developed). At 20, 30 Sv the Kuroshio current flowing along the south coast of Japan indicated a serpentine path following the cold water mass. However, this serpentine path (huge cold water mass) changed into a linear pattern after continuing for $3^1/2$ years with 20 Sv and after 6 years with 30 Sv winds. Thus in this experiment a normal wind stress is applied to a steady ocean but in all cases the formation of shallow water wave was rapid (within two months) and died down very slowly. With 50 Sv winds no shallow water wave developed and only a linear pattern continued. However, in all these cases, when a linear pattern exists, the Kuroshio current flows around Hachijoshima. The above results indicate that the same phenomenon is responsible for the formation of shallow water waves as at Shikoku strait (Kyushu and Izu Ridge can be considered its two extremities). However, this explanation leaves some problems that should be studied hereafter. At present in this experiment, we studied the response of the Kuroshio currents to the steady wind stresses, but it is also necessary to study the response to periodically changing wind stresses. Similarly, it is necessary to review the sensitivity of the Kuroshio current to mixing forces under each of the above cases (with no regard to their cause).

So far we have mentioned the various aspects of the problem and the salient features can be summarized as follows:

The shallow water wave pattern obtained during the isostatic model could be considered a shallow water wave pattern under linear propagation, particularly at 20 Sv.

With the isostatic model, some sort of warm water mass exists in a region corresponding to southwesterly islands and these islands are greatly affected by those currents. This could also affect the tides in Kuroshio within the Shikoku Sea.

The Kuroshio shallow water waves seen in the isostatic model may be explained on the basis of Rosberry waves developed south of Kyushu as a result of the jutting land.

These conclusions were extremely useful for future numerical model calculations.

4) Related topics (16:10-18:00)

Evaluation of energy in the Kuroshio region (numerical experiment)—Emura.

The biomodality of the Kuroshio current (analytical theory)—Masuda.

Results of current measurement off Cape Shio—Fukazawa.

Small shallow water waves on the east Kyushu coast—Sekimoto.

5) Review (18:00-19:00)

The discussions centered round the course of the Kuroshio current during linear propagation. In the numerical model only the course traversing south of Hachijoshima was seen under a linear flow, but it is yet to be seen to what extent such a course is followed by the actual Kuroshio current.

(Reported by J.-H. Yoon)

**Studies on Basic Measurement Techniques Related to Oceanographic Data**

The 3rd Study Meeting

The third study meeting on the above topic was held as part of the special research project (1) as per the following program:

Place: Hakuen Building, 3rd Floor, Seminar Room, Doshisha University.

Date: December 14, 1981 (Monday) and December 15, 1981 (Tuesday).

Topics discussed:

1) Radiation of sound wave with limited pulse width in free space—Toshiaki Nakamura and Akira Nakamura (Osaka University).

2) Parametric receptacle array—Kamakura and Ikeya (Nagoya University).

3) Measurement of current speed and waves using a parametric sound source—Takeuchi (Dentsu University).

**Studies on Basic Measurement Techniques Related to Oceanographic Data**

The 4th Study Meeting

Fourth study meeting of the group on "Studies regarding basic measurement techniques related to Oceanographic Information" was held as per the following program:

Place: Room No. 5315, West II Building, Denki Tsushin University (Electrical Communication University) (1-5-1 Chofugaoka, Chofu-city, Tokyo).

Date: January 26, 1982 (Tuesday) and January 27, 1982 (Wednesday).

Topics discussed:

1) Waveform control of large amplitude sound pulses—Nakamura and Nakamura (Osaka University).

2) Simulation of propagation of sound waves with finite amplitude—Watanabe (Doshisha University).

3) Studies regarding the course of propagating a low frequency sound beam under sea—Kamakura and Ikeya (Nagoya University).

4) Measurement of oceanic current speeds using the Doppler dispersion of sound waves—Takeuchi and Endo (Dentsu University).

5) The sound impedance of materials at the sea bottom and analysis of reflected wave—Tsuchiya and Sugimori (Tokai University).

6) Basic studies of different sensors using underwater sound receivers—Kumamoto (Dentsu University).

7) Studies on the transmission of underwater information using a coded sound pulse train—Matsuda (Osaka Dentsu University).

## Conference on Tomorrow's Geoscience—A Report

This conference was held at the seminar hall of Hachiego University for two days on January 24 and 25, 1982. The objective of the conference was to provide a self education opportunity to young scientists and provide a platform for all scientists to exchange their views. Eminent scientists in the field of the geosciences were invited to deliver lectures followed by open discussions.

In all 59 persons from 18 national and private universities, three Museums, a Geological Survey Department, etc., engaged in the field of geopaleontology, geophysics, geochemistry, etc., participated in the conference. The topics for discussion were selected from marine micro-paleontology, marine geology, marine chemistry, seabed physics, etc., and the theme was future study programs in these fields. Both speakers and audience participated actively in the discussions on the contents of the addresses and the problems yet to be solved in each field. This conference, along with the senior scientists conference held on the 24th provided an opportunity to establish a good communication between the scientists, irrespective of their field of research or age.

In the concluding session, all points were summarized and many people expressed their satisfaction regarding the Geoscience Symposium and the themes of this conference. Most of them expressed their desire that a similar symposium should be held next year as well. The same people who coordinated this conference were requested to make arrangements for next year.

Lecture program:

January 24

Film show

Akiba Fumio (Petroleum Resource Development & Technology Research Institute):

Analysis of the Paleoenvironments of the upper part of new layers using sea weeds.

Fujioka Kantaro (Institute of Oceanography, Tokyo University):

Eruption history of an island arc as seen from the ocean sediments with the northeast Japanese arc as example.

Koizumi (Osaka University):

The state of the art of paleoenvironmental analysis using sea weeds, Bacillariaceae or diatonic, and the problems involved.

Open discussions:

January 25

Kato (Tokai University):

Geochemistry of ocean sediments as seen from gap water.

Masuzawa (Nagoya University):

Fluctuations in the oxidation-reduction cycle of waters in the Japanese seabed as seen from gap water and sediments.

Yamano (Earthquake Research Center, Tokyo University):

Measurement of heat flow from the earth crust in the ocean and its analysis.

Kodama (Earthquake Research Center, Tokyo University):

History of the formation of the Philippine Sea as estimated from paleomagnetic data.

Matsumoto (Institute of Oceanography, Tokyo University):

Abnormalities in the gravitational force and the structure of ocean floor.

Takigami (Faculty of Science, Tokyo University):

$^{40}$AR-$^{39}$AR age measurements of marine rocks.

**Summary and Discussions Regarding Future Studies**

Conference on Satellite Application—A Report

Date and time: January 27, 1982 (Wednesday)—10:00-17:00.

Place: Conference Room of the Institute of Oceanography.

Participants: Teramoto, Segawa, Taira, Fukazawa, Kawabe (Institute of Oceanography, Tokyo University); Takagi (Biological Research Laboratory, Tokyo University); Toba (Science Faculty, Tohoku University); Nishi (Science, Kyoto University); Umaya (Applied Research Laboratory, Kyushu University); Suzuki (Dentsu University); Yokoyama (Iwate University); Hirozawa (Space Research Center); Sugimori (Dept. of Oceanography, Tokai University); Isuchide (Directorate of Science and Technology).

Discussions:

First the results of studies conducted so far were presented.

1) Taira, Toba, and Nishi mentioned the necessity, usefulness, and problems encountered while using the IR Image data from the viewpoint of satellite data users.

2) Sugimori and Yokoyama indicated the present condition of processing the data in a visible region and IR Image, their applications, and the problems encountered during studies.

3) Segawa explained the use of DCS (Argos plan), particularly the signal reception or data processing and indicated the possibility of using the same for studies and their effectiveness.

4) Takagi's talk was centered around the system he developed based on his experience on data reception from NOAA satellites, etc.

5) Hirozawa explained the properties of a microwave dispersion meter based on his experiment.

6) Suzuki illustrated the characteristics of MR (Microwave radiation meter) signal reception and the effect of surrounding conditions (e.g., humidity) using the results of his experiment.

These were followed by discussions regarding the steps to be taken for propagating the use of satellite data in oceanographic studies. It was decided, as a result of discussions, that the analysis of IR Image, etc., and its effectiveness, etc., should be studied as the first step. For this purpose it is necessary to train young scientists who can be engaged in such studies. Similarly, the system hardware group should also be strengthened so that the university administrators (Ministry of Education) can provide even more funds for data collection.

**Meeting of the Committee on Sediment Trap Application**

Date and time: February 13, 1982, 10 a.m.

Place: Conference Room in the Institute of Oceanography, Tokyo University.

Participants: Azuma, Hattori, Handa, Saino, Masuzawa, Takayanagi, Tsunogai and Tanimura.

Program:

1) Inaguration

2) Self-introduction

3) Program followed so far (Tsunogai): The progress made by this committee and the aims of study were explained

4) Problems encountered and solutions: The aspects being studied by each member were first presented and then discussed: i) problems encountered while quantitatively analyzing the granular masses using a sediment trap; ii) problems encountered during the sediment trap experiment, including the buoy system; iii) plans regarding the preparation of sediment traps

5) The next meeting was fixed to be around June.

(Reported by S. Tsunogai)

**Reports by Administrative Office**

The Third Meeting of the Managing Committee:

Date: February 12, 1982 (Friday).

Place: Institute of Oceanography, Tokyo University.

Representatives of each committee reported the program and results of their studies during this year; with 20 minutes allotted for each group.

Next year a conference is proposed to publish results on the research project and general research work (tentative schedule: two days, June 29 (Tuesday) and 30 (Wednesday), 1982 at the Institute of Oceanography) and it was also decided to publish a book on the interim results obtained.

Thorough discussions were held on the method adopted for this special research project and the program to be followed next year.

## ANNOUNCEMENTS

### Symposium on the Structure and Movements of the Philippine Sea

Date and time: March 25, 1982 (Thursday), 10:30-17:00
March 26, 1982 (Friday), 09:30-17:30.
Place: Institute of Oceanography, Tokyo University.

Program:

First day: March 25 (Thursday), 10:30-17:00

Welcome

Chairman: Kobayashi (Institute of Oceanography, Tokyo University)

Keynote address:

1) Structure and topography of the Daito Ridge—Tokuyama and Nakata (Institute of Oceanography).

2) Abnormalities in the earth's magnetism observed in space up to 500 m around the north-south Daito Islands—Iguchi, Isezaki, Nishimura, Hyoto and Yasukawa (Faculty of Science, Kobe University).

3) Geology of the surface layer of the north-south Daito Islands and the formation of the Dromites—Konishi and Obara (Faculty of Science, Kanezawa University).

Lunch break.

Chairman: Tokuoka (Faculty of Science, Shimane University)

Keynote address:

4) Selection of topics—Matsuda (Faculty of Science, Kobe University).

Discussions: Mizuno (Directorate, Geological Survey), Kobayashi (Institute of Oceanography, Tokyo University), Misawa (Institute of Oceanography, Tokyo University), Aoki (Institute of Oceanography, Tokyo University).

Break.

Chairman: Motosaki (Directorate, Geological Survey)

Keynote address:

5) Expansion of coastal region—Fujii and Ouchi (Faculty of Science, Kobe University).

6) Structure of the deep sea region along the south sea trough and stress fields—Fukao (Faculty of Science, Nagoya University) and Kamogawa (Distress Center).

7) Selection of topics—Sugi (Geological Research Laboratory, Tokyo University) and Yamano (Geological Research Laboratory, Tokyo University).

Discussions: Tomoda (Institute of Oceanography, Tokyo University).

Fujita (Faculty of Science, Osaka Municipal University).

Second day: March 26 (Friday), 09:30-17:30

Chairman: Murauchi (Faculty of Science, Chiba University)

Keynote address:

8) Distribution of earthquakes in the northern central Philippine Sea—Yamazaki (Faculty of Science, Nagoya University).

9) A novel viewpoint of the plateau creep theory as applied to southwest Japan—the geo-stress phenomena—Shiono (Science, Osaka Municipal University).

10) Structure of the south sea trough as observed from reflected earthquake recordings—Tamano, Ikebe, Toba and Aoki (Petroleum Resource Development Corporation).

11) Structure and topology of the region around southwest Japan—Okuda (Petroleum Corporation).

Lunch break.

Chairman: Minamijumo (Earthquake Research Institute, Tokyo University)

12) Igneous rock formation in the fore arc and formation of sedimentary rocks—Miyake and Hisatomi (Science, Kyoto University).

Keynote address:

13) Selection of topics—Kagami (Institute of Oceanography, Tokyo University), Sakurai (Hydrographical Department of the Marine Safety Board), Taira (Science, Kochi University).

Break.

Chairman: Shigi (Science, Kyoto University) and Kagami (Institute of Oceanography, Tokyo University)

14) Conclusion.

Requests for the Newsletter should indicate the number of copies desired and sent to the Administrative Office.

## Guidelines for Authors

Newsletter welcomes any enterprising article.

The contents are generally classified as:

| | |
|---|---|
| Highlights: | Quick (intermediate) results of this Special Research Project |
| Bulletin: | Domestic or international news reports closely related to this Special Research Project |
| Reports: | Reports regarding the activities of each section or committee |
| Announcements: | Announcements regarding proposed meetings or conferences of each group |
| Bibliography: | Listing of papers published by the members on this Special Research Project or other reports. |

Manuscripts should be written on standard 400 character size paper (handwritten material also accepted). Figures should be drawn in such a way that they can be reproduced as they are.

Manuscripts may be sent to this office through the respective group leader.

Manuscripts should reach this office at least 20 days prior to the scheduled date of publication.

The future publication schedule is May 10, 1982 and June 25, 1982.

**Note:**

When reproducing or citing the material published in this Newsletter, due acknowledgment should be given. Authors' names, wherever mentioned, should also be quoted. A copy of the publication containing such a reproduction may be sent to the Editorial Section of the Administrative Office.

| | |
|---|---|
| Published by: | Administrative Office,<br>Special Research Project "The Ocean Characteristics and Their Changes" |
| Published at: | Administrative Office,<br>Physical Oceanography Department,<br>Institute of Oceanography,<br>Tokyo University,<br>1-15-1 Minami-dai, Nakano-ku,<br>Tokyo 164<br>Tel.: 03-376-1251, Ext. 251. |

Special Research Project

# The Ocean Characteristics and Their Changes

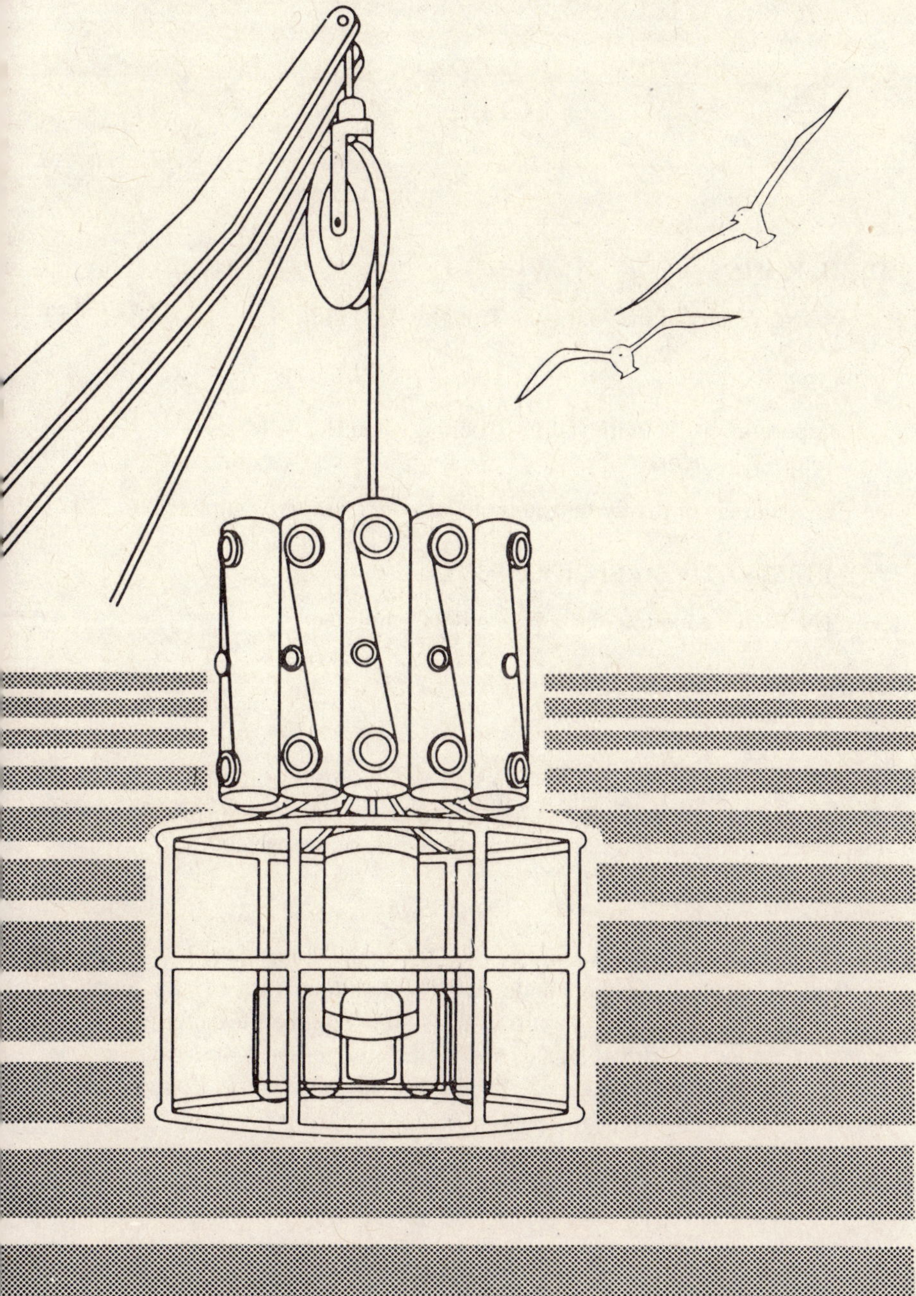

NEWSLETTER NO. 6, JUNE, 1982

## Contents



Researchers engaged in this Special Research Project, who wish to submit their papers in English and acknowledge receipt of a research grant for this special study, should indicate the above title of the project.

### Cover Page

The cover page shows a model of CTD equipment required to obtain the vertical distribution of the conductivity and temperature of sea water by measuring C, T along each depth. In this equipment the measured values are transformed into electrical signals and sent to the research vessel via coaxial cable (Courtesy Prof. Kitagawa, Institute of Oceanography, Tokyo University).

## HIGHLIGHTS

# Warm Waters of the Marianas Trough

*Yoshio Horibe*

(Institute of Oceanography, Tokyo University)

(Received: May 13, 1982)

During its voyage, the *Hakucho-maru* (KH-82-1 short voyage—Cepheus Expedition) surveyed the Marianas trough and testified chemically to the fact that hot water springs from the seabed at a depth of 3,600 m, which corresponds to the Life valley of the Marianas trough.

The main locations so far known for the hot springs coming out of the deep seabed are the Red Sea and the East Pacific Rise. From the viewpoint of plate techtonics, both these areas lie in the dispersion type plateau region where a new seabed is continuously being formed as a result of volcanic activity. The Marianas trough is in the Back Arc Basin on the west side of a trench down the Pacific plateau and is geologically different from the above two locations. However, hot water masses with some periodicity are observed even in the Marianas trough, which is testimony to the continuous expansion of the ocean bed. During 1981, Lonsdale of the Scripps Institution of Oceanography observed abnormalities in water temperatures near the seabed at ten points which lie along the central volcanic mountain peaks of the Marianas trough, using a CTD apparatus fixed to a deep tow. Since then scientists, mainly American and Japanese, have planned to survey the seabed in the Marianas trough using a submarine *Alvin.* The chemical evidence we obtained this time of the presence of hot springs would greatly support the execution of the above research plan.

One useful proof of the presence of hot springs at the deep seabed would be to detect the presence of $^3$He in the sea water. Helium can be found with mass numbers 3 and 4 in isotope form, i.e., $^3$He and $^4$He. The ratio of these two isotopes in the atmosphere ($^3$He/$^4$He) is $1.40 \times 10^{-6}$ and is almost the same in sea water. However, at the interface of the open plateau, i.e., in the region where the mantle material is being discharged from the deep seabed, the concentration of $^3$He is higher. Fig. 1 shows a cross section of $^3$He concentration distribution in the sea water around the East Pacific Rise at 15°S. In the earth's atmosphere the amount of $^3$He formed by mechanisms other than the disintergration of tritium ($^3$H) is extremely small. Generally, since $^3$H does

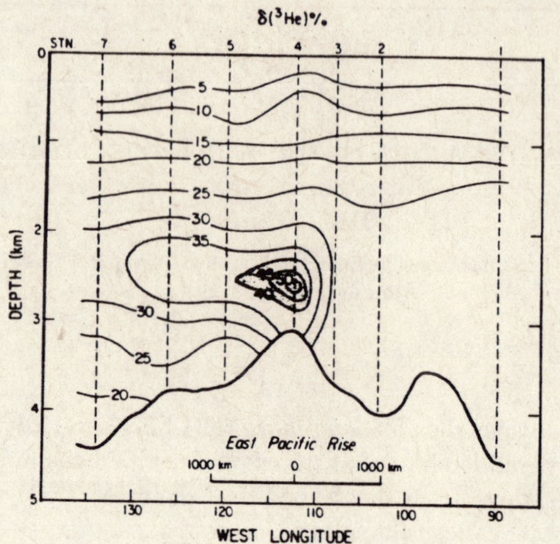


Fig. 1. Distribution of $^3$He around the East Pacific Rise of 15°S. The numbers along the horizontal axis represent west longitude (from H. Craig and J.E. Lupton: *The Sea,* Vol. 7, pp. 391–428, 1981).

not exist at the deep seabed, the presence of $^3$He with an abnormally high concentration at the deep sea, therefore, is due to the mantle.

The isotope ratio ($^3$He/$^4$He) in the atmosphere is considered standard and the concentration of $^3$He in a given sample is expressed in terms of deviation from the standard value (as %), i.e., δ $^3$He. Thus

$$\delta\,{}^3\mathrm{He} = (R/R_{air} - 1) \times 100.$$

Here $R$ and $R_{air}$ are the ratio of helium isotopes in sample and air, respectively.

Although $^3$He indicates that mantle material is being discharged from the deep seabed, it is presently measured in laboratories only. It cannot, however, be used to explore the locations of hot springs progressively as is done at present. In this voyage the waters were sampled on a research vessel and the results used to determine locations with higher $^3$He concentrations. As indicated by the studies in the East Pacific Rise, the measurements can be made quickly on a research vessel and we could also find materials which correlate with δ$^3$He. These are Mn and $CH_4$. There could be other materials as well, however, since the concentration of many substances is already higher in sea water, it is not possible to notice the increase in their concentration due to discharge. Silicon is one such example. During this voyage also, the members took the sea water samples collected to their respective laboratories and analyzed them to find peculiar substances. Let us see what the result will be.

Anyhow, during this survey of the Marianas trough, we could not carry out the measurements on the vessels itself, considering the technical limitations of Japanese researchers.

Fortunately, last November, I was invited to join the voyage of the Plut IV of the Scripps Oceanographic Research Laboratory and had the opportunity to take part in the voyage around the East Pacific Rise at 21°N. The sea water collected by sampler or the hot water samples collected by the submarine *Alvin* were chemically analyzed on the research vessel *Melville* itself. Particularly, the trace of hot water springing from the seabed was obtained using gas chromatographs for the dissolved $CH_4$. We, therefore, discussed with Dr. Craig, the chief scientist of the *Melville,* and sought his permission to use the services of Mr. Kim, who analyzed the $CH_4$ on board the vessel surveying the Marianas trough and decided to find out the trace of hot springs by $CH_4$ analysis on the vessel itself.

We know from Lonsdale's report that temperature abnormalities existed

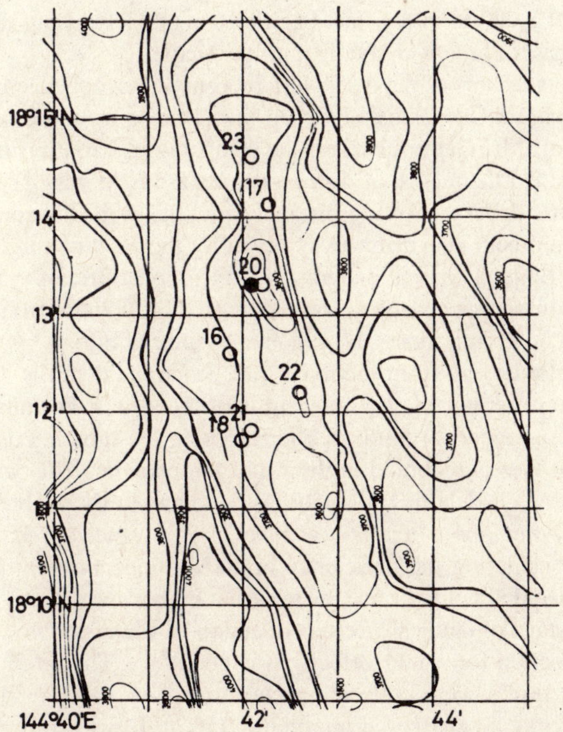


Fig. 2. The seabed topography around the center of the Marianas trough. Points marked ○ indicate water sampling points. At the ○ mark of Stn. 20, the water was sampled at a depth of 600 m from the seabed, whereas at the ● point on its west side, water was collected between 600 m and 1,100 m.

at about ten places along the opening of the Marianas trough with its center at 18°2′N, 144°42′E. We first tried to obtain the topography of the seabed of this area to make the basis for future studies. Using the 12 KHz PDR of the *Hakucho-maru*, we measured the depth along the east-west direction at 0.5 miles (900 m) intervals. Mr. Iguchi and Morinaga of Kobe University measured the depths and prepared a topographic chart. Fig. 2 shows the seabed diagram of a region extending five miles east-west and seven miles north-south. The points of abnormal water temperature, according to Lonsdale's report, are around the mountain peaks at the mouth of the trough but a clear detection of a rift valley was difficult with the usual PDR equipment.

We therefore installed CTD equipment 10 m from the seabed and decided to observe abnormal temperatures. A binger was attached 5 m below CTD fish, to measure the distance from the seabed. However, since the seabed is a rocky area with sharp peaks, it was extremely dangerous. Last November, when we observed the seabed while collecting hot water samples at the hot spring mouth at 2,600 m in the East Pacific Rise of 21°N, using the *Alvin,* we found it to consist of layers of lava in a pillow shape.

Very small amounts of deposits were present on the concave region. Prof. Hirata of Hiroshima University observed the distance from the seabed using PDR equipment, Mr. Gamo checked the temperature variation of CTD and Fukazawa obtained the record of data on the YHP-45. At vessel speeds above 1 knot (including drift) measuring distance from the seabed becomes difficult as the CTD equipment also drifts away. Even for the *Hakucho-maru,* which is equipped with Baus raster and 2 sterns in 2 axes, it is difficult to move in the desired direction at a speed of 1 knot. It started its voyage from the northernmost tip in the mouth along the axis but in spite of efforts by the navigating crew, it drifted away from the axial line. Since the depth is 3,600 m, the water temperature remains fairly constant up to 1/1000°C. Fortunately, at station 20 of Fig. 2, about 20 minutes after crossing the mouth axis, an abnormality of 1/100°C was recorded in the water temperature. This was the same point noticed by Lonsdale in 1981 with his deep tow to show abnormal water temperatures. After about ten hours travel, we advanced 3 km along the mouth axis. Considering the deficiency in preparations and the capability of vessel, it was decided not to cruise the vessel at higher speeds.

The third stage of studies was the collection of water samples. At the first station 16 correction for sound velocity was necessary. Therefore for the first 1,000 m, the samples were collected at an interval of 100 m and for further depths, they were collected at an interval of 250 m. However, between a distance of 500 m from the seabed, sampling was done more closely for a proper estimation of $CH_4$. Although the ship was anchored at a given position, it took almost 1.5 hrs for the sampling apparatus to collect the samples, and during this time, the vessel had drifted by more than 1 km. We therefore requested the crew to maintain the vessel at a given point for at least 1.5 hrs by proper operations. Samplings were completed at stations 17, 20, 22, and 23 and the samples were collected from the shallow waters along the mouth axis.

As shown in Fig. 1, the highest value of $\delta\,^3$He in the East Pacific Rise was observed about 200 m from the seabed. Therefore, samples were collected at close intervals to a distance of 500 m from the seabed (i.e., at every 50-60 m) and $CH_4$ was analyzed by gas chromatography. If we look at the results of the $CH_4$ analysis obtained for stations 16–21 we find that the $CH_4$ concentration is much lower than expected but shows a tendency to increase after a distance of 500 m. At stations 20, 22, and 23, where some abnormality in water temperature was observed, the samples were collected at close intervals between 500–1,000 m from the seabed. As shown by the cross sectional diagram of Fig. 3, the highest value of $CH_4$ concentration was obtained (75 μcc/kg sea water) at a distance of 600–700 m from the seabed. This value is almost 20 times higher than the normal concentration of $CH_4$ in sea water. The presence of $CH_4$ with such high concentrations in the deep sea is obviously due to the hot waters springing from the seabed rather than due to biological sources as we can see from the example of the East Pacific Rise.

Currently we are preparing the sea water samples used to find the $CH_4$ concentration, for an analysis of $^3$He. According to the preliminary reports $\delta\,^3$He = + 27% and $CH_4$ = 8.2 μcc/kg were the results for samples from Station 20 (depth 3,615 m) very close to the seabed (depth 3,585 m). The highest

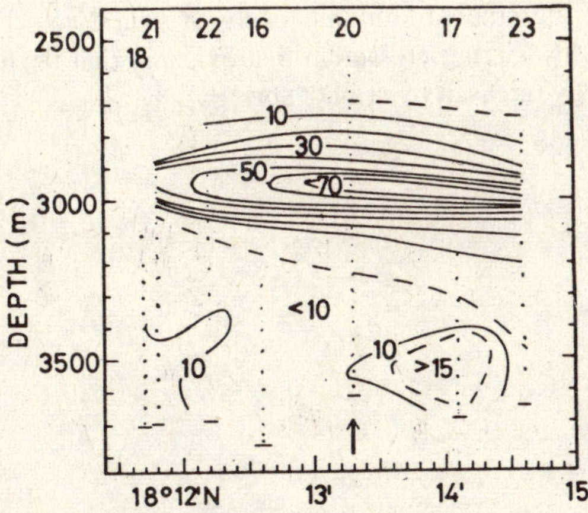


Fig. 3. Cross section of vertical distribution of $CH_4$.

concentration of $CH_4$ was 75.2 μcc/kg at a depth of 2,957 m at Station 20; and we are eager to know the location at which the $\delta\ ^3He$ concentration will be the highest.

One of the most important facts of the Marianas trough expedition was that the $CH_4$ was immediately analyzed on the vessel itself. Since we were expecting the highest concentration of $CH_4$ at about 200 m above the seabed the samples were collected up to 500 m in the first half. After knowing the analysis results, the sampling points or sampling layers were changed in the later half, which enabled us to point out the layer with highest $CH_4$ concentration. Had we continued to take the samples in the same way as first half and analyzed the samples in our laboratories at the Institute then probably this finding would not have been possible. Mr. Kim was very busy analyzing $CH_4$ and plotting the results.

It took three days to collect the water samples. During this period we had seven CTD measurements and eight water samples (Station 19 is beyond the range of Fig. 2). Prof. Kanamori and Prof. Nozaki took responsibility for collecting samples. All these eight points were within a range of three miles in the north-south direction. The vessel was manoeuvred in such a way that waters could be sampled at any desired point and we would like to thank all the research officers who carried all the study equipment on the *Hakucho-maru.*

During the next stage of study, probably a better profile of the seabed will be necessary. Then we would also like to take photographs of the seabed and the like to know what exists and where. This would help us collect directly the samples of hot spring water using a deep-diving submarine. It is necessary to learn the chemical composition and physical properties of this hot spring water and then compare the same with the hot spring water of the East Pacific Rise, which is being studied simultaneously. It would also be interesting to learn the type of life (living creatures) that exists in that peculiar atmosphere based on the energy released by the hot spring.

REPORTS

# Report on the Activities of the Committee on Heavy Metals

*Satoru Kanamori*

(Hydrological Research Laboratory, Nagoya University)

(Received: January 21, 1982)

The chemical analysis of various trace elements existing in sea water has been of special interest to the chemical oceanographers of Japan for a long time. It was Japanese scientists who were first involved in the great research project incorporating studies on the distribution of various trace elements in all the seas of the world. However, during these 15 years the concentration of heavy metals in sea water has slowly reduced over the time. This change is indicated in Table 1 and we find that almost all the heavy metal concentrations have been reduced to 1/10. With the progress of time, analytical methods of high accuracy are being developed and, therefore, it is possible to detect even the smaller values. However, another major reason for this trend is the techniques and uncontaminated equipment for the experiments. One such example is the contamination of the free sampling apparatus of Patterson (CIT), well-known for his studies on lead; another example is the clean room. The situation in the case of lead is particularly bad and since there is not much difference in the lead content in air and sea water of the same volume, the use of a clean room is indispensable. A similar situation exists in other elements as well.

**Table 1. Changes in the reported values of concentration of heavy metals in sea water (μg/kg)**

| Element | 1965* | 1975** | 1980 |
|---|---|---|---|
| Al | 10 | 2 | — |
| Mn | 2 | 0.2 | 0.02-0.1 |
| Fe | 10 | 2 | 0.2-0.5 |
| Co | 0.1 | 0.05 | 0.005 |
| Ni | 2 | 1.7 | 0.2-0.6 |
| Cu | 3 | 0.5 | 0.02-0.3 |
| Zn | 10 | 4.9 | 0.01-0.6 |
| Pb | 0.03 | 0.03 | 0.02- 0.06 |
| Cd | 0.11 | 0.1 | 0.01-0.1 |

*J.P. Riley, G. Skirrow: Chemical Oceanography, Academic Press, London, 1965.

**J.P. Riley, G. Skirrow: Chemical Oceanography, 2nd Ed., Academic Press, London, 1975.

During the autumn of 1980, before this Special Research Project took shape, a committee was formed to study the prototype of a contamination-free water sampler (coordinated by Yoshio Horibe) and the members were Yoshio Horibe, Yukio Kodama, Satoru Kanamori, Hiroyuki Hirata, Muroi, Masayo and Suji Nakamura. During the execution of this project, we also decided to study the clean laboratory on board, which is indispensable for further experiments. With the progress of this special research project, a prototype could also be developed. Then a small committee on clean room (with Yoshio Horibe as group leader) installed a 250 cm (length) × 180 cm (width) × 185 cm (height) clean room on the Tokyo University's research vessel, the *Hakucho-maru*. This clean room was found to behave like static class O (in 1 cubic feet water, there are no particles larger than 0.3 μm).

While preparing a prototype of contamination-free water samples, we referred to the CIT sampler, the Niskin-type, bacteria-free water sampler, the GOFLO sampler, etc.

The CIT sampler is, of course, the sampler with the highest degree of cleanliness and has a capacity of 10 l. However, it weighs 200 kg and since it is suspended by wire, it cannot be used on the No. 1 winch of the *Hakucho-maru.* The operation is also complicated and costs as much as 5 million yens. We thought that we should have a lighter sampler that could be transported easily.

With the GOFLO sampler (with teflon coating), the contamination is avoided by closing the nozzle for the passage of the surface water layer. A good result is reported to have been achieved using this technique, but then the sample cannot be called absolutely uncontaminated. When the inner parts are contaminated, its detection and removal are both difficult in this apparatus. The Niskin-type, bacteria-free water sampler (1.5 l) collects the water sample in a polyethylene bag. It was excepted that with this method it would be easy to get an uncontaminated structure, but it was difficult to expand the size (to 10 l) due to mechanical reasons.

Thus the above study revealed that it is not possible to satisfy the requirements of a contamination-free sample using the existing samplers or their improved versions and it is therefore necessary to develop a new sampler method. As a part of this proposal, we studied the bellow-type, universal compression-type sampler equipment. Using materials easily available a prototype was prepared and tested. The results indicated that it may be possible to make an apparatus with a 10 l capacity as desired. However, there were many problems to be solved in order to complete the development of this apparatus. Thus:

1) The material used should be very strong, the bellows must compress to a sufficient extent, and at the same, the water sample must not become contaminated.

2) The apparatus should be lowered into the sea in the compressed condition, then the valve opened and water sample allowed to be absorbed. After that the valve is closed again and the apparatus is brought to the vessel. It is necessary to develop mechanisms that would achieve these operations. A proper material must be selected so that the water sample does not become contaminated and, at the same time, a mechanism is to be incorporated whereby samples from a region away from the sampler or wire could be obtained.

To solve the above problems, tests were repeated to find the suitable material. As a result, a water sampler shown schematically in Fig. 1 was designed and fabricated.

The main body of the apparatus was made of stainless steel. Valves are made from teflon, the outer pipe from acrylic and the bellows-type water collector is made from a good quality polyethylene. The unit is designed so that the water sample is not much in contact with polyethylene or teflon. The broad functioning of this unit is explained in Fig. 1. Thus it is necessary to execute three operations with a proper time interval between them. The front end of the test water carrying pipe is cut to open at that end. The apparatus extends in length under its weight and then pulls the washer, closing the cock.

Design responsibility was taken by Hirata, prototype preparation of the bellows-type water collector was handled by Kanamori, and the material test-

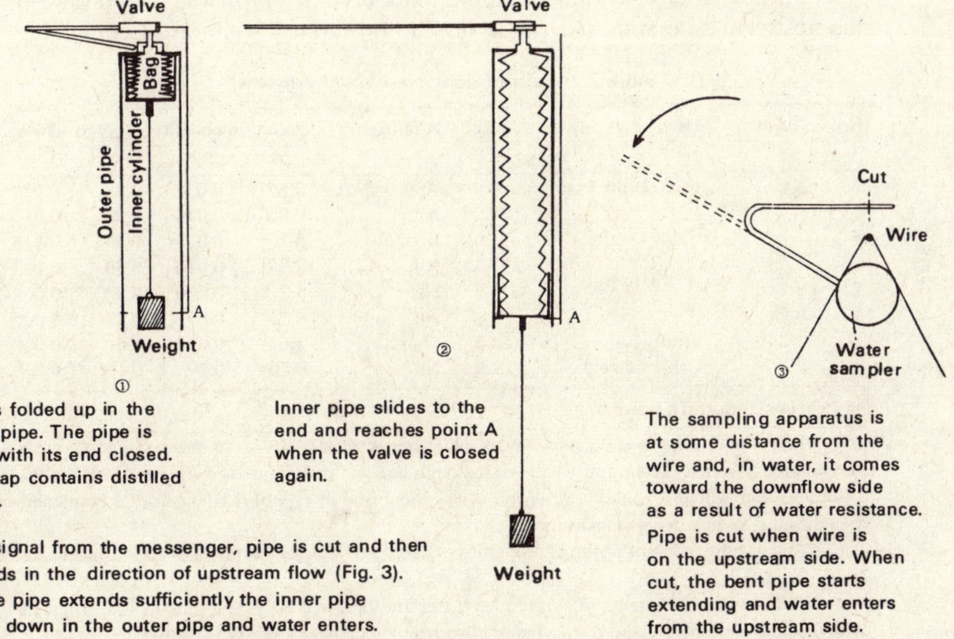


Fig. 1. Schematic diagram of contamination-free water sampler.

ing was entrusted to Muroi. All members participated in discussions related to the design drawing or selection of material, etc. On this occasion a small committee was regularly formed to study heavy metals (Chairman: Kanamori) within the general group and all members of the study group were adopted on this committee.

Some of the components used in the fabrication of the prototype were ordered outside and the mobile part was fabricated in the workshop of the Faculty of Science, Hiroshima University. Since we anticipated some changes in design at the last moment, the mobile part of the prototype was made from brass. With some minor changes the prototype was completed in the later half of October and a pool test was conducted.

The bellows-type water collector was made by a blow molding technique. Since this calls for a material with sufficient strength and ease of stretching, we decided to use polyethylene after studying its physical properties and chemical stability. Results of our test are indicated in Table 2. We noted from the results that sample No. 4 or No. 6 shows the desired properties. Using good quality materials, we made the containers to hold water samples which can be subsequently analyzed by the members.

November 7-13, we went on Tokyo University's research vessel *Tansei-maru* on its KT-81-17 cruise, collected the water samples with the prototype developed and tested the same on board. The waters were collected at different depths after confirming that the water collector pipe was turned toward the upstream side and after observing the operating condition of equipment

**Table 2. Results of elution test of polyethylene***

| No. | Maker | Uses | M.I.** | Additives | Elution quantity ($ng/10^3\ cm^2$) | | | |
|---|---|---|---|---|---|---|---|---|
| | | | | | Cu | Cd | Pb | Ti |
| 1 | A Co. | Film, Blow form | 0.3 | Antioxident | 0.65 | 0.032 | 0.38 | 0.023 |
| 2 | " | Film | 0.7 | Nil | 0.96 | 0.038 | 0.32 | 0.013 |
| 3 | " | Cable coating | 1.0 | Bonding | 3.2 | 0.070 | 0.49 | 0.013 |
| 4 | " | " | " | Nil | 0.39 | 0.016 | 0.30 | 0.013 |
| 5 | " | " | 2.0 | Nil | 5.2 | 0.080 | 0.33 | 0.000 |
| 6 | B Co. | " | 1.3 | Nil | 0.51 | 0.37 | 0.22 | 0.000 |
| 7 | " | Film | 1.5 | ? | 5.6 | 0.037 | 1.6 | 0.000 |
| 8 | " | Cable coating | 2.0 | Nil | 0.70 | 0.30 | 0.16 | 0.000 |
| Normal contents in sea water (ng/kg)*** | | | | | 300 | 30 | 50 | 15 |

*Preparatory tests were conducted for other heavy metals but no excessive elutions were particularly obtained. Testing method: washed with acetone then immersed in (1 + 1) $HNO_3$ for 24 hours, after which it was rinsed. Again it was immersed in (1 +100) $HNO_3$ for 24 hours and then the elution quantity was measured.

**This is the index of physical properties of fused polyethylene. MI = 1.0-1.5 was adequate for our applications.

***The normal contents of surface sea water are indicated only as a reference. Compared with this, the elution obtained from a material with a high degree of purity is below 1%.

**Table 3. Results of the analysis of samples collected by the (prototype) contamination-free sampling apparatus**

| Stn. | Depth (m) | Cu (μg/kg) | Zn (μg/kg) | Cd (μg/kg) | Pb (μg/kg) | Hg (μg/kg) | As (μg/kg) | Sb (μg/kg) | Ti (μg/kg) |
|---|---|---|---|---|---|---|---|---|---|
| 7 | 0 | 0.136 | 2.2 | 0.00342 | 0.0166 | 0.0024 | 1.4 | 0.20 | 2.8 |
| 7 | 200 | 0.159 | 1.3 | — | 0.0159 | 0.0026 | — | — | 2.6 |
| 7 | 1500 | — | 5.9 | — | — | 0.0013 | 1.9 | 0.09 | 3.3 |
| 6 | 1000 | 0.235 | 1.2 | 0.0923 | 0.0072 | 0.0014 | 2.0 | 0.10 | 1.2 |

near the water surface. In order to compare this sample with conventionally collected samples, a Niskin-type sampler was fixed about 10 m above and below the contamination-free sampler developed at this time and samples were collected simultaneously. Also, in order to know the contamination caused by the rope or messenger (Brass made), a rope and stainless steel messenger combination was used and samples were collected. The test was repeated several times to see if there were any apparent problems caused by repetition. The sea-water samples collected by the equipment were carried to the Institute of Oceanography in the bellows-type container and kept in the clean room. They were then handed over to members of the Heavy Metals Group and analyzed. The results of analysis are still awaited but some of the available results are shown in Table 3. In spite of the fact that the moving part was made of brass and a lead block was used as weight, the amount of Cu, Zn, Pb, etc. is much lower than that obtained by the conventional method. By using all metal parts made of stainless steel we hope to achieve the anticipated performance. Presently six contamination-free samples are being manufactured without changing the basic design but improving those parts which were found to be weak during repetitive trials and slightly modifying the apparatus to make it easy to handle. We hope they can be used during the KH-82-1 cruise of the *Hakucho-maru.*

There are still some problems which remain to be solved in this sampling apparatus. For example, it uses a number of springs which we would like to improve subsequently. Also the dead volume of the bellows-type container is still very large and therefore some improvements in its shape and operating method are desirable.

N.B.: The activities of the study group on "Distribution and behavior of heavy metals in ocean" are closely related to the activities of the "Heavy metals" study group. Prof. Kanamori's first attempt in this direction was to establish a clean room and a multifaced back-up for the preparation of a contamination-free sampler. Accordingly, this report forms the main part of the activities of the Kanamori group during this year. Currently, contamination-free sampling apparatuses are in operation on the KH-82-1 cruise of the *Hakucho-maru* and the activities of the Kanamori group continue.

# Proceedings of the Symposium on Oceanographic Measurements

Day and time: April 13, 1982 (Tuesday), 09:30–17:00.
Place: No. 1, Conference Room, Denki Tsushin University.
Participants: 53.

This symposium was arranged to present the results of two topics of the planned studies on measurement aspects carried out under this special research project for fiscal 1981 and four other topics of general studies. There were a total of 20 research papers.

We first thought of summarizing the contents of the symposium in general. However, the planned and general studies are combined and since we felt it is easier to introduce each topic separately, we decided in that way.

## 1. On-board Measurement Method and Movement of Nontowing Bodies

1.1. Mechanical construction of the buoy system and problems in fluid dynamics (Higashi, Watanabe and Nakamura, Faculty of Engineering, Tokyo University).

A system to measure the ocean characteristics was proposed which could be installed on ships or airplanes and which could be towed over a wide range in the sea. A substance something like a kite was quoted as a nontowing object. As a result of the analysis of the dynamic characteristics and experimental observations, it was concluded that a rectangular plate with a low aspect ratio leads to a stable operation.

A differential equation for the continuous motion of a nontowing rope object was also presented and methods to obtain (steady state) stable conditions were discussed.

## 2. Measurement of Current Speed, Water Temperature, Salinity Pressure, Potential Difference, etc.

2.1. Measurement of distance and speed using the image sensor (Mihashi, Mochizuki, Denki Tsushin University).

For an indirect measurement of current speed, a method using an image sensor as space filter was proposed. This method does not involve any mechanically moving parts and it is very easy to select the proper filter parameters, depending on the object of measurement. Also it is possible to focus the object automatically in spite of the changes in its distance. The experiment was carried out in space but it was ascertained that the method could be used over a wide range.

2.2. The drop and revolution characteristics of an electromagnetic current speed profiler (Kinjo, Faculty of Science, Hokkaido University).

The structure of a free falling-type electromagnetic current meter now being developed to obtain the vertical profile of horizontal flow (velocity) was discussed. Also the possible range of measurement for its drop-revolution characteristics was pointed out. Subsequently factors such as drift velocity or the sinking distance after the equipment is dropped into sea, which considerably cause errors in measurements, were studied and it was concluded that they are not significant errors caused in the equipment being developed.

2.3. Measurement of ocean temperature and salinity using microwave techniques (Suzuki, Arai, Watanabe and Ishikawa, Dentsu University).

An indirect method to measure the temperature and salinity of sea water near the surface was proposed, which uses the naturally released energy of X and L band microwaves. In this case the L band electromagnetic waves indicate the condition of particles at certain depths whereas X band waves can express the condition of surface particles only. As a result, the former strongly affects the temperature as well as salinity and the latter only the temperature. The results of actual experiments were almost the same as expected.

2.4. Response of ion sensor to sea water (Fujiwara, Faculty of Science, Chiba University).

A method using an Ion Selective Electrode (ISE) was proposed for the *in situ* measurement of salinity composition and concentration.

It is based on the fact that as voltage is applied to ISE in series its internal resistance increases. Thus another electrode is inserted into a container which already has an ISE and by applying the voltage to that electrode, the subsequent voltage at the ISE output is measured. As a result we obtained a quantitative response wave that is independent of ionic concentration and an ionic conductivity response wave which is affected by ionic concentration.

2.5. Impedance characteristics of an electromagnetic strain transmitter under vigorous vibrations (Kumamoto, Dentsu University).

A method using the mechanical load characteristics of an ultrasonic converter (source), which is used for submarine communication or measurements, has been proposed to measure pressure. Here use is made of the fact that the impedance of an ultrasonic converter varies with pressure. The paper also discussed the analysis of impedance and measurement error as well as the impedance measuring circuit.

2.6. Prototype of equipment to measure the electric potential at the seabed (Tomoda, Institute of Oceanography, Tokyo University).

The prototype to measure electric potential difference was proposed in order to study abnormal variations in electrical conductivity at the seabed. This structure had two arms, each 10 m long, and could measure a potential gradient of 10 μV/m or above. However, some fish are sensitive to a gradient of 0.01 μV/cm and the analysis of that mechanism would be extremely interesting.

## 3. Information Processing and Transmission

3.1. Passive telemetry using a radar reflector with variable reflectivity (Ogami, Biological Research Institute, Tokyo University and Hasebe, Engineering Physics Department, Japan University).

The reflectivity of a radar reflector is varied depending on the volume of information to be transmitted. In the telemeter system proposed here, this reflector is illuminated with radar to collect the information. It is possible to classify the reflector response depending on the frequency range. The equipment is not affected by the allocation of electrical waves and thus can be used widely for the collection of oceanographic data and information transmission.

3.2. Prototype of Local Users Terminal and results of experiment (Segawa, Institute of Oceanography, Tokyo University).

Various systems have been developed which use a satellite to collect oceanographic data. Here a new development regarding the receiving equipment (Local Users Terminal) which can use the Argos system has been discussed. The object satellite is a NOAA No. 6, No. 7, frequencies concerned are 136.6 MHz, 137.9 MHz, and a turnstile antenna is used. These days, position determination was attempted with the system and fairly satisfactory results were obtained.

3.3. Underwater information transmission using codified sound waves (Matsuda, Osaka Dentsu University).

A method is described here which uses codified sound pulse trains to transmit information underwater. Here, taking into consideration the facts that: a) speed of information propagation is slow, b) the medium of transmission is non-uniform, and c) it is considerably affected by factors such as temperature, noise, and topography. A new code called the M-C code was proposed for information transmission in place of the Golay code which is quite noisy. As a result, the S/N ratio for the M-C code was found to be better. For the same gain, the bit length was shorter and the speed of propagation was faster. As against these merits, the hardware for this code becomes a little more complicated.

## 4. Measurement of Marine Variations Using Sound Waves

4.1. Measurement of velocity vector space distribution using a numeric regeneration of wave field (Yamazaki and Tamura, Department of Engineering, Tokyo University).

In order to explore the total phenomena of oceanic currents we should know the space distribution of the velocity vector and its time variation. A numerical regenerating technique has been proposed here for this purpose. It is based on records of reflected acoustic waves and their regeneration. Here 1) the sound waves reflected from sea water and the substances moving within are recorded with a transducer array, 2) these are subjected to proper data processing and the current velocity vector distribution is calculated. Presently the measurements are carried out in space but the results are quite satisfactory.

4.2. Measurement of wave height distribution of sea waves using acoustic techniques (Ando, Faculty of Engineering, Tokyo University).

The wave motion of the ocean can either be expressed as a convex-concave pattern or in terms of the space-time frequency spectrum. In both methods, array-type receivers are installed at the seabed. The observation of the wave pattern in the first case involves reconstruction from a simulation of wave transition, whereas, in the second case, a block reflection from the sea wave is used. Both the numerical analysis and experiments in space have brought these methods to a practical stage.

4.3. Observation of temperature profile using the vertical propagation characteristic of sound waves (Takeuchi, Dentsu University).

Sound waves propagating through sea water are greatly affected by the temperature, salinity, and pressure of sea water, but particularly by temperature. Accordingly if we measure the propagation time of the sea wave accurately we can estimate the temperature profile. Here we have studied the relation between the propagation time of the sound wave that is reflected vertically from the seabed and various parameters of temperature. Also a temperature profile was created in a water tank and the experimental results were found to tally with the theoretical findings.

4.4. Basic studies regarding the measurement of vorticity in the ocean using ultrasonic waves (Motooka, Chiba Institute of Technology and Okujima, Department of Precision Engineering, Tokyo Institute of Technology).

In order to measure the strength of the vortex generated in the ocean, three ultrasonic pulse receivers are used to form a triangle surrounding the vortex center. In this method the time of propagation for a sound wave cycle is measured in clockwise and anticlockwise directions and thereby the structure of the vortex is discovered. This technique can also be used to measure the angular velocity of the vortex, its vorticity, and the strength of the vortex when the center of the ultrasonic receiver system is away from the vortex center.

4.5. Measurements of current velocity and waves using a parametric sound source (Endo and Takeuchi, Dentsu University).

A method is proposed whereby a parametric array can be used simultaneously to measure the current velocity on board and the vessel speed with respect to the floor. Since a primary, very high frequency wave is absorbed and attenuated by sea water to a large extent, it cannot traverse a long distance. It is therefore used to measure flow velocity. On the other hand, a secondary low frequency wave can traverse a long distance and therefore can be used to measure velocity with respect to the floor. If the array is installed at the sea bottom, this secondary wave can be used to measure surface waves. This paper also analyzed the conditions of the parametric array and the corresponding sound wave propagation characteristics and indicated the practical conditions that can be achieved.

## 5. Sound Source and Propagation Characteristics of Sound Waves

5.1. Radiation of large amplitude sound wave pulse in free space and the waveform control (Nakamura and Nakamura, Osaka University).

When sound pulses are used for various acoustic measurements, it frequently becomes necessary to use sound waves with high amplitude and a wider frequency spectrum. Also, if the waveform of the sound wave that would be incident on the object of measurement could be controlled, it improves both the ease of data processing and precision. This paper discussed the analytical approach to achieve such a high amplitude sound source from a structure consisting of piston, cylindrical pipe, and index phone. During actual experiments currently being carried out in space, the results of analysis were found to be in good agreement with actual observations.

5.2. Propagation of sound waves in the vicinity of water clutter (Okushima and Otsuki Nario, Department of Precision Engineering, Tokyo Institute of Technology).

It is possible to obtain the water temperature profile over a wide region using the propagation time of sound waves in sea water. Here the observed data of the water temperature profile in the Kuroshio region is used to calculate the path of sound waves and positions of sound wave receivers were indicated which could measure the propagation time of a direct wave and the adjoining two propagation channels. The range of observation included a horizontal distance of 20 km, and the receivers were placed at depths of 500 m and 700 m. The results indicated that sound velocity is minimum around the depth of 450 m. However, studies regarding sound channels were also conducted, assuming the minimum sound velocity at a depth of 1,000 m.

5.3. Computer simulation of nonlinear sound wave propagation (Watanabe, Faculty of Engineering, Doshisha University).

When a powerful sound source is used, the effect of the nonlinearity of the medium cannot be ignored. If we could estimate the propagation characteristics of sound waves quantitatively, considering the nonlinearity effect, it would be a major step in acoustic measurement. Here we studied the nonlinear distortion in a large amplitude sound wave due to the difference in the velocity of propagation of each section in a sound pressure wave and the effect of linear attenuation due to the absorption by the medium or dispersion during traverse, etc., independently. Results of this analysis are in extremely good agreement with the results of measurements of sound waves propagation through a pipe under an atmospheric medium.

5.4. Bending of sound propagation channel due to oceanic currents (Kamakura and Ikeya, Faculty of Engineering, Nagoya University).

The propagation channel or sound waves, having a wave length much greater than the localized variation of temperatures in sea water, are supposed to be affected strongly by the oceanic currents alone. The relation between the propagation channel and oceanic currents under such conditions were calculated using Snail's principle for a two-layered flow.

The properties in this case were expressed in a comparatively simpler form.

## 6. Measurement of the Soil Properties of the Ocean Floor

6.1. Acoustic impedance measurement of the soft sediments at the ocean floor and analysis of reflected wave (Tsuchiya and Sugimori, Department of Oceanography, Tokai University).

The sound reflection properties were studied using a three-layer sedimentation (sea water, surface layer, lower sediment layer) model. As a result we could estimate the concentration of the sediment surface layer, its density and layer thickness from the frequency response characteristics of the coefficient of reflection of complex sound waves. If we could estimate the conditions of different types of sediment layers at the seabed and determine the characteristic values beforehand, then it would be helpful for the actual oceanographic measurements.

So far we have only outlined various research papers discussed. The details are contained in "The Proceedings of the Symposium on Oceanographic Measurements" under the auspices of the Special Research Project "The Ocean Characteristics and Their Changes" (70 pages) published on the same date. Those desirous of obtaining a copy of the above publication may write to the Administrative Office, Coordinating Section of the Special Research Project (the address is available on page 133).

(Coordinator: Hitoshi Mochizuki)

## ADMINISTRATIVE REPORT

The eighth meeting of the executive committee was held on March 30, 1982 (Tuesday) in the C Wing Conference Room of the Institute of Oceanography, Tokyo University.

A research convention was decided to be held for two days on July 5 (Monday) and 6 (Tuesday) to announce the interim results of studies. (The details may be found in the Announcement column. In the last issue, the dates were announced as June 29 and 30 but have been changed as mentioned above.)

The following subcommittees were formed from the committee on the application of large equipment.

Subcommittee for on-site filtering equipment: Chairman: Yoshio Horibe (Institute of Oceanography, Tokyo University).

Subcommittee for mud mixer: Chairman: Yoshio Horibe (Institute of Oceanography, Tokyo University).

The ninth meeting of the executive committee was held on April 22, 1982 (Thursday) in the A wing Seminar Room of the Institute of Oceanography, Tokyo University.

It was decided to prepare an interim report of this special research project and the editorial committee was formed:

Members: Yokichi Takayanagi (Faculty of Science, Tohoku University). Toshihiko Teramoto (Institute of Oceanography, Tokyo University). Yoshio Horibe (Institute of Oceanography, Tokyo University). Hitoshi Mochizuki (Dentsu University).

The abstracts of reports to be presented at the research conference (July 5, 6) are to be printed for distribution to participants on those days.

## ANNOUNCEMENTS

### Research Convention on the Special Research Project "The Ocean Characteristics and Their Changes"

The subject of study has now entered the second year and in order to provide guidelines for future studies, it has been decided to hold a symposium to discuss interim results of studies as per the following program. Through this symposium, it is hoped, an interaction between different working groups will be established and the nexus will be strengthened which can eventually accelerate the total research. Hence, all are requested to attend.

Date and time: July 5, 1982 (Monday), 09:30–17:00.
July 6, 1982 (Tuesday), 09:30–17:00.

Place: Conference Room, Institute of Oceanography, Tokyo University.

Topics to be discussed: Representatives from specific research projects, general studies (continued from last year) will discuss the results of studies conducted so far and plans for future research.

## Guidelines for Authors

This Special Research Project has now entered the second year.

The state of the art of this study and the results, etc. are published in Newsletter. We request all readers to contribute their views and articles which will establish closer ties between different groups so that this research project can advance as a whole.

The contents are classified as:

| | |
|---|---|
| Highlights: | Quick (intermediate) results of this Special Research Project |
| Bulletin: | Domestic or international news reports closely related to this Special Research Project |
| Reports: | Reports regarding the activities of each section or working group |
| Announcements: | Announcements regarding the proposed meetings or conferences of each group |
| Bibliography: | Listing of papers published by the members on this Special Research Project or other reports |

Manuscripts should be written on standard 400 character size paper (handwritten material also accepted). Figures should be drawn in such a way that they can be reproduced as they are

It is proposed to publish this Newsletter every six weeks.

**Note:**

When reproducing or citing the material published in this Newsletter, due acknowledgment should be given. Authors' names, wherever mentioned, should also be quoted. A copy of the publication containing such a reproduction may be sent to the Editorial Section of the Administrative Office.

| | |
|---|---|
| Published by: | Administrative Office, Special Research Project "The Ocean Characteristics and Their Changes" |
| Published at: | Administrative Office,<br>Physical Oceanography Department,<br>Institute of Oceanography,<br>Tokyo University,<br>1-15-1 Minami-dai, Nakano-ku,<br>Tokyo 164<br>Tel.: 03-376-1251, Ext. 251. |

Special Research Project

# The Ocean Characteristics and Their Changes

NEWSLETTER NO. 7, JULY, 1982

# Contents



Researchers engaged in this Special Research Project, who wish to submit their papers in English and acknowledge receipt of research grant for this special study, should indicate the above-mentioned title of the project.

Cover Page

The figure shows a prototype of a hydraulic hand boring machine type KURD-1 in operation, made to excavate shallow coral reef. This is operated by four persons. One operating the drill handle, one adjusting the wire, one observing and recording the progress of excavation, and one supervising the hydraulic engine and water pump. The prototype was made in 1978 and tested at Guam as well as Tokai islands last year. Since 1981 it has been included as equipment for the Special Research Project "The Quaternary environments of the ocean floor". Height about 3 m (Courtesy Mr. Futakuchi, Faculty of Science, Kanazawa University).

HIGHLIGHTS

# Analysis of Quaternary Period Environments Using Coral Reef Boring (1)

*Kenji Konishi and Katsuhito Futakuchi*

(Faculty of Science, Kanazawa University)

Recently coral reef boring samples are being used for the qualitative analysis of changes in the Quaternary period environments of the ocean floor. This is because coral reef sediments are "bio-rocks" containing mainly $CaCO_3$ and a lot of the porolithic remains of benthonic creatures with skeletons formed from aragonite or high magnesium calcite. Among other reasons we can mention that 1) analysis of paleoenvironments is easy because of lot of fossils are available, 2) the geochemical analysis and measurement of ore of such hard composition (such as the isotope ratio of oxygen, carbon) is comparatively easy, 3) coral reefs are quite suitable for "dating" with the radioactive carbon measurement method which uses beta rays and also uranium-type radioactive materials which use alpha and gamma rays based on the radioactive nonequilibrium condition, 4) the sedimentation speed of a coral reef is quite large; about 2,000-3,000 Bubunof (1 Bubunof corresponds to 1 mm sedimentation in 1,000 years). This means it can give higher resolution on a time scale. If we use skeletal dating techniques for reef organisms with a higher sedimentation speed (e.g., scleractinia), then it is possible to read the records from yearly variation to seasonal variations.

Particularly in the case of coral reef growths, such as those on the northern side of Japan or in a region which is always subject to seismic variation, the above method is optimum for analyzing records obtained in extremely shallow water over an interglacial or subglacial period (within a glacial period) during which the seas had a relatively higher level irrespective of the discontinuity of records. In order to obtain an accurate picture of paleoenvironmental changes, it is necessary to supplement these observations with the analysis of deep sea core samples where a continuous record can be expected. Obviously this record helps more to determine causes behind historical environmental phenomena such as the glaciatic surface changes or localized crust changes. During the fourth period also, a large part of coral reefs contains a calcareous material. To collect these samples, a good boring machine is necessary. Due to the lack of such machines, this study has been delayed in all parts of the world including Japan. For the last 20 years, in our laboratory, we have been studying the paleoenvironmental analysis of coral reefs found in the southwestern islands of Japan, particularly those in Oto-gun of the Kagoshima prefecture. Since 1977, however, we started the paleontological analysis of newly formed coral reefs using a boring sampler (Konishi, 1979).

A simple hand-boring machine, which is shown schematically on the cover of this issue, was manufactured in 1979 with the help of research grant and is designated the KURD-1. This machine is also used in this special project "Quaternary Period Environments of the Ocean Floor." As can be seen in Fig. 1, this consists of a rotary hydraulic drill head (0–100 rpm, 15 kg) suspended with wire from a tripod whose legs are 4 m long. Below that there is a water-swivel, drill rod (dia. 70 mm), double tube core barrel (dia. 76 mm), and a drill bit (tungsten carbide with 12 tips) all of which are made to follow a drill guide column that is fixed separately by a pipe. A large, 70 mm dia. rod is used to avoid choking from the fragments. Inclined drilling is also possible as long as the angle is limited to 10°. The important parts (not shown in the figure) include: hydraulic pump (Robin EY-44-40; 10 ps/3600 rpm:

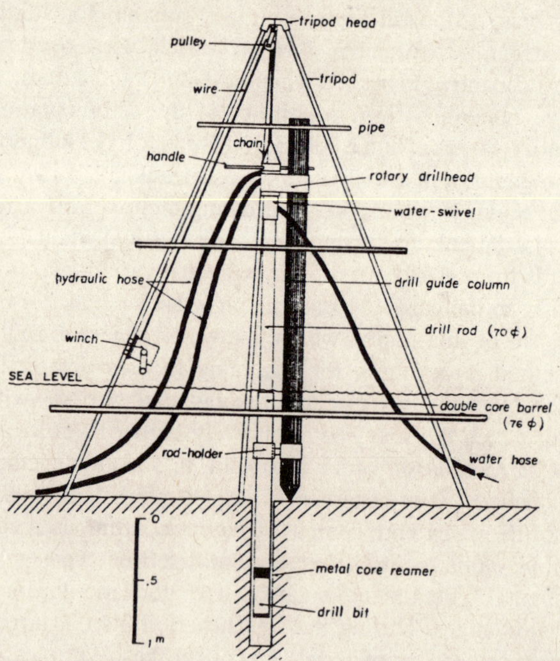


Fig. 1. Prototype of simple boring machine used to excavate coral reef (KURD-1).

maximum torque 2.15 kg/2600 rpm with a clockwise and anticlockwise changeover switch); fuel tank (75 liter, with wheels for shifting 80–150 kg), and water pump (marine pump LP-20W, 25 kg). The unit can be installed on board or on land (sometimes on the operation platform). There are some drawbacks to the system, such as, it requires lot of time to shift the equipment as it is fixed with pipe but this is because the machine has been fabricated with safety as a prime requisite.

During the 12 days between September 30, 1981 and October 11, 1981, we conducted five boring trials in shallow layers at the northwest coast of the Kikaigashima islands in Nakakuma (129°50′09″E, 28°20′08″N). The results are still being processed but interim findings are reported below and corroborate the earlier results.

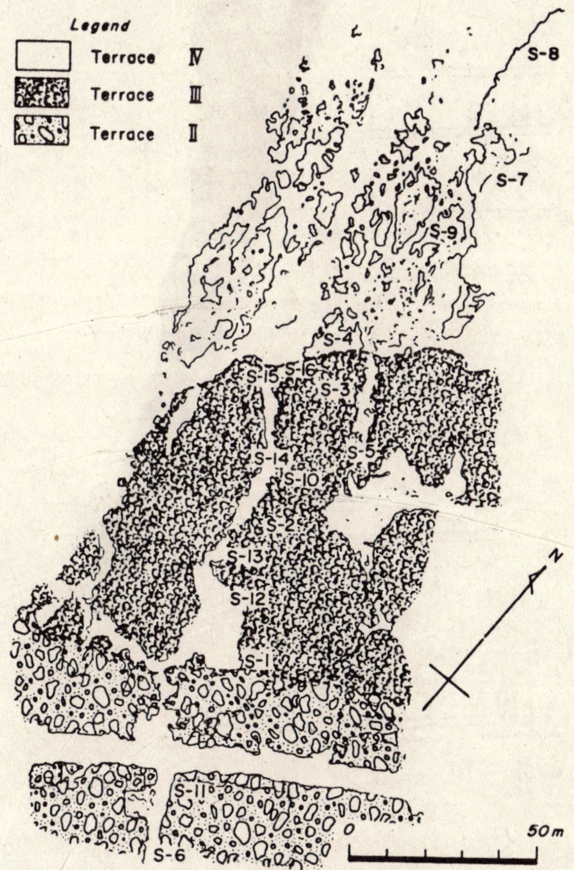


Fig. 2. The plane view of Kikaigashima Nakakuma region indicating the boring points for coral reef samples (S-1 to S-11 are spur region, S-12 to S-16 are trench region).

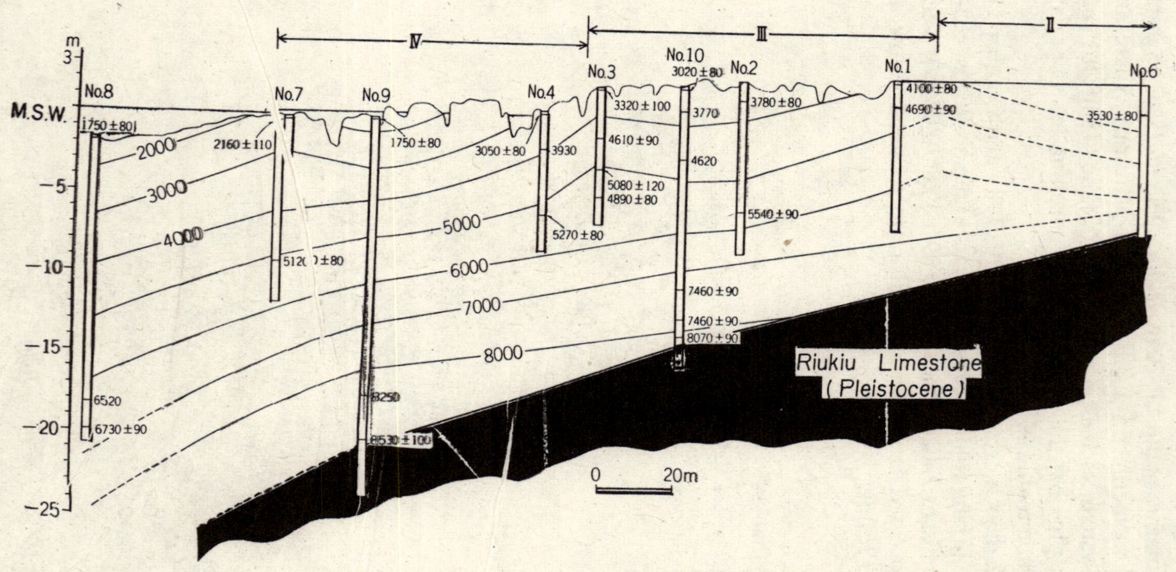


Fig. 3. Measured age values by the radioactive carbon method for boring samples from the new coral reef in the Nakakuma region (spur region) and the isochron lines (interval—1,000 years) indicating the extent of growth of the coral reef. The roman numerals at top indicate the respective sections.

The excavation points were concentrated in a trench region instead of a conventional spur region (Figs. 2 and 4, points S-12–S-16). This was to ascertain the isochron along the trench which contradicts the isochron [lines of the same period (interval 1,000 years Fig. 3: Konishi and Tanaka, 1982)] determined along the cross section normal to the river line in the spur. Also we wanted to study, if possible, whether trench or spur systems are causes of the antecedent valley defined by the erosional landform of a modern coral reef in the earth's crust. By arranging the trench and spur, it is first possible to learn the growth of the coral reef in the last 8,000 years and the variations in environment dynamically and in three dimensions.

Let us now discuss the results of each bore.

| | Height above sea level (below high tides), m | Length of excavation, m | Length of recovery, m | Recovery ratio, % |
|---|---|---|---|---|
| S-12 | -0.25 | 6.00 | 4.34 | 72.3 |
| S-13 | -0.66 | 5.00 | 3.19 | 63.8 |
| S-14 | -0.67 | 4.80 | 2.73 | 56.9 |
| S-15 | -1.23 | 3.50 | 1.77 | 50.6 |
| S-16 | -0.91 | 5.00 | 3.47 | 69.4 |
| Total | | 24.30 | 15.40 | 63.4 (mean) |

The depth of the excavation in all cases is shallower than planned. This was also found during the test excavations in Agana Bay at Guam in 1980. However, as the depth exceeds 6 m, further excavation becomes almost impossible with the KURD-1 and, therefore, we had to abandon it. Compared to the double tube core barrel with an inner diameter of 66 mm, which is used to collect a large quantity of samples to be used for a variety of studies, the hydraulic pressure of the pump was lower and therefore its efficiency also decreased. As a countermeasure either the diameter can be reduced or the pump should be used for shallow excavation only. It would be an objective for further studies to increase the power of the entire system. For the reasons mentioned above and also for fuel economy, it is desirable to replace the hydraulic pump with a diesel engine.

Figure 4 shows the seabed profile between S-12 and S-16 as well as the bar chart (of excavation). Fig. 5 shows the rock divisions of the core sample from S-14 in great detail, divided into top and bottom. It was understood from the studies that autochthonous scleractinia are the main part of the coral phase (corresponding to the upper phase of a three-part division). Compared with the spur region in general, this part had a well-developed algal crust consisting of porolithon, lithophyllum (particularly in S-14 and S-16). Further

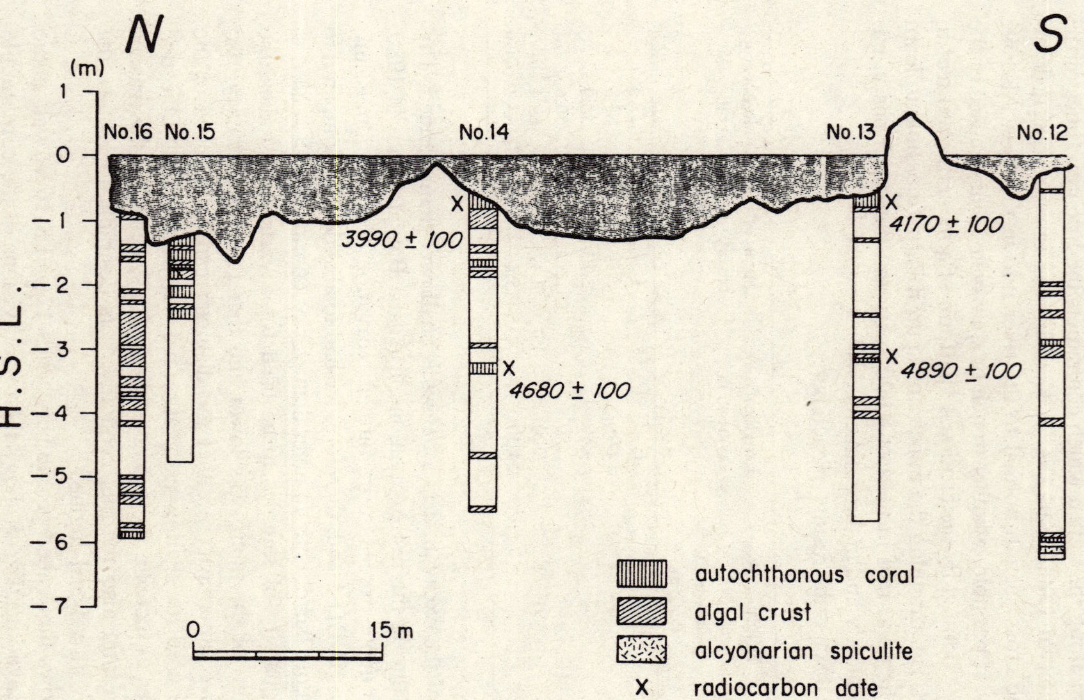


Fig. 4. Profile of spur region, locations of excavation point and the bar chart. The bar chart shows the divisions of different rocks and radioactive carbon dating.

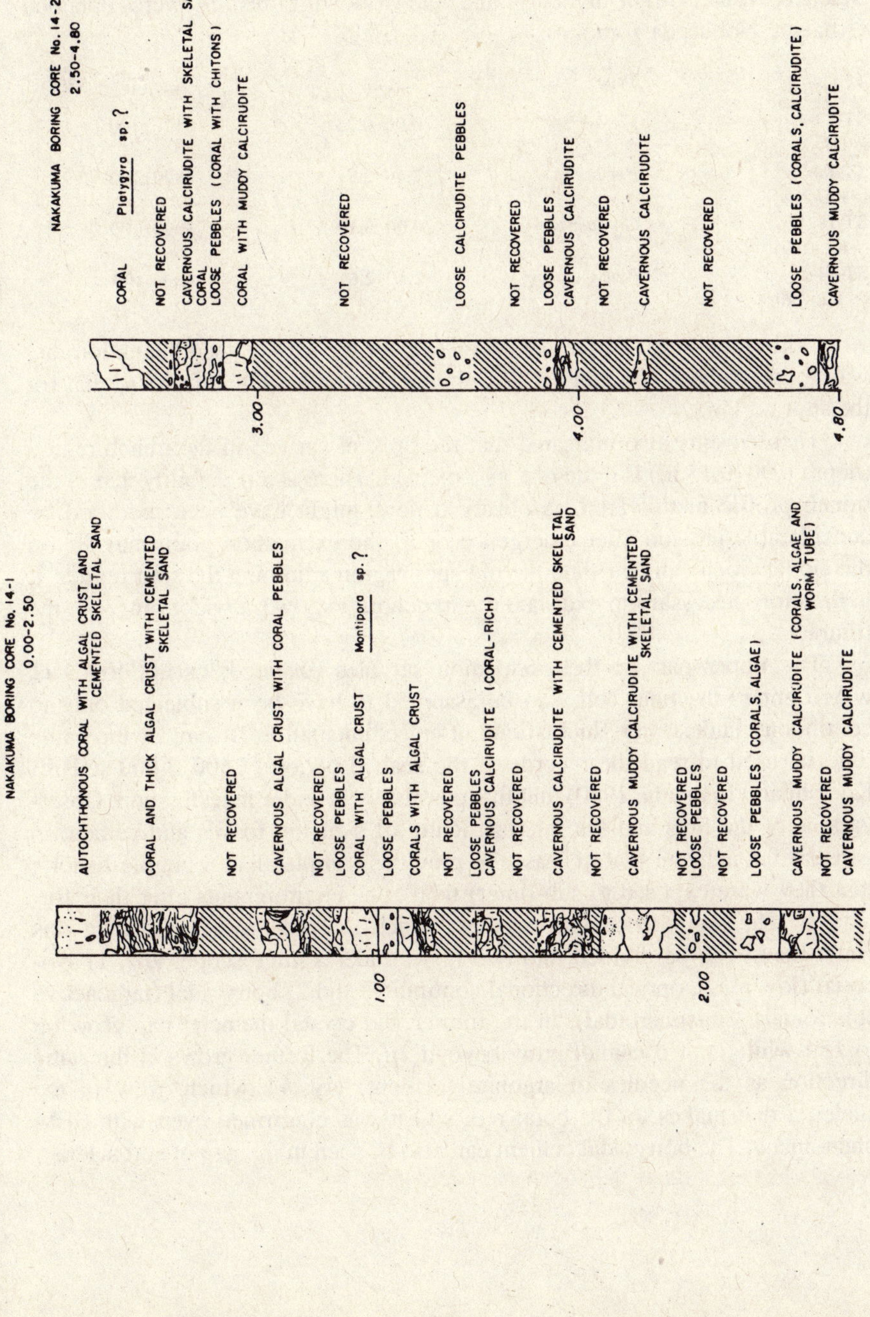

NAKAKUMA BORING CORE No. 14-1
0.00-2.50

AUTOCHTHONOUS CORAL WITH ALGAL CRUST AND CEMENTED SKELETAL SAND
CORAL AND THICK ALGAL CRUST WITH CEMENTED SKELETAL SAND
NOT RECOVERED
CAVERNOUS ALGAL CRUST WITH CORAL PEBBLES
NOT RECOVERED
LOOSE PEBBLES
CORAL WITH ALGAL CRUST *Montipora* sp.?
1.00
LOOSE PEBBLES
CORALS WITH ALGAL CRUST
NOT RECOVERED
LOOSE PEBBLES
CAVERNOUS CALCIRUDITE (CORAL-RICH)
NOT RECOVERED
CAVERNOUS CALCIRUDITE WITH CEMENTED SKELETAL SAND
NOT RECOVERED
CAVERNOUS MUDDY CALCIRUDITE WITH CEMENTED SKELETAL SAND
NOT RECOVERED
LOOSE PEBBLES
2.00
NOT RECOVERED
LOOSE PEBBLES (CORALS, ALGAE)
CAVERNOUS MUDDY CALCIRUDITE (CORALS, ALGAE AND WORM TUBE)
NOT RECOVERED
CAVERNOUS MUDDY CALCIRUDITE

NAKAKUMA BORING CORE No. 14-2
2.50-4.80

CORAL *Platygyra* sp.?
NOT RECOVERED
CAVERNOUS CALCIRUDITE WITH SKELETAL SAND
CORAL
LOOSE PEBBLES (CORAL WITH CHITONS)
CORAL WITH MUDDY CALCIRUDITE
3.00
NOT RECOVERED
LOOSE CALCIRUDITE PEBBLES
NOT RECOVERED
LOOSE PEBBLES
CAVERNOUS CALCIRUDITE
4.00
NOT RECOVERED
CAVERNOUS CALCIRUDITE
NOT RECOVERED
LOOSE PEBBLES (CORALS, CALCIRUDITE)
4.80
CAVERNOUS MUDDY CALCIRUDITE

Fig. 5. Classification of core sample from S-14.

studies of coral phase confirmed the generation of soft coral spikes (Konishi, 1981). As such this supports the university of the distribution of these rocks in coral environments.

The four samples of autochthonous scleractinia were subjected to radioactive carbon dating in a low-level radiation test at the Faculty of Science, Kanazawa University and the following results were obtained (Analyst: Nobusada Tomoko).

| Sample code | Type | Depth | Age ($T^{1/2}$ = 5568) |
|---|---|---|---|
| ST-13-1 (=KL-287) | *Favites* sp. | 0.00–0.13 | 4170±100 |
| ST-13-2 (=KL-288) | *Porites* sp. | 2.40–2.51 | 4890±100 |
| ST-14-1 (=KL-289) | *Goniastrea* sp. | 0.00–0.19 | 3990±100 |
| ST-14-2 (= KL-290) | *Platygyra* sp. | 2.50–2.67 | 4680±100 |

Within the range of the period measured, the average sedimentation rate was found to be 3,500 and 3,900 Bubunof, which is quite similar to that for the spur region.

These results also indicated that the present surface in the trench region (depth 0.00–0.15 m) is quite old, as expected. There is a possibility that a) the trench profile in this area, as observed now, might have been modified by considerable erosion after emergence or b) the excavation point may lie on the slope which can be called the old spur region adjoining the spur region. It is therefore necessary to excavate the trench region over a wider range, in the future.

The upper part of the excavation samples obtained, except for S-12 which enters the tidal belt, can be assumed to have been subjected only to continuous underwater fluctuations after sedimentation. It can at the same time be used to read the records of the seabed period (1,500–3,000 y. B.P., Konishi and Matsuda, 1980), but this possibility is under investigation. Observations of the ores and the microstructure of skeleton fossils and cement in gaps so far indicate that at least the previous samples had a simple history, that they were exposed to sub-(inter) tidal zone environments after their formation. The argonite cement needles, which are the characteristics of this water mass, can be divided into two types under a microscope, viz., 1) syntaxial (low mass, optical directional continuity) and 2) botryoidal (radioactive spheroidal, semispheroidal). In the former, the crystal diameter can grow up to 15μ while later it cannot grow beyond 2μ. The former grows in the same direction as the needles of argonite (cement) crystals which grow in the skeleton that makes up the coral reef and it was confirmed even with SEM phenomena. The botryoidal cement can also be seen in the gap of soft scleractinia which are formed by high magnesium calcites. Now, the micrite-type high magnesium calcite, when seen under a microscope, shows an internal structure that is like imperfect pellets. This is formed as a sea water based cement, unlike the clastic limestones mud in gaps (MacIntyre, 1979; Goto, 1980 MS). However, under SEM phenomena, it was found to be an aggregate of Scalenohedral idiomorphic crystals of the order of 2μ. As such, the above-mentioned uniform microstructure like the pellet microstructure can be used as a standard for judging the cementation action of sea water.

We would like to study the sea level fluctuations in the later Holocene epoch in more details. This should provide important data for the comparison of samples from the same period which are exposed to a tidal region and a permeable region and for studies about the dolomitization action of mixed waters. During 1982-83, we plan to excavate the reef forest bed whereby we can expect maintaining the continuous record.

## Acknowledgment

The studies about SEM phenomena were conducted as a sponsored study with the help of visiting researchers (No. 81122). It is our duty to express our thanks to Prof. Nasu, Director, Institute of Oceanography and to Prof. Kobayashi, Dr. Fujioka and Dr. Nishida for their guidance: 1st year Master's student, 4th year Master's student, 4th year Master's student, 4th year Master's student. We also received assistance from M/s Ohera, Takahashi, Yoshioka, and others during the field studies.

## Bibliography

Goto, T. 1980. MS Cementation and diagenetic alteration of the Holocene raised reef formation. M.Sc. Thesis 1980-36, Fac. Sci., Kanazawa Univ.

Konishi, K. 1981. Alcyonarian Spiculite: Limestone of soft corals. *Proc. Fourth Internl. Coral Reef Symp.* (Manila) 1, 17-23.

Konishi, K. and S. Matsuda. 1980. Relative fall of sea level within the past 3,000 years. *Trans. Proc. Paleont. Soc. Japan,* N.S. 117, 243-246.

Konishi, K. 1979. The sea level fluctuations during the Holocene epoch as seen from the solid structure of postglacial coral reef. General Research Report (B) of 1977-78 Mombusho Science Awards.

MacIntyre, I. 1979. Distribution of submarine cements in a modern Caribbean fringing reef, Galeta Point, Panama. *Jour. Sed. Petrol.* 47(2), 503-516.

RESEARCH REPORTS

# *Hakucho-maru* Cruise KH-82-1 (Cepheus Expedition)

*Yoshio Horibe*

(Institute of Oceanography, University of Tokyo)

(Received: June 28, 1982)

The main aim of this cruise was chemical studies on deep waters in the northwestern Pacific Ocean which covers the area 30°N and southward. As such the cruise was important to the "Mixed Sea Waters" and "Heavy Metals in Sea Water" groups of this special research project. During its course KH-80-2 between April–June, 1980 (Cygnus Expedition) the *Hakucho-maru* has covered an area north of 30°N for similar studies. These two cruises have therefore provided important data useful for chemical studies on the deep-sea waters of the northwest Pacific region, particularly the East Houshu and South seas.

## Chemical Studies

The solid line of Fig. 1 indicates the route of the Cepheus Expedition time and the dotted line indicates that of the Cygnus Expedition.

Leg 1: Departure from Tokyo on January 22, 1982 and arrival in Ponape on February 15. Total 25 days. CTD observations were carried out to the bottom layer at points marked 0. At each measuring point except CE-7, the waters from each layer were collected with a 23 liter capacity Niskin type sampler. The water sampler was used every 100 m for depths less than 1,000 m, and every 250 m for depths more than 1,000 m. Since a pinger is attached (Benthos Co., Model 2216) at the wire tip, the maximum depth is restricted to 10 m above the ocean floor. Using the water samples, salinity and nutrient salts were analyzed onboard and results were disseminated to concerned scientists. At CE-5 and CE-8, a large capacity sampler was used and about 230 liter of water was drawn at each of 15 layers. This 230 liter water sample was used to analyze natural radioactive elements such as $^{14}$C, U-Th, etc. as well as artificial radioactive elements like $^{90}$Sr, $^{137}$Cs, Pu, etc. The researchers onboard conducted the processing, separation, etc. and the respective elements were brought back to the laboratory.

The CE-5 and CE-8 points, where large quantities of water were

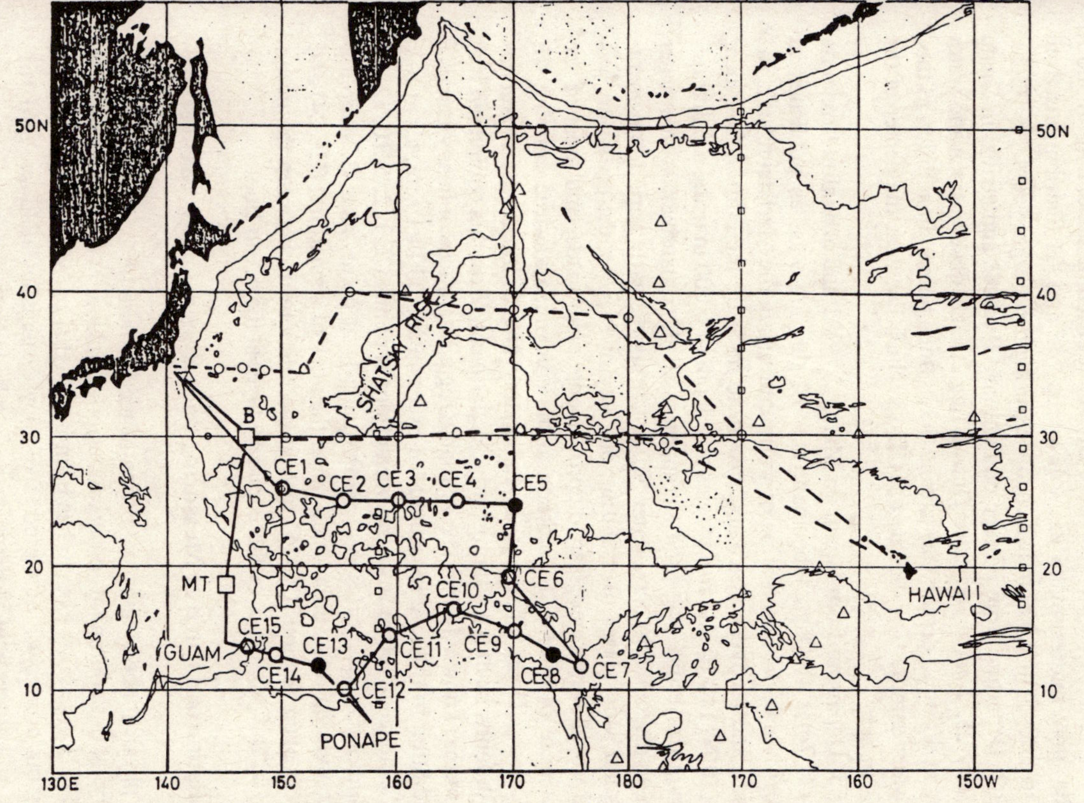


Fig. 1. □MT indicates points of the Mariana trough. There are eight points in this □(Ref. Newsletter No. 6, page 115, Fig. 2): B is a proposed dumping ground for radioactive waste (30°N, 147°E). With this as a center, a detailed profile of the seabed is available for the surrounding 50 miles. Hence this is very convenient for studies on the seabed profile. Depth 6,000–6,200 m.

sampled, correspond to points 227 and 229 of the GEOSECS plan executed during 1970s. The 1973 results of $^{14}C$ and $^{3}H$ distributions are already available for these points. The point CY11 discussed earlier [Newsletter No. 2, p. 25 (GEOSECS 226, 30°34′N, 170°36′E)] has a different sea condition and it would be quite interesting to know how the $^{14}C$ and $^{3}H$ distributions from surface to a depth of 1,000 m has changed during these eight years.

Leg 2: Departure from Ponape on February 19, 1982 and arrival in Guam on February 25. Total seven days. During Leg 2, CTD observations and water sampling at different depths was carried out at three points. In addition to this, a large water sample was obtained at CE-13 which lies near the center of the Marianas trench.

Leg 3: Departure from Guam on March 2, 1982 and arrival in Tokyo on March 17. Total 16 days. The main task of this leg was, as published in Newsletter No. 6, the observation of warm currents in the Mariana trough, the installation and recovery of a buoy system to measure the current flow near point B (30°N, 147°E) and fish studies at a depth of 6,000 m using a fish trap.

During this cruise, we used six stainless steel, contamination-free water samplers as discussed in Newsletter No. 6. During the early period, the operation was not satisfactory. But, during our stay at Ponape, the operation was observed under water, necessary improvements were made, and most of the problems were overcome. In addition to the contamination-free sampler, we also installed a clean laboratory in Laboratory No. 2 aboard the *Hakucho-maru* with funds from the administrative department. With this equipment it is felt that further knowledge will be available regarding traces of heavy metals.

In addition to this, we measured Rn excess using 23 liter samples at 6–7 levels in a range of 300 m from the ocean floor at points CE 1, 2, 6, 10, 12, 14 and an additional two points near B. Particularly at points near B, sampling and CTD observations were carried out at the peak of a small hill (CE-24) and the deepest plain section (CE-25) to find the relationship between the ocean floor profile and properties of the boundary layer at the ocean floor.

## Installation of a Current Measuring Buoy System and Its Recovery

The physical oceanography section of the Institute of Oceanography has been carrying out continuous observations of deep layer currents near point B since 1977. Here a buoy system has been installed about 90 km west of point B. During our cruise, we recovered three buoy systems installed on April 16–18, 1981 and three new systems were installed. Both the installation and recovery operations were carried out smoothly. Considering the situation in 1975, when the study was first planned, we felt much improvement. Table 1 shows the positions of the buoy system installed and recovered as well as the

**Table 1. Installation and recovery of buoy system**

| Point | Position | Distance of current meter from ocean floor (m) |
|---|---|---|
| | Recovery system | |
| TA | 30°00′N, 145°45′E | 50, 200, 400, 600, 800, 1800 |
| TC | 30°00′N, 145°00′E | 1000, 2000 |
| TF | 31°00′N, 145°45′E | 100, 500 |
| | Installed system | |
| TA | 30°00′N, 145°45′E | 50, 200, 400, 600, 800, 1800 |
| TF | 30°00′N, 145°00′E | 30, 230 |
| TH | 29°29.8′N, 145°20′E | 30, 230 |

distance of the current meter (Belgen Model No. 5) from the ocean floor.

Among these, the position of TA point has been monitored continuously since December 4, 1977. Adding the data obtained by our recovery, we have available long term data (1,260 continuous days) about the current velocity. Point TH was established for the first time but all three points represent the apexes of a triangle with each side being 80 km long. As mentioned in Newsletter No. 1, such an arrangement is made to find out whether the average current velocity is affected by the seabed profile as we go near it. The current at TH is expected to have a westward component.

## Conclusion

The course of this cruise was luckily quite smooth. The winds were rather strong but since they were in a favorable direction, they helped increase the vessel speed, which sometimes reached 13 knots. As a result, on such a lengthy cruise, some room was available for time adjustments and we were comfortable.

The large water sampler used at present has reached a state of perfection. The safety and preparation of the scientists, as well as handling by the vessel, were smooth and without any problem. We could collect large samples even under a wind pressure of 15 m.

There were many problems in the sampler fabricated by copying the tripod core sampler of the Woods Hole Oceanographic Institution. This was probably because we did not properly understand the merits of the original model.

As one of the main organizers of this cruise, I feel that we had never before had a cruise where all the tasks were completed just as planned. I take this opportunity to thank all the researchers accompanying our cruise for their kind cooperation as well as the crew members of the *Hakucho-maru.*

# The *Tansei-maru* Cruise (KT-82-Part 4)

*Prof. Keisuke Taira*

(Institute of Oceanography, University of Tokyo)
(Received: July 2, 1982)

## 1. Current Observations with a Float

The above cruise was conducted between May 8–14, 1982. The current meters were installed at SR1, SR2, SR3, OW and SG1 as shown in Fig. 1. This cruise is a part of the research activities under the special research project "Variations in the Kuroshio and effect of the earth profile (Chief Coordinator: Prof. Toshihiko Teramoto)."

Points SR2 and SG1 are the same points used by Prof. Tomota in his Japanese-French joint survey of Japan Bay. These flow meters are to be

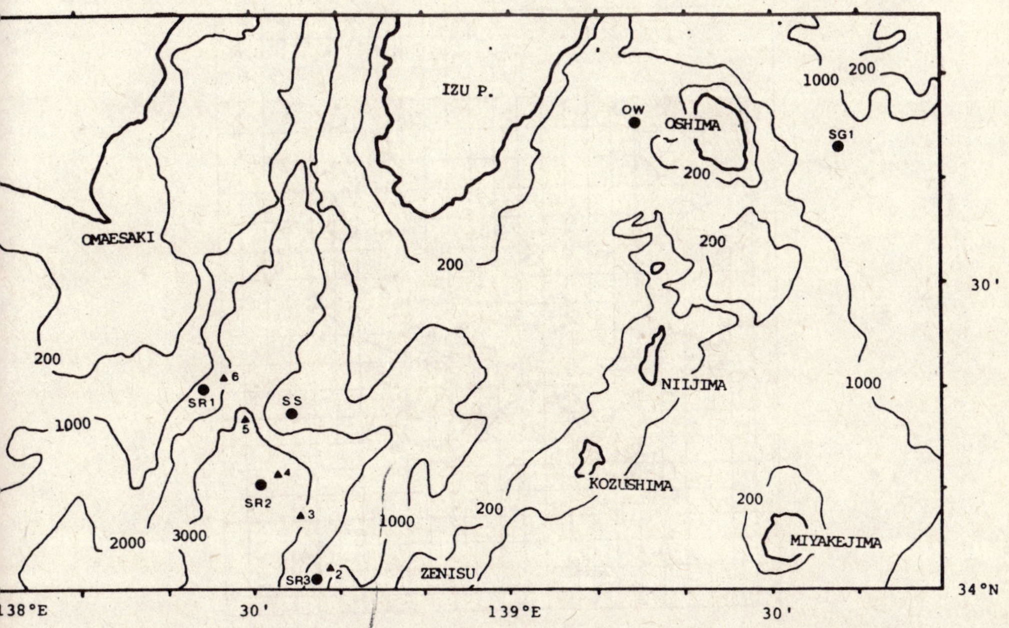


Fig. 1. Observation points of the *Tansei-maru* cruise (KT-82-4). ● current meters installed (SR1, SR2, SR3, OW, SG1), ▲ Accoustic dropsonde (points 2–6).

recovered this December by the *Tansei-maru.* As mentioned elsewhere, the cross sectional currents off Toi-misaki and Shiono-misaki were measured during this special research project. Fig. 2 shows the ocean current diagram based on our observations during May 7-20, (*Kaiyosokuho,* 1982, No. 10; Hydrographic Survey of Maritime Safety Agency). In our studies we observed the Kuroshio current velocities along its axis at the following six places. Thus three cross sectional points off Toi, three cross sectional points off Shiono-misaki, Hachijojima west (33°N, 139°E, continuously monitored since May, 1978), Funka trough (SR 1-3 in Fig. 1), Oshima west lane (OW in Fig. 1), and Sagami trough (SG 1 in Fig. 1). This is the first time in Japan that an attempt has been made to simultaneously measure the Kuroshio current at 12 points on its course, even the gulf currents are not measured this way. This should hopefully make the characteristics of changes in the Kuroshio current quite clear.

## 2. Measurement of the Vertical Profile of the Kuroshio Current Velocity Using Acoustic Dropsonde

We have developed an acoustic dropsonde to directly measure the vertical profile of the horizontal current velocity and the flow mass of ocean currents [Toshihiko Teramoto: Shiken Kenkyu (1) Hokoku, March 1982;

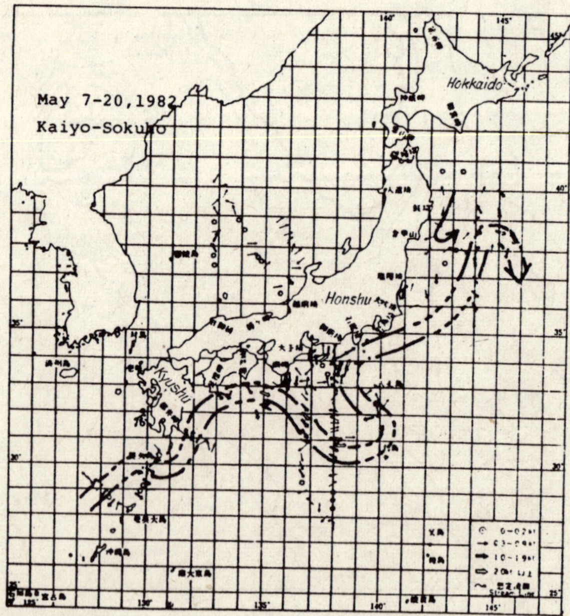


Fig. 2. Ocean current diagram (*Kaiyosokuho,* 1982, 10, Hydrographic Section).

Kitagawa, et al., Acoustic Dropsonde Development; Proceedings of 1982 Spring Symposium of the Oceanographic Association of Japan]. Here a sound source is allowed to drop freely and its distance under water is used from the front and rear tips of the vessel. If the vessel position is determined by Lolan C and the direction by compass then we can know the position of the sound source under water. However, it is not possible to say whether it (sound source) is on the left or right side of the vessel. The sound source can be balanced with the help of a buoy and weight so that, after reaching the bottom, it drops the weight and continues to float. We propose to fix a depth meter to the sound source. The possibility of an acoustic measurement technique was confirmed last year during experiments from the observation tower (Bosai Center) and the harbor. This equipment was first used during this cruise at the five points marked by ▲ in Fig. 1.

Looking at Fig. 2, the current axis seems to be at 34°N, 138°30′E, where a strong flow, over two knots has been measured with GEK. It is a northeast flow which means the current axis lies at points 3-4 (Fig. 1). CTD observations were carried out at SR 1-3, SS which should be confirmed by subsequent analysis. This also means that the measuring line of the dropsonde was cutting the Kuroshio current axis at right angles. The current axis at this cross section follows the axis of the Funka trough. However, as it flows in northeasterly, it enters the Oshima channel and as the maximum depth of this channel is only 500 m, the deeper waters would be stopped to form a dam (Taira and Teramoto, 1981). Now if (the Kuroshio) current flows eastward as shown in Fig. 2, the depth of water from Zenisu to Oshima is around 200 m. Thus only the waters above the 200 m level can cross the ridge. Another estimation says that about $30 \times 10^6$ $m^3 s^{-1}$ of the flow must cross the ridge. The waters crossing the west Oshima channel are calculated to be $2 \times 10^6$ $m^3$ $s^{-1}$. Direct measurements by acoustic dropsonde should dispel all doubts in this respect.

The analysis of data has, unfortunately, just begun and we cannot say anything at this point. Fig. 3 shows the vessel's position and the direction at station 4. This figure also shows the time lapse after dropping the sonde. The ship sailed in northeastward at 1.5 knots, indicating that it was caught in the Kuroshio current axis. Fig. 4 shows the position of the accoustic dropsonde, assuming a dropping speed of 70 cm $s^{-1}$ (estimated value 60-80 cm/s). The position at an interval of 200 m was recorded. The actual position is within ±25 m from that indicated by the straight line. The accuracy of the Lolan C, shown in Fig. 3 is ±20 m. We assumed that the sonde lay on the right side of the stern. (The positional curve corresponding to the left side of the stern is not realistic as the currents in shallow waters are weak, whereas those in deep water below 1,000 m are as strong as 3 m/sec. These two curves do not match.) It is clear from Fig. 4, that the Kuroshio vertical profile is quite complicated. The thickness (range) of the northeast current which causes ship drift spans few 10s of meters 50–60 m and an average current up to 200 m depth is north-northeast. Between 200 and 600 m it is northwest whereas between 600 m and 1,400 m it flows southward. Fig. 5 shows the vertical distribution of the east-west and south-north components of current velocity. We still have to wait for a comparison with results from a dynamic computation using CTD data.

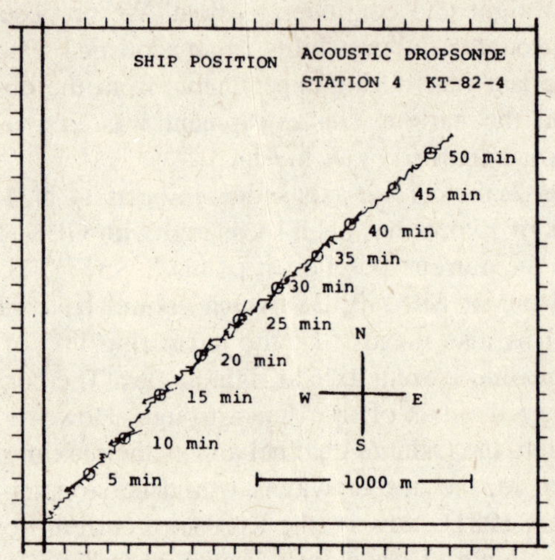


Fig. 3. Position of the *Tansei-maru* during acoustic dropsonde observations. Time of sonde after dropping was recorded. It was flown in a northeast direction at 1.5 knots.

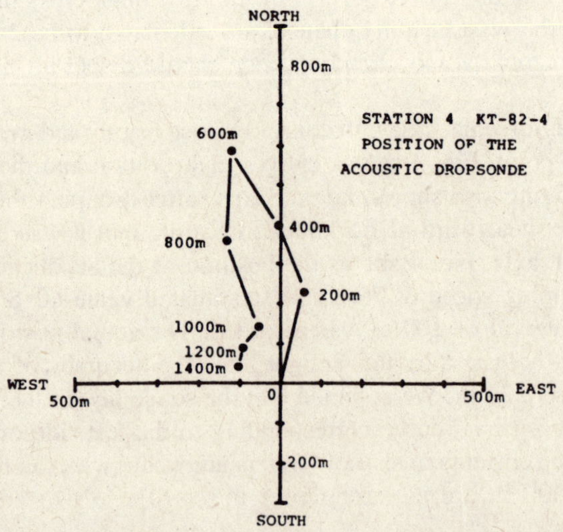


Fig. 4. Position of dropsonde at various depths. The point at which dropping started is considered the origin.

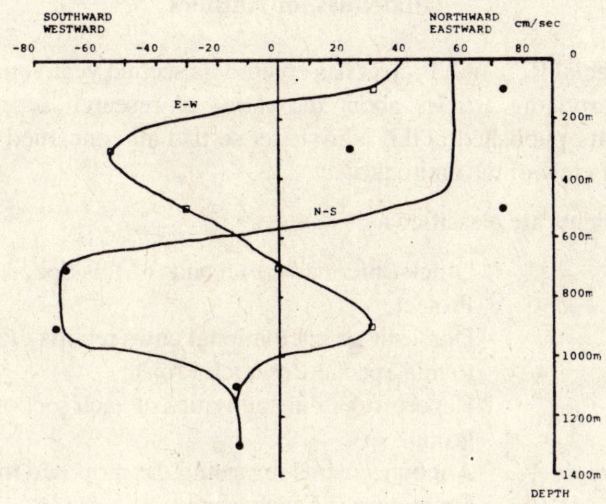


Fig. 5. Vertical distribution of Kuroshio current velocity as determined from the locus of the dropsonde.

During the research plans (2) by Taira and Fukazawa in 1982, it was decided to procure CTD equipment. This will not only make measuring the dropsonde depth possible, but also the density and velocity fields.

## 3. Conclusion

At point SS of Fig. 1, a sound source was installed at 1,000 m and the tests regarding sound propagation were carried out. These results will be discussed separately by Prof. Okujima.

Prof. Kitagawa was responsible for collecting data from the acoustic dropsonde, whereas the data was analyzed by Prof. Fukazawa.

We conclude this report by expressing our gratitude to the crew members of the *Tansei-maru* and the research members for their kind cooperation.

## Guidelines for Authors

This Special Research Project has entered its second year.

All enterprising articles about the status of research activity, interim results, etc. are published in this Newsletter so that all concerned persons can have a broad view of the entire project.

The contents are classified as:

| | |
|---|---|
| Highlights: | Quick (intermediate) results of this Special Research Project |
| Bulletin: | Domestic or international news reports closely related to this Special Research Project |
| Reports: | Reports regarding activities of each section or working group |
| Announcements: | Announcements regarding the proposed meetings or conferences of each group |
| Bibliography: | Listing of papers published by the members on this Special Research Project |

Manuscripts should be written on standard 400 character size paper (handwritten material also accepted). The figures should be drawn in such a way that they can be reproduced as they are.

It is proposed to publish this Newsletter every six weeks.

**Note:**

When reproducing or citing the material published in this Newsletter, due acknowledgment should be given. Authors' names, wherever mentioned, should also be quoted. A copy of the publication containing such a reproduction may be sent to the Editorial Section of the Administrative Office.

| | |
|---|---|
| Published by: | Administrative Office,<br>Special Research Project "The Ocean Characteristics and Their Changes" |
| Published at: | Administrative Office,<br>Physical Oceanography Department,<br>Institute of Oceanography,<br>Tokyo University,<br>1-15-1 Minami-dai, Nakano-ku,<br>Tokyo 164<br>Tel.: 03-376-1251, Ext. 251. |

Special Research Project

# The Ocean Characteristics and Their Changes

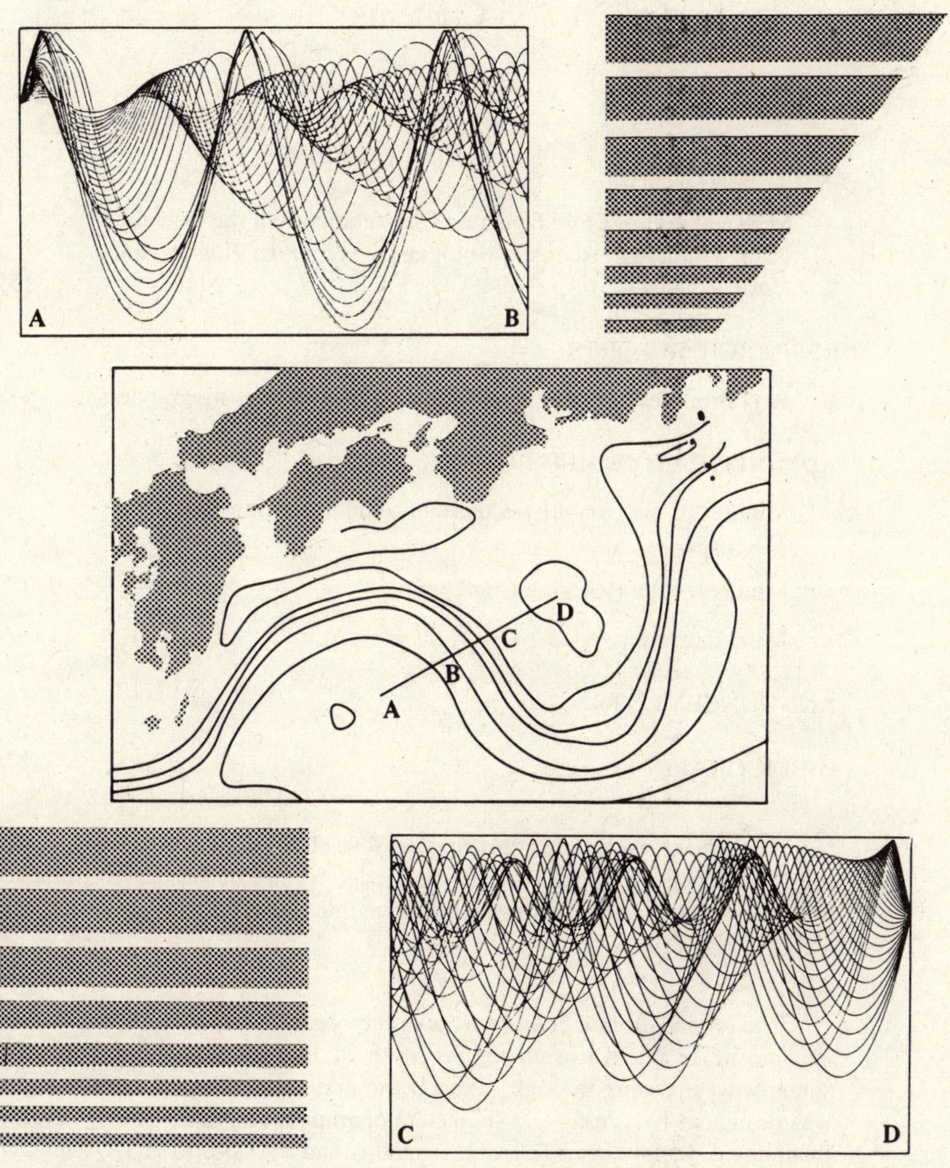


NEWSLETTER NO. 8, SEPTEMBER, 1982

# Contents



Researchers engaged in this Special Research Project, who wish to submit their papers in English and acknowledge receipt of a research grant for this special study, should indicate the above title of the project.

Cover Page

The propagation of sound waves in the ocean as estimated from the density profile in the Kuroshio region south of Honshu are shown here. The sound waves passing through A and D are particularly important. The figure was designed by Prof. Okushima (Department of Precision Engineering, Institute of Technology, Tokyo University) and was prepared by Kilagawa (Institute of Oceanography, Tokyo University).

## HIGHLIGHTS

# Measurement of Fluid Dynamic Characteristics of the Ocean Using Numerical Acoustic Holograms

*Yasutaka Tamura and Hiroo Yamazaki*

(Faculty of Engineering, University of Tokyo)
(Received: September 6, 1982)

### 1. Introduction

Optical or electromagnetic waves are greatly attenuated in the ocean. Hence sound waves are generally used in marine surveys, measurements, and communication. The oceanic structure occupy three-dimensional space and during many submarine surveys or measurements it is convenient to project the spatial distribution of information diagrammatically.

Studies have been made regarding the techniques whereby the position or condition of the object can be photographed using sound waves, particularly in the fields of medical and oceanographic studies. Recently some techniques using acoustic holography have been developed for acoustic photographic systems [8]. This paper discusses primarily the practical experiences of authors on the use of acoustic holography in oceanographic studies and simultaneously discusses various techniques developed.

### 2. Acoustic Holography

2.1. *Holography*

Holography was first invented in the optical wave field [7] and then various applications were studied. Holography is a technique for recording and reproducing images and involves two-stage processing, viz. 1) record the wavefront and 2) reproduce the wavefront. Nowadays, holography is not restricted to optical waves, but the image can also be built up with electromagnetic and sound waves.

In the case of optical waves, a photographic dry plate is commonly used to record the wavefront. This photographic plate is highly sensitive and a plane recording is possible. However, since it is not sensitive to the phase of incident light, a coherent laser beam is used to obtain a hologram that records both the amplitude and phase of the wave front and the actual recording is obtained in the form of interference with the reference light wave.

As against this, no external reference wave is necessary in the case of acoustic waves and the sound field can be detected directly. Similarly, the transmission, reception, and recording of a large band of frequencies is easily possible in this case. A long time record of a variable wavefront (in amplitude or phase) is also possible. Sometimes a holographic record is even obtained with a non-standard sound source [9]. However, there are also some problems. Thus, a high sensitivity plane detector for sound waves does not exist yet and, as a result, either we have to depend on a single detector or an array of multiple detectors to record the wavefront. The wavelengths of sound waves used in submarine holography are comparatively larger, between a few mm to a few cm, and, therefore, image quality is not so good. It is easily affected by factors such as water temperature, variation in composition, medium movement, etc.

Various methods have been proposed for recording the sound field and reproducing the image. In oceanographic applications it may, however, be desirable to use a scanner or array of receivers to detect the sound field and record it in a digital form so that the image can be built using a computer. This is the numerical acoustic holographic technique for image reproduction.

### 2.2. *Principles of Acoustic Holography*

The method uses reflected and dispersed waves wherein the equipment is comparatively smaller and still distant objects or objects spread over a wide region can be easily observed.

Acoustic holography consists of two-stage processing: wavefront recording and wavefront reproduction (image reproduction). The object is exposed to sound waves and the reflected or dispersed sound field is recorded in digital form.

The recorded hologram can be considered to have sampled the combined spherical sound wave propagated from fronts, each point of the object as a point reflection source, using the detector. Each spherical wavefront would be received with a certain delay as it follows the acoustic path: transmitter → object → receiver. When the transmitted signal has a narrow band width, the above delay can be considered to be the phase rotation.

Exactly the reverse procedure is adopted to reconstruct the acoustic image from the recorded data. Information about a certain point in the object region can be determined as follows. The delay or phase rotation for each problem point on the hologram is calculated and then the delay or phase rotation is carried out in such a way that it will annul the previous effect. This parameter is added to all points on the hologram. The problem points can be assumed as the source of reflected sound waves and the spherical wavefronts originating from these points are added on the hologram surface with the same phase. The reflected waves from other points are added in different phases and become cancelled. This operation corresponds to illuminating different points (to a different extent) while reconstructing the hologram in optical holography. Now, focusing the phased array or delay and sum beam former can be assumed to have been done by the computer.

The signal so obtained represents a sound field reflected from the problem point. Generally this operation is carried out only for necessary points in the object region and, by reconstructing the space distribution of the strengths of the reflected sound field, the position and shape of objects is expressed.

The signal indicating the reconstructed sound field contains more information than amplitude alone. We can, in fact, obtain images of phase distribution or the frequency dependence of the reflection ratio. Also, if we could determine the space distribution of the Doppler displacement, it may be possible to picturize the state of movements of the object.

### 2.3. *Merits of Acoustic Holographic Measurements*

Measurements or surveys using acoustic holographic techniques (particularly those which reconstruct the object numerically) claim the following merits:

1) By selecting a larger hologram surface, it is possible to improve the directional resolution.

2) Information regarding the entire object area can be recorded simultaneously. Therefore, any point can be focused by computer. Accordingly, it is possible to collect information on a wide region in a very short time.

3) Flexible signal processing is possible so that images can be reconstructed by computers. Correction in the deviation of the transfer element's properties or changes in signal processing, depending on the condition of object and purpose of measurement, can be carried out easily. Now, since image resolution obtained by acoustic holography is better than that obtained by optical holography, further image processing such as improvements in picture quality using a computer, characteristic refraction, etc. can be assumed to be without defects.

## 3. Application of Acoustic Holography in Oceanographic Studies

### 3.1. *Application to Submarine Image Processing*

Acoustic holography techniques using numerical reconstruction can find the following applications in oceanographic studies. One is its use in holographic sonar [6] which aims to improve the acoustic television camera or directional resolution. Here a comparatively high-frequency sound source is used to observe objects in a highly turbid sea. In another application, a low-frequency sound source with low attenuation can be used and, by forming an array with a large mouth, one can think of scanning a larger area with a high resolution (mouth formed side locking sonar, etc.). In either case, the improved directional resolution as a result of the holographic mouth formation is used to picturize the position and condition of undersea objects. This can then act as a submarine projection equipment.

3.2. *Problems in Oceanography*

Following problems are faced when using acoustic holography techniques in oceanographic studies.

1) Sound path bends or deviations in propagation characteristics under the ocean.

2) If we widen the hologram opening to improve directional resolution, many array elements become necessary, which further increases the size of the electronic circuitry.

3) Unwanted reflection waves or noise emanating from the seabed or surface.

4) Array movement.

5) Corrosion due to sea water or the adhesion of living organisms.

It is expected that the first problem can be solved by acoustic calculations or corrections; whereas, the second may be solved as high density parallel processing circuits become a reality with developments in integrated circuit technology. Some announcements have been made to reduce the number of array elements and memory capacity. When the effect of an unwanted signal or movement in the detecting plane is too large to be ignored in numerical computations, some pre-processing then becomes necessary to eliminate unwanted signals while recording the hologram.

3.3. *Measurement of Velocity Vector Distribution Using Acoustic Holography*

Scientists have been studying [1-4] a measuring technique employing acoustic holography which could be used to calculate the velocity vector distribution of mobile objects. This method is explained here as it can also be used in studying sea water movements.

The fluid dynamic structure of the ocean spans three-dimensional space over a wide region. A localized measurement would be insufficient to survey this structure as it is desirable to explore all aspects of the flow. This then calls for measurements of the spatial distribution of the velocity vector and its variation with time. The authors proposed to obtain this information with acoustic holography techniques.

It is possible to observe the total area under study and, at the same time, use numerical acoustic holography methods. Also mobile (dynamic) objects also can be observed and the beam of any shape can be formed. Our study aims to "visualize the flow" quantitatively and prepare a "velocity vector measuring sonar" using the above properties.

3.3.1. Measuring sequence

The measurement of velocity distribution is made in the following two stages as shown in Fig. 1.

1) Sound waves are propagated in the ocean and their spatial and temporal pattern reflected from sea water as well as other mobile reflecting objects (scatterers) is detected by transducer array elements and recorded after digitizing.

2) The recorded hologram is subjected to the proper processing after which the velocity vector field is calculated and displayed.

Among various reflecting objects (scatterers) we can mention floating material, air bubbles of the impedance change caused by variations in the

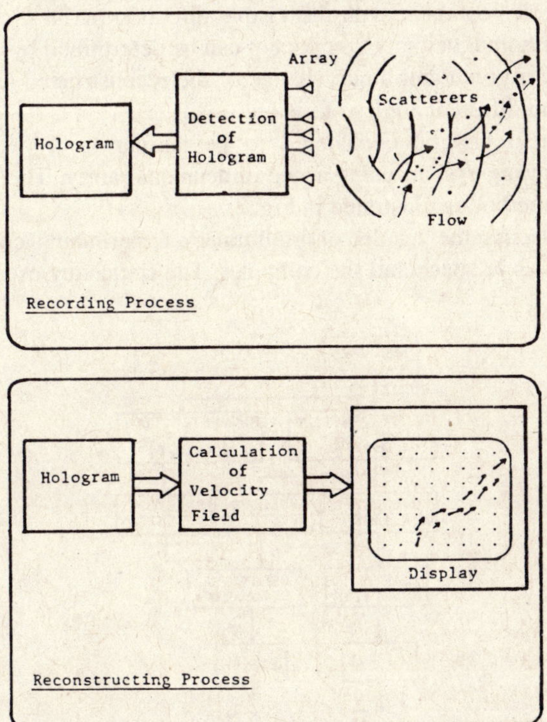


Fig. 1. Measurement of velocity distribution using acoustic holography.

concentration or composition of the sea water.

Various methods can be thought of for the calculation of the velocity vector depending on the methods used to change the frequency of transmitted sound waves or the structure of array elements, etc. So far, a number of methods have been proposed and the possibility of measuring the velocity vector distribution has been confirmed in a model experiment using ultrasonic waves in space. Those methods and experimental results are discussed below.

3.3.2. Method 1

Here a sinusoidal sound wave is either propagated continuously or as repetitions of burst waves. The wave-form data, recorded as a wave intercepting the reflected wave, is by overlapping Doppler beat signals corresponding to the movement of objects. This data is actually a varying reflected sound field recorded on the hologram time scale. Using this data we can reconstruct the time-based signal indicating variations in the sound field reflected from a point at which the velocity vector is to be determined. This signal is equivalent to the output of Doppler sonar focused toward the problem point. Using a central frequency of the reconstructed signal, it is possible to determine the velocity component in the visible direction (axial direction) at that point. The horizontal velocity component can be determined by estimating the time delay for a given amplitude change of the reconstructed signal between two points with different angles of vision.

This system can be considered to be similar to a two-dimensional velocity measuring system using a one-dimensional array. The sequence followed in this method is illustrated in Fig. 2.

Let us discuss the results of preliminary experiments conducted with ultrasonic waves in space and the computer. The frequency used was 40 kHz

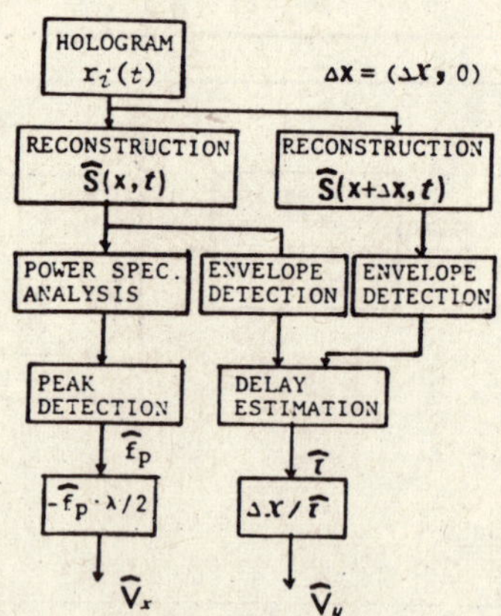


Fig. 2. Sequence followed in velocity estimation method 1.

(wavelength about 8 mm). A linear receiving array was made by spacing eight microphones at an interval of four wavelengths and one fixed frequency transmitter was kept at the center of that array during the experiment. Changes in the amplitude of the received signal are simultaneously sampled for all eight channels and recorded as a 12 bit information. Subsequent estimations

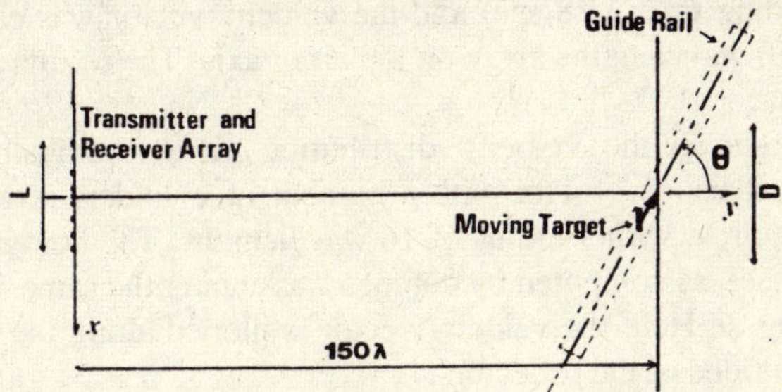


Fig. 3. Arrangement of a velocity vector measuring experiment using ultrasonic waves in space.

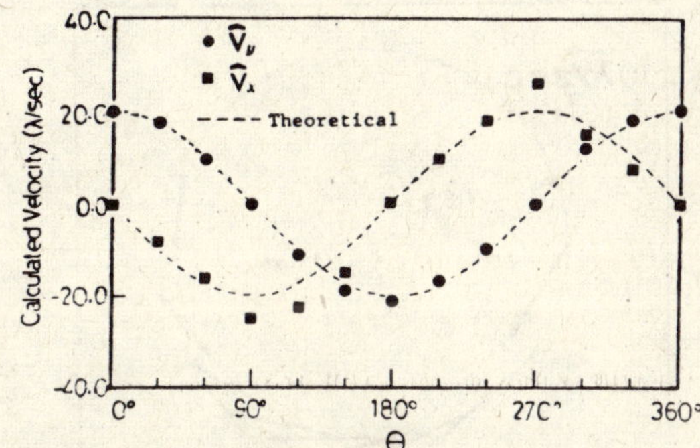


Fig. 4. Angle of guide rail θ and measured values of velocity vector.

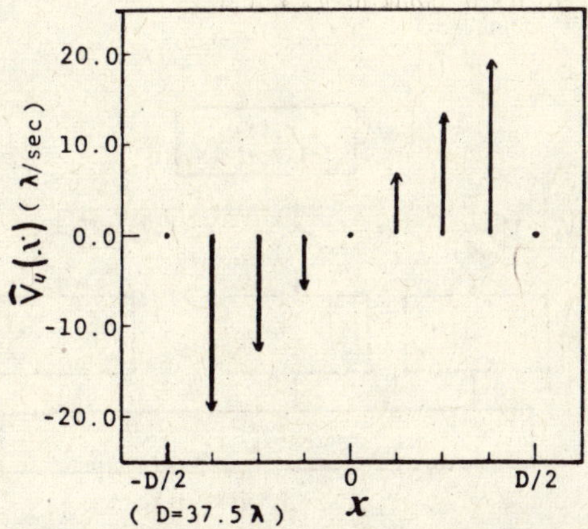


Fig. 5. Measured velocity distribution along the visible directions for multiple rotating and revolving bodies.

of velocity components were done with a large computer.

With the arrangement shown in Fig. 3, an aluminum sheet was used as a reflector and was allowed to move linearly (with a velocity of about 20 wavelengths/sec) and an experiment was conducted to measure the velocity vector. Here the waveforms were recorded by changing the angle of movement θ (recording time 1.28 sec) and the velocity vector was estimated at a point about 150 wavelengths away on the array axis. The results are shown in Fig. 4.

Figure 5 shows the velocity distribution along the visible direction measured at a distance of 1 m with a number of cylinders arranged on the periphery of a circle with a radius of 16 wavelengths. The hologram (recording time 2.56 sec), as computed by computer assuming the same arrangement, is shown in Fig. 6. Here the velocity vector is plotted along the circle representing the positions of the reflectors.

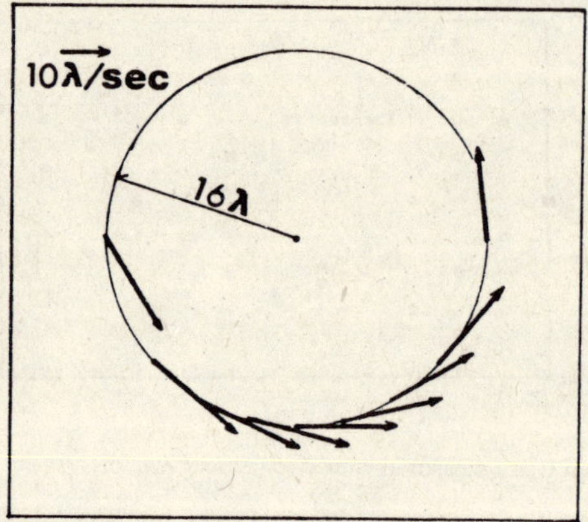


Fig. 6. Results of computer simulation.

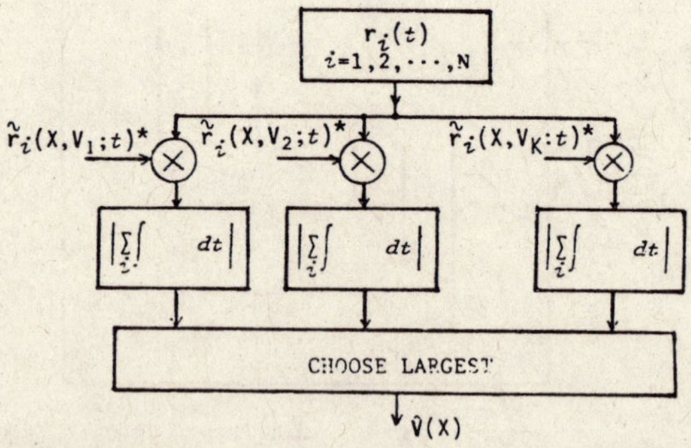


Fig. 7. Sequence observed in method 2.

3.3.3. Method 2

It is possible to scan the focus point by various methods using a computer. If we could ascertain that focal point movement matches with the actual target movement, then the scanned velocity vector at that time would be the estimated value of vector velocity of the object. The following method can be considered to confirm the matching of movements. Thus the focus of a numerical image system is scanned with a certain speed and the output (of system) is integrated with respect to time. This procedure is repeated by varying the (scanning) speed within the necessary speed range and with necessary quantification. If the speeds match, then the phase variation (difference) of output becomes zero and the absolute value of integration reaches maximum. A general schematic diagram of this method is indicated in Fig. 7. In this figure $r_i$ $(t)$ $(i = 1, \ldots, N)$ is the waveform detected at time $t$ for each of the $N$ array elements. Here the point $x$ under consideration is assumed to be a point mass moving with velocity $V$. The complex conjugate of the obtained waveform $\tilde{r}_i$ $(x, V; t)$ is multiplied by the detected waveform and, by carrying out addition, integration within the hologram plane and by observing time, we can intergrate the scan of velocity $v$ around the point $x$ with respect to time.

Compared to method 1, method 2 can be used as a measuring system

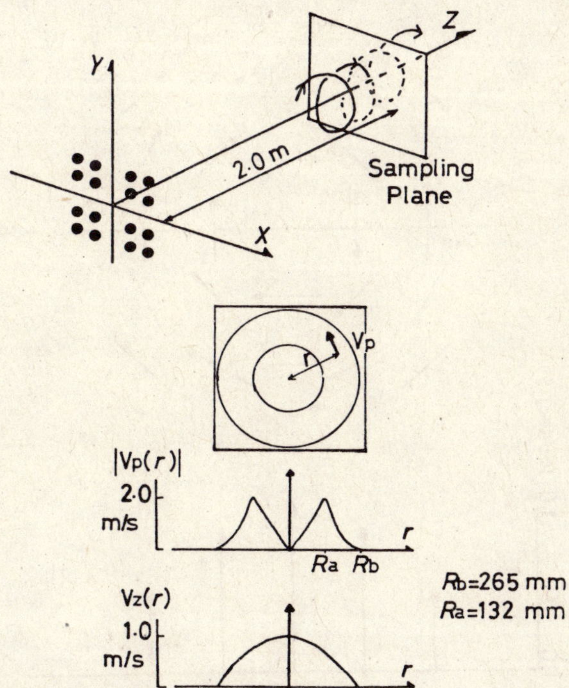


Fig. 8. Arrangement of computer experiment.

with a much higher space resolution and can also be applied to a three-dimensional velocity measurement using a two-dimensional array. On the other hand, since the number for which calculations are repeated is quite large, a much longer time is required to estimate the velocity.

Let us now present the results of calculations. Let us assume a submarine ultrasonic wave of 300 kHz (assume a wavelength of 5 mm). Then 16 × 16 receivers are set up on a lattice arrangement with an interval of 2 cm. A burst wave with a pulse width of 0.3 m/sec is propagated every 1 m/sec from the center (of above arrangement) and a hologram is computed, assuming turbulent multiple point radius of 27 cm and at a distance of 2 m as shown in Fig. 8. The recording time is 64 m sec. The vortex has a center along the array axis ($z$-axis in figure). The absolute value of the velocity component $|V_p|$ along the plane at right angles to the array axis and $V_z$ the velocity component along the axis become functions of the distance from axis "$r$" alone and were assumed to maintain the waveform mentioned in Fig. 8. The focus was scanned only for directions and by using the central frequency of the reconstruction signal to estimate velocity in the visible direction, the computer time was reduced significantly. Fig. 9 shows the result of estimating a three-dimensional distribution of velocity vector.

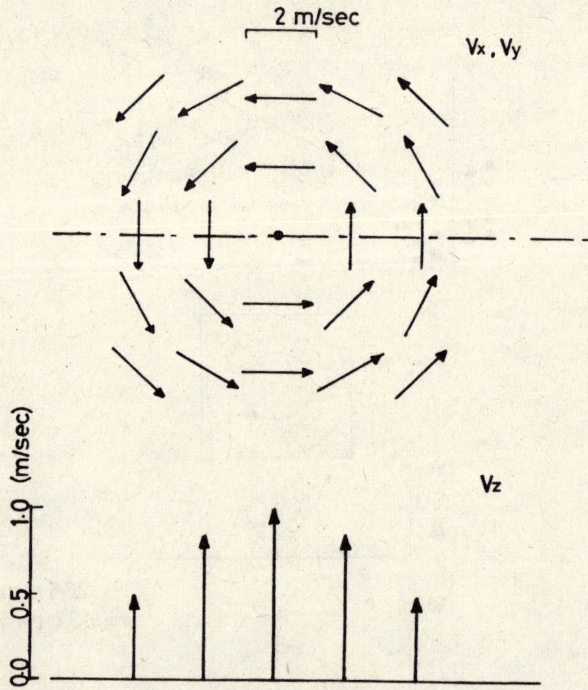


Fig. 9. A three-dimensional velocity distribution reconstructed from data obtained from a computer experiment.

3.3.4. Use of LPM modulation

In both the above methods for velocity measurement, the distance over which a measurement is possible becomes restricted if the reflector bodies are scattered over a wider range and if a continuous wave or repeated burst waves are used for measurements. Thus at 300 kHz, when we want to measure the flow with a maximum velocity of 1 m/sec in a visible direction, it is necessary to use a burst wave repetition frequency above 800 Hz and the distance over which the measurement is possible turns out to be only about 1 m. This problem occurs since the quantum of the Doppler shift of the reflector bodies is measured with the central frequency of the Doppler beat.

These two methods for estimating the velocity can also be used, however, with a different type of frequency modulation. Thus if a LPM modulation (Linear Period Modulation), as proposed by Altes and Skiner [5], is used with first transmission wave, the quantum of Doppler shift can be measured in terms of the phase rotation of detected output. In this case, the burst repetition frequency can be reduced and we can also extend the distance range over which measurement would be possible, save the memory capacity, and simplify computations.

The experiment using the LPM method was conducted in space. Two transmitters were used as shown in Fig. 10. Here λ is the wavelength of sound

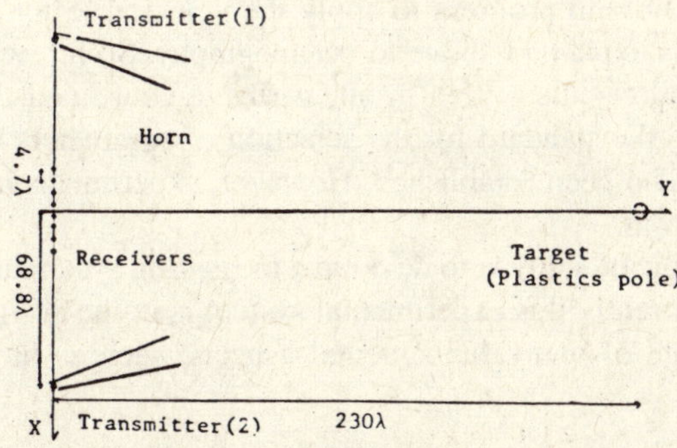


Fig. 10. Experimental setup for LPM sound waves.

Fig. 11. Waveforms of LP modulated signal: a—40-48 kHz, b—32-40 kHz.

waves at 40 kHz. From each transmitter, mutually noncorrelated LPM sound waves are radiated in the frequency range of 40–48 kHz and 32–40 kHz instantaneously as shown in Fig. 11. To reconstruct the reflected sound field, a delay and addition was used. Whereas, to estimate the velocity, method two was used. To calculate focus scanning, the waveforms from each receiver, mutually correlated to transmitter signals, are delayed to a certain extent and the phase change corresponding to the scanning velocity vector is added. In this way computation time can be reduced.

When a cylindrical object with a 4 cm radius was allowed to move with a velocity of 0 [1] and 20 λ/sec [2–4], the estimated values of velocity vector were as shown in Fig. 12. In this experiment, however, the transmitter and receivers were interchanged manually. Also, since waveform recordings were repeated synchronously with the movement of the object, the accuracy of the measurement has dropped as a result of the effect of air swinging in the laboratory, etc.

When two or more transmitters can be placed with a wider interval (or if multiple receiving arrays can be used) as shown in Fig. 10, the velocity vector can be measured in a very short waveform recording time. In the example of this experiment, the velocity vector can be estimated with a recording time of only 6 m/sec.

3.3.5. Future programs

Studies are now in progress to apply the velocity vector distribution estimation methods explained above to oceanography. So far, with a model experiment using ultrasonic waves in air, we have confirmed that the method works fine. Also the standard for the selection of parameters for the measuring system has also been established. However, experiments in the ocean are yet to be carried out.

The next topic of study is to develop a measuring system that can be used in the ocean. Currently this experimental system is in the design stage but we lack a knowledge of many fundamental aspects, such as how to apply the

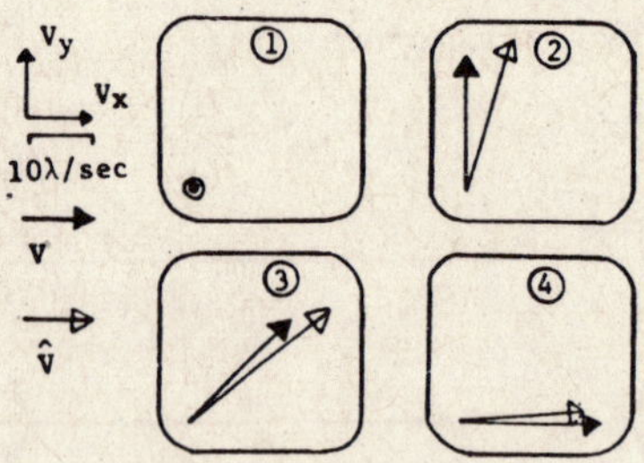


Fig. 12. The actual velocity $V$ and estimated velocity $\hat{V}$.

acoustic holography mentioned in section 3.2 to oceanographic studies.

Thus, the first step is knowledge about reflector bodies in the ocean. It is necessary to know the actual reflector, its size, its distribution condition, strength of reflection, etc. Second, we should know the span of the measurement area and the range of velocity. In oceanography, the guidelines for a system design change depend on the range of the area to be measured and the range of velocity distribution to be measured. The present authors have no experience or adequate knowledge regarding experiments in the ocean. We therefore shall be happy to receive suggestions for these studies.

We are planning to measure the velocity distribution up to a few m/sec in a range of few meters to few tens of meters (distance) by collecting reflections from a floating body using a sound source of a few hundred kHz. For distances beyond that range, comparatively low frequencies can be used and the scale of the reflector bodies should also be larger. Depending on the circumstances, it may be necessary to use an artificial reflector or a sound generator. But even in those cases, the measurement would be possible using the merits of acoustic holography. The application of methods proposed by the authors may not be restricted to the measurement of velocity alone but can also be used to observe marine life.

## Bibliography


1. Tamura and Yamazaki. 1982. *Proceedings of the Symposium on Measurement Automation and Control,* 18-1, 44/51.
2. Tamura. 1982. Doctoral Thesis in Faculty of Engineering, University of Tokyo.
3. Tamura and Yamazaki. 1979. *Proc. IMEKO Tokyo Flow Symposium,* 245/250.
4. Yamazaki and Tamura. 1982. *Proceedings of the Seminar on Oceanographic Measurements,* 30/33.
5. Altes and Skiner. 1977. *J. Acoust. Soc. Amer.,* 61-4, 1019/1030.
6. Nitadori, K. 1975. Acoustical Holography, Vol. 6 (Plenum Press), 507.
7. Okoshi. 1977. Holography. Denshi Tsushin Gakkai.
8. Sato and Ueda. 1970. Measurement and Control, 9-4, 251/262.
9. Joba. 1977. Measurement and Control, 16-5, 49/55.

# Report of the Sixth Summer Seminar of Young Oceanographers

## 1. Introduction

The Young Oceanographers Association is a group formed about five years ago mainly by post-graduate students of physical oceanography. The aims and objectives of the association are to provide a platform for young oceanographers who are interested in physical oceanography or those who are engaged in oceanographic studies to exchange their information so as to learn more about the subject. For this purpose a summer seminar is arranged every year so that discussions at any level may be possible.

The sixth summer seminar of young oceanographers was held between August 4 and 7, 1982 at Kawawata Seminar Center of Tohoku University. A total of 29 members participated in the discussions.

## 2. Program

August 4: 20:00

Scope of Program, self-introduction.

August 5: 09:30–12:30

Paper 1: "Studies on the circulation of Izu Bay using a double-layer model." In his lecture, Prof. Okumura (Department of Oceanography, Tokai University) introduced his studies about the effect of the earth's rotation of the tidal currents in Izu Bay.

Paper 2: "Status of studies regarding the mixed layers in Japan." Prof. Hanawa (Faculty of Science, Tohoku University) introduced the CRP plan of Japan and mentioned some of the characteristics of mixed layers in the seas surrounding Japan.

13:30–17:30

Paper 3: "Weather on seas." Special lecture by Mr. Endo (Japan Meteorological Research Institute). He explained a number of long-term oceanographic phenomena corresponding to weather conditions in the atmosphere and dynamic analysis of these phenomena available today.

Paper 4: "Long-term variations in the tropical belt of Pacific Ocean."

Prof. Kutsuwada (Institute of Oceanography, University of Tokyo) introduced El Niño's studies and also explained some aspects of the annual variation in the tropical region of the western Pacific Ocean as well as their dynamic structures.

Unscheduled: ("Kelvin waves following environmental currents"). Kubokawa (Faculty of Science, Tohoku University) explained the dispersion characteristics due to a shift from the hemisphere equilibrium flow as well as their non-linearity and accordingly explained Kelvin waves.

19:30

"Outline of studies and introduction of problems handled." Each participant introduced problems of his choice and explains how he solved a particular oceanographic phenomenon theoretically.

August 6

09:30-12:30

Paper 5: "Measurement of energy dissipation ε on the upper layer of the ocean." Koga (Faculty of Science, Hokkaido University) explained the method to determine ε from observations and introduced the basic considerations of the method.

Paper 6: "Surface atmospheric pressure in the north Pacific." Prof. Ishihara (Department of Oceanography, Tokai University) explained annual variations in the atmospheric pressure in the north Pacific and the annual variation of the Kuroshio currents.

Paper 7: "Two-dimensional *prium* experiment." Yoshida (Faculty of Science, University of Tokyo) introduced some of the aspects of his studies regarding the structural composition of mixed currents which can be seen in a two-dimensional *prium* experiment using hydrogen air bubbles.

13:30-15:20

Paper 8: "Physical oceanographic impressions." Special lecture by Prof. Toba (Faculty of Science, Tohoku University). He narrated the nature of studies he followed and explained the qualities required of young scientists and students.

15:30-16:30

Recreation (Soft ball).

16:30-18:30

General body meeting of the Young Oceanographers Association.

The activities of the association over the last five years, including the last executive committee meeting, were explained to members and then guidelines for future activities were discussed. Also University of Tokyo was decided to be the administrative school for next year (October 1982-September 1983).

18:30-late night
Party.

August 7
09:30—Closing ceremony.

### 3. Participating Organizations and Number of Participants

Faculty of Science, Hokkaido University (1)
Faculty of Science, Tohoku University (7)
Meteorological Research Institute (1)
Faculty of Science, University of Tokyo (5)
Institute of Oceanography, University of Tokyo (2)
Department of Oceanography, Tokai University (11)
Faculty of Science, Kyoto University (1)
Department of Cultural Studies, Kyoto University (1)
Thus a total of 29 persons participated from eight organizations.

### 4. Other Activities

The details of papers discussed during this summer seminar of the Young Oceanographers Association are recorded in Vol. 3 of the Journal of the Association of Young Oceanographers. Those desirous of obtaining a copy of Vol. 3 or who want to know more about the Young Oceanographers Association can write to Atsushi Kubokawa at the following address.

For further developments, the meetings shall not be restricted to persons connected with physical oceanography or academic field alone but people from all walks are welcome.

*Executive Committee, 1982 Summer Seminar of Young Oceanographers Association*

Seminar Room of Physical Oceanography, Geophysics Department, Faculty of Science, Tohoku University.

Aobe, Aramaki Aza, Sendai City 〒 980. Tel: 0222-22-1800 (Ext. 3343).

## ADMINISTRATIVE REPORT

### Tenth Executive Meeting

The tenth meeting of the Executive Committee was held on May 27, 1982 (Thursday) at the Institute of Oceanography, University of Tokyo.

Administrative progress of the Research Conference to be held on July 5 and 6 and also the preparations for the managing committee were discussed.

It was decided to prepare a "Plan outline for research activities during the fiscal year 1982" and representatives of all groups under the plan research as well as general research were requested to send drafts.

The editorial committee for "Interim research report" informed that the first meeting of the committee was held at the Institute of Oceanography, University of Tokyo on May 15 (Saturday) and the work is in progress.

The second meeting will be held on May 27 (Thursday), in continuation of the Executive Committee meeting, whereas the third meeting is planned for June 21 (Monday).

### First Research Conference for 1982–83

The first research conference during 1982-83 was held on July 5, 6, 1982 at the Institute of Oceanography, University of Tokyo.

Representatives of the plan, research and general research groups, which have been working on the project since last year, reported the results of studies conducted so far and mentioned their future activities.

The detailed reports were published in 1982. Research report [1] was distributed to concerned persons.

The Fourth Steering Committee meeting was held on July 6, 1982 (Tuesday) at the Institute of Oceanography, University of Tokyo.

The activities and necessity of continuing the Large Equipment Application Committee as well as that of the small committees for 1982-83 were read and it was decided to continue with the same structure. Clean laboratory, field survey equipment, non-mixing type water sampler equipment, etc. shall be entrusted as common equipment to the same committee.

Large Equipment Application Coordination Committee
- Mass spectrometry
- Mooring apparatus

Common equipment (clean laboratory, field survey equipment, non-mixing type water sampler)

Committees

Heavy metals

Numerical experiments

Measuring techniques

Satellite applications

Sediment trap application

The next meeting (fifth) of the managing committee is planned for October 5 (Tuesday).

**Eleventh Executive Meeting**

The eleventh meeting of the Executive Committee was held on August 2, 1982 (Monday) at A wing, Institute of Oceanography, University of Tokyo.

Items proposed to be discussed in the fifth Steering Committee meeting were presented.

1) Plan to be submitted for an application to receive the research grant for the next fiscal year.

2) Handling of general research groups.

3) Guidelines for future plan.

4) Publication of Interim Research Report.

5) Future schedule of this special research project.

Printing the following publications is over and copies have been distributed to the concerned persons. From this year, it is decided to number the publications of this special research project as mentioned below.

Research Composition for Fiscal Year 1982 and Plan Outlay

| The Ocean Characteristics and Their Changes<br>57—1 |
|---|

Research report [1] for studies conducted in fiscal year 1982. (Proceedings of the first conference held in 1982.)

| The Ocean Characteristics and Their Changes<br>57—2 |
|---|

## ANNOUNCEMENTS

Sixth Symposium of Young Chemical Oceanographers: In continuation of last year's activities (Newsletter No. 2, pp. 31-35), the above symposium is planned to be held in the manner indicated herein. This symposium, with its residential nature, provides a platform for discussing any subject without restrictions and presents an opportunity to develop a new perspective or thinking which would not be possible with normal symposiums. Every year about 30 young scientists in the 20-30 age group take part in this symposium.

This time the discussions are proposed to be centered round the keynote address and the other topics raised by young scientists. Persons desirous of participating in the symposium (suggestions for topics of discussions are also welcome) or persons with any enquiries can contact the coordinator.

### Agenda

1. Time and Place: October 22 (Friday) 15:00 to October 23 (Saturday) 12:00
2. Place: University Seminar Hall 1987—1, Hachioji city, Tokyo.
3. Fee: About 5000 yen (includes lodging/boarding)
4. Address (Proposed).

### Keynote addresses

Prof. Ueda (Earthquake Research Center, University of Tokyo). Recent topics in plate techtonics.

Prof. Endo (Meteorological Research Laboratory). Mean general circulation of ocean.

Prof. Niizuma (Faculty of Science, Shizuoka University). Variation in biomass during the reversal of earth's magnetic field.

Others (not decided).

5. Coordinator: Gamo (Institute of Oceanography, University of Tokyo, Department of Inorganic Chemistry). 1-15-1 Minami-dai, Nakano-ku, Tokyo 164. (Tel. 03-376-1251, Ext. 274).

Designs for cover page are invited.

The special project now enters the second phase and at this juncture, we would like to change the cover layout of Newsletter as well. Your designs/sketches for the new cover page may be sent to the Administrative Office.

## Bibliography

Taira, K. 1982. Changes in ocean (characteristics) and acoustic oceanography. The Japan-American joint study. *Journal of the Association of Acoustic Oceanography*, 9(1), 6–8.

Mitsuhashi, W. and H. Mochizuki. 1982. Indirect speed measurement using an image sensor. *Proceedings of the Electronic Communication Association, J.*, 65-C (7), 578–579.

Persons who have published their papers or reports on this Special Research Project at other places are requested to inform the Administrative Office.

## Guidelines for Authors

Newsletter welcomes enterprising articles from all persons. Present activities under the project, results obtained etc. can be published here so that the connections between various study groups can be closer and can help the entire project to go smoothly towards completion.

The contents are generally classified as:

| | |
|---|---|
| Highlights: | Quick (intermediate) results of this Special Research Project |
| Bulletin: | Domestic or international news reports closely related to this Special Research Project |
| Reports: | Reports regarding the activities of each section or working group |
| Announcements: | Announcements regarding the proposed meetings or conferences of each group |
| Bibliography: | Listing of papers published by the members on this Special Research Project or other reports |

Manuscripts should be written on standard 400 character size paper (handwritten material also accepted). The figures should be drawn in such a way that they can be reproduced as they are.

It is proposed to publish Newsletter every 6 weeks.

**Note:**

When reproducing or citing the material published in this Newsletter, due acknowledgment should be given. Authors' names, wherever mentioned, should also be quoted. A copy of the publication containing such a reproduction should be sent to the Editorial Section of the Administrative Office.

| | |
|---|---|
| Published by: | Administrative Office,<br>Special Research Project "The Ocean Characteristics and Their Changes" |
| Published at: | Administrative Office,<br>Physical Oceanography Department,<br>Institute of Oceanography,<br>Tokyo University<br>1-15-1 Minami-dai, Nakano-ku,<br>Tokyo 164<br>Tel. 03-376-1251, Ext. 251. |
| Editorial staff: | Taeko Kitano, Noriyo Kimura. |

Special Research Project

# The Ocean Characteristics and Their Changes

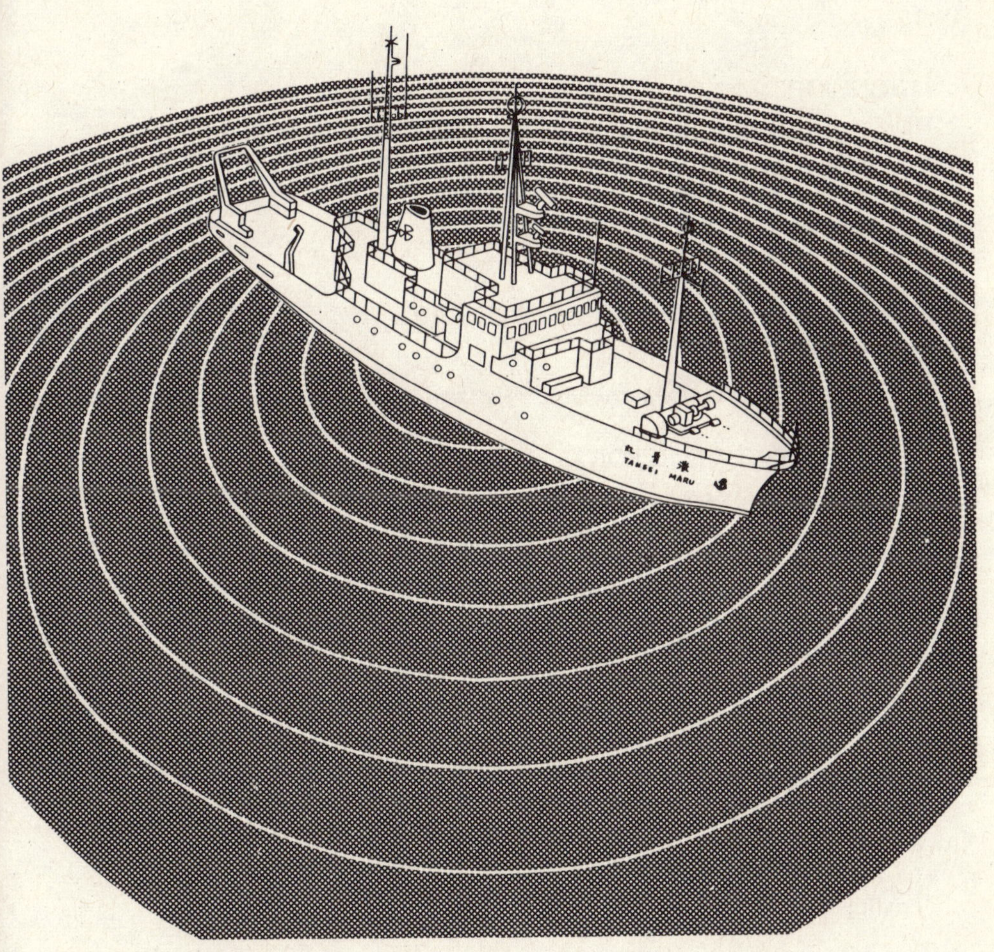


NEWSLETTER NO. 9, NOVEMBER, 1982

# Contents



Researchers engaged in this Special Research Project, who wish to submit their papers in English and acknowledge receipt of a research grant for this special study, should indicate the above title of the project.

Cover Page

This is a schematic outline of the oceanographic research vessel the *Tansei-maru* (second generation) built recently in October, 1982 as per the design of Kitagawa (Institute of Oceanography, Tokyo University) to whom the credit goes.

## HIGHLIGHTS

# Free-fall Type Electromagnetic Velocity Profiler (EMVP)

## Preparatory Experiments in Shallow Seas

*Seiichi Kanari*

(Faculty of Science, Hokkaido University)
(Received: October 25, 1982)

Since 1981, we have been conducting prototype experiments on a free-fall type electromagnetic velocity profiler with the aim of using it in shallow seas. The basic studies are over and data processing is in progress so as to be completed before November. At the end of August, the amplifier section of the velocity probe was ready and at that stage we conducted some preparatory experiments using an analogue tape recorder without waiting for the completion of the A-D converter system. The aim of this experiment was to study whether there is any irregularity in the output of the salt water bridge or whether there is noise as a result of revolution, etc. Therefore, we did not adjust the weight to make the unit fall with a certain fall velocity (50 cm/sec) nor did we mutually compare the results with another velocity meter.

The experiment was conducted off the Suma-ura at a depth of 7 m. However, as a measure of safety, the bottom contact weight was suspended 2.5 m below the instrument capsule (Fig. 1) and hence the effective fall distance was not more than 4 m.

Only the east-west component of the compass (*Emx*) was measured. Also the AC component of the salt water bridge output, following its rotation, was recorded as analogue output on a two-channel small cassette recorder. The output was again reconstructed on a two pen recorder after recovery.

An example of a reconstructed record is shown in Fig. 2. Here the upper part corresponds to the east-west component (*Emx*) of compass output. Since the profiler rotates as the weight drops (or rises), the signal draws a sinusoidal wave form following this rotation. At a point marked (▲), in the center, the weight suspended from the capsule touches the bottom and by automatic release starts to rise. For this reason the capsule starts rotating in the reverse direction. The lower part of this figure corresponds to the output from the salt

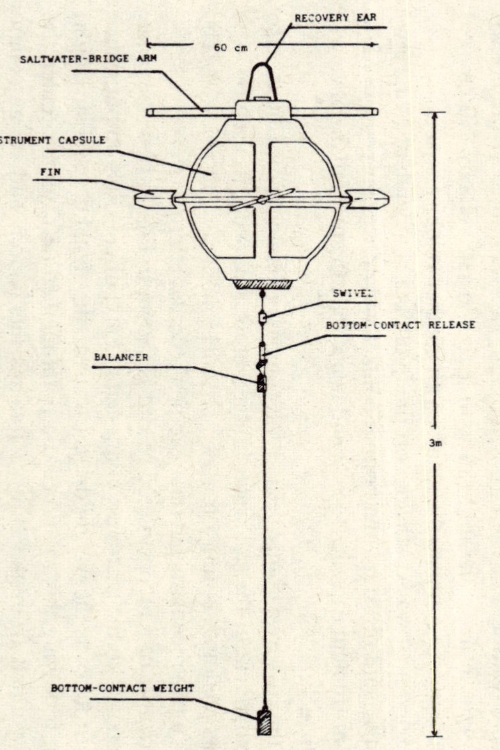


Fig. 1. Plan View of EMVP.

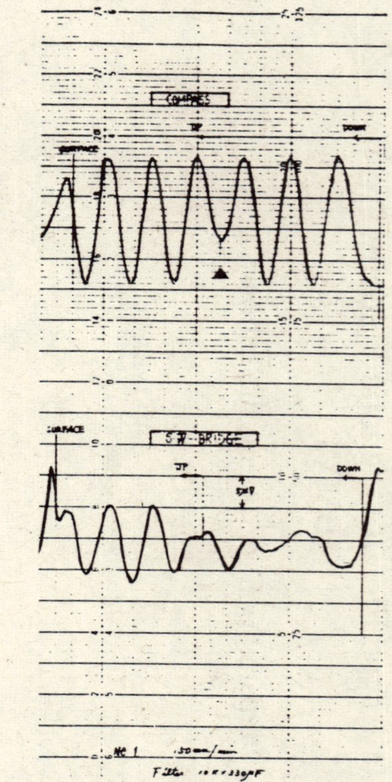


Fig. 2. Analog reconstruction record of EMVP. Upper part shows east-west component of compass output *(Emx)* whereas the lower part shows output signal of salt water bridge duly filtered and amplified (46 dB).

water bridge recorded simultaneously. A slight distortion is observed in the down profile but the up profile is a very clear waveform. The cause for the distorted waveform can be the effect of currents following the capsule. This aspect may, however, require further investigation and here we shall discuss only the analysis of the up profile waveform, which is devoid of distortion.

For analysis, we used the same method employed in the data processing system currently in progress. However, since data were read from the pen records visually, it includes a sampling point error of ± 0.5 mm. The waveforms shown by solid lines in Fig. 3 are the compass output of the up profile (east-west component) and the output of the salt water bridge (lower part). The dotted line in the upper part shows the north-south component of compass output (*Emy:* has a phase difference of 90° with respect to the east-west component). The output signal of the salt water bridge, at a point where both the above compass output signals become zero, is used to compute the flow velocity. These points are shown on the waveform of the salt water bridge output by O (*Ey*) and • (*Ex*) marks. The sign of the salt water bridge output reverses just before the bridge arm, which is also a velocity sensor, and rotates through 180°. This point (of sign reversal) is decided by the region where the sum of two compass outputs (*Ex* and *Ey*) becomes negative. Thus the desired output of the salt water bridge at sampling points outside the symbol (⊥+⊥) shown below the compass output, can be obtained by -1.

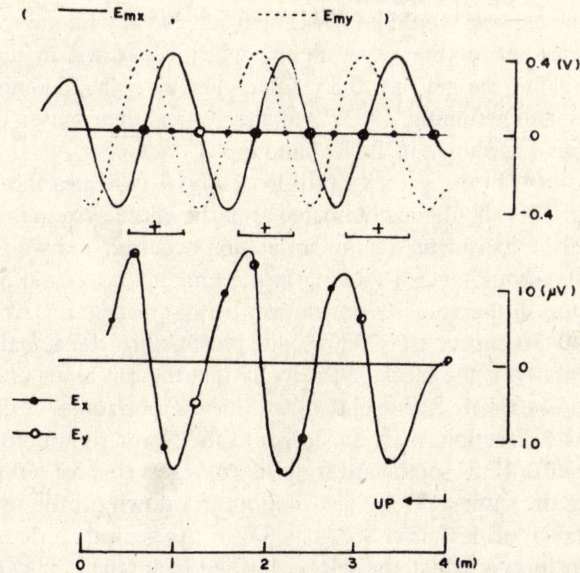


Fig. 3. The compass output (upper part) and salt water bridge output (lower part) when EMVP is floating upward. (—•—) indicates Ex sampling point and (—o—) indicates *Ey* sampling point.

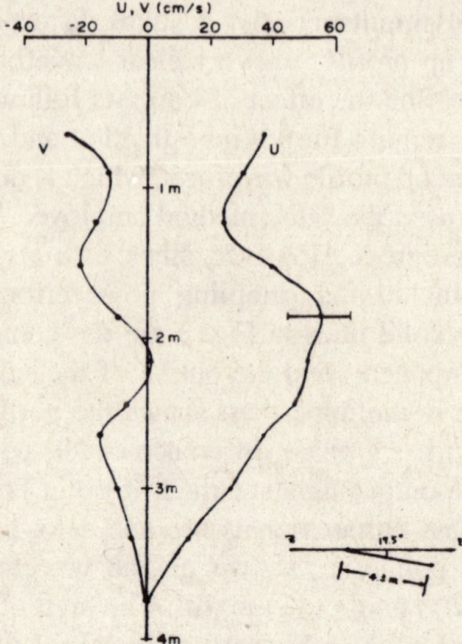


Fig. 4. Vertical profile of velocities expressed at a standard of 3.75 m. (—•—) indicates the error range in velocity calculations as a result of reading errors. Below right: The variation in direction of EMVP during its transverse from sea bottom to the surface and the distances.

The time between dropping the capsule to its rise to the surface is 32 seconds. The effective dropping distance is 4 m. Accordingly, the mean fall velocity of the capsule would be $w$ = 25 cm/sec. Now, with the magnetic field strength at the site of the experiments, when converted in terms of field strength of alunite, we get $Fz$ = 0.355 Gauss and $Fy$ = 0.315 Gauss. Using the above values, and assuming $Z$ = 3.75 m, the vertical distribution of velocity $U$ and $V$ was found as shown in Table 1 and Fig. 4.

The mark (•) in the velocity profile of Fig. 4 indicates the error arising from reading. This should be eliminated after the entire system is completed.

During the experiment, a large undulation occurred and we perceived the existence of a strong current looking at the cruising condition of the vessel. The undulation in the velocity profile can be assumed to reflect that. Also a velocity of 40–50 cm/sec truly represents, presumably, the actual flow conditions. Incidentally, if the measured velocity distribution is integrated over the average time of capsule traverse, to determine the horizontal deflection of the capsule and its direction, then, as shown in the lower right corner of Fig. 4, the capsule shifts 17.5° southward from its east-west line for a distance of 4.3 m. Assuming the same velocity distribution, if a down profile is constructed, then the distance of deviation becomes 8.6 m. As such there is no contradiction in the position at which the fall begins and later when the capsule returns to the surface after its travel.

**Table 1. Induced voltages *Ex* and *Ey* sampled from the output of the salt water bridge and the corresponding computed values of *U, Vo*. Suma-ura, August 12, 1981. Up profile (filtered), $w = -25$ (cm/s), $Fz = 0.355$ (Gauss), $Fy = 0.315$ (Gauss)**

| *Z*(m) | *Ex* (mv) | *Ey* (mv) | *U-U** (cm/s) | *V-V** (cm/s) | *U* (cm/s) | *V* (cm/s) |
|---|---|---|---|---|---|---|
| 0.65 | 8.50 | -8.00 | 34.70 | 10.47 | 37.15 | -25.31 |
| 0.90 | 6.90 | -11.20 | 28.16 | 23.54 | 30.61 | -12.24 |
| 1.23 | 5.30 | -10.10 | 21.63 | 19.04 | 24.08 | -16.73 |
| 1.52 | 9.20 | -9.00 | 37.55 | 14.55 | 40.00 | -21.22 |
| 1.85 | 13.00 | -11.80 | 63.07 | 25.98 | 55.52 | -9.79 |
| 2.15 | 12.00 | -14.50 | 48.98 | 37.01 | 51.43 | 1.22 |
| 2.44 | 11.00 | -12.50 | 44.90 | 28.84 | 47.35 | -6.94 |
| 2.65 | 8.50 | -10.50 | 34.70 | 20.68 | 37.15 | -15.10 |
| 3.00 | 6.00 | -11.80 | 24.49 | 25.98 | 26.94 | -9.79 |
| 3.33 | 2.70 | -13.00 | 11.02 | 30.88 | 13.47 | -4.89 |
| 3.75 | -0.60 | -14.20 | -2.44 | 35.78 | 0.00 | 0.00 |

It became quite clear from these experiments that the polar noise of the salt water bridge, noise following the rotation of the bridge arm or the noise arising from the amplifier circuit, do not pose any practical problem. Also the objectives of the experiments are achieved fairly satisfactorily using the data processing method mentioned above. Studies on the problem of distortion in the down profile are now in progress. If it is caused by the turbulence following the capsule, then it can easily be resolved by reversing the top and bottom of the capsule during fall.

Now prototype preparation, after taking into account the above findings, is in progress and is expected to be completed by the end of November. Immediately after its completion, a comparative experiment is proposed to be conducted using the normal velocity meter.

# The *Keiten-maru* Cruise

*Masao Fukasawa*

The committee on "Variations in the Kuroshio current and the effect of the earth's profile" (Chief coordinator: Prof. Teramoto) conducted the cruise south of Honshu during August 19 to September 2, 1982 to back up the long-term measurements. For this purpose they used the *Keiten-maru* of the Fisheries Department of Kagoshima University (Captain: Watami). During the cruise we met with a typhoon but escaped safely. However, the newly instituted measuring points and the previous measuring point were disturbed and of 11 points, recovery was possible only at eight. The main research team included Teramoto, Taira, Fukazawa, Kawabe, and Kitagawa from Institute of Oceanography, University of Tokyo and Ichikawa, Maeda, Sakurai, and Yamashiro from the Department of Fisheries Kagoshima University.

So far this team has measured the current at three places in the Kuroshio belt. Among them the data obtained during the six months from April to October, 1981 off Shiono-misaki have been analyzed and published in the proceedings of the conference of the Oceanographic Association. The details of that project will be discussed elsewhere. However, one of the salient features is that there was not much difference in the variability of the deep layers during the 200-day observations off Shiono-misaki as well as off Cape Hatteras. These two belong to the Pacific Kuroshio belt and the Atlantic gulf region respectively. Considering the fact that the seabed profile in these two cases differs considerably, we can conclude that it is an important parameter. As a next step to resolve this aspect, it would be natural to obtain the space distribution of variability in the Kuroshio belt and the distribution of variability within a given frequency range. The points selected this time take this into consideration and there will be no substitute.

The figure shows the region south of Kyushu and a broad profile of the seabed. It also indicates the measuring points set up this time. The position depth and observation depth of each point is indicated in the table. The SB points which lie in Shikoku Basin are the new points where measurement has begun with this cruise. The SB points were set up in order to compare the condition of flow with that at point B, which lies east of the Izu Ridge but is on the same latitude. Points CS and CT, which lie in the Kuroshio lower

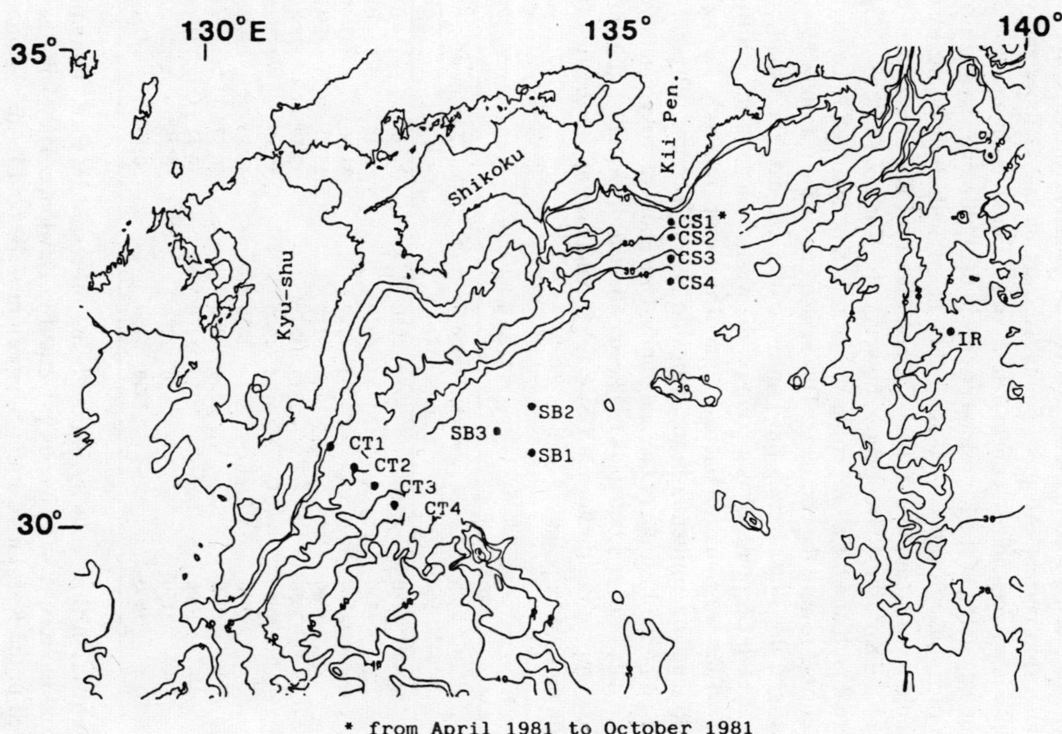


The path traversed by the *Keiten-maru* between August 19 — September 2, 1982 and the locations of the mooring system.

stream, were set up in order to compare the variation in flow velocity at these points with that in the outer ocean. The flow was measured at each SB point at depths of 400 m and 450 m. This is because recently the resolution of the Anderaa flow meter has worsened and the above arrangement only means that two meters are being used in the same layer. CT points lie off Cape Toi in the lower Kuroshio current. All these points were set up last October and with this cruise it was possible to collect data from nine months. In a sense, Cape Toi is the beginning of the Kuroshio currents in the Philippine Sea and a comparison of results of deep as well as shallow layers with those on the eastern side should be of great interest. Each CS point lies off Cape Shio in the lower stream of the Kuroshio current. Now, however, the Kuroshio follows a slightly different path and the above points therefore can be assumed to be slightly off the current toward the continent. This is also one of the reasons why CS 1 was not used this time. During this cruise we obtained data for a total of 14 months at each CS point. IR is a point on the western slope of the Izu Ridge and lies on the west side of Hachijoshima. This points presently can be assumed to enter intermittently in the Kuroshio current's course. IR point has been set up by the Institute of Oceanography, University of Tokyo and over a long time data of over 1,000 days has been collected. It would be quite interesting to compare the findings at CS, CT, SB points set up for the first time now with those at IR point.

**Details of each buoy point**

| station | depth (m) | observation depth (m) | location | |
|---|---|---|---|---|
| | | | lat. | lon. |
| SB1 | 4480 | 4000, 4050 | 31°15′ | 134°07′ |
| SB2 | 4450 | 3970, 4020 | 30°45′ | 134°07′ |
| SB3 | 4540 | 4060, 4110 | 31°00′ | 133°48′ |
| CS2 | 2640 | 1420, 2220 | 33°05′ | 135°47′ |
| CS3 | 3650 | 2230, 3230 | 32°52′ | 135°46′ |
| CS4 | 4680 | 3140, 4260 | 32°40′ | 135°48′ |
| IR | 1800 | 1670 | 32°59′ | 138°59′ |
| CT1 | 1070 | 590, 890 | 30°57′ | 131°31′ |
| CT2 | 1700 | 770, 1270 | 30°41′ | 131°48′ |
| CT3 | 3150 | 2120, 2720 | 30°28′ | 132°02′ |
| CT4 | 3860 | 2330, 3430 | 30°15′ | 132°17′ |

The data on flow velocity is now being processed by Prof. Teramoto's group and the results will be soon presented in Newletter. They will also be presented in the Kuroshio Current Symposium to be held by the Institute of Oceanography, University of Tokyo around January 25, 26 next year. Interested persons are earnestly requested to attend the same.

Before we conclude, we would like to express our gratitude to Captain Watami and other crew members who worked very hard and also to all those who helped us before and after the actual cruise in different ways.

BULLETIN

# Diary of the *Keiten-maru* Cruise

*Masao Fukasawa*

(Institute of Oceanography, University of Tokyo)

Since the launching of this Special Research Project, the Teramoto group has set up mooring systems off Cape Toi at Kyushu and off Shiono-misaki at Kii Peninsula in order to observe the deep layer currents over a prolonged period. So far the equipment has been recovered (and reinstalled) twice. Once in April, 1981 during the cruise of the *Hakucho-maru* of the Institute of Oceanography, University of Tokyo and then in October, 1981 during the cruise of the *Keiten-maru* of the Fisheries Department of Kagoshima University. In our last cruise we planned to recover and install the equipment at three points off Cape Shio, four points off Cape Toi and one point on the west side of Hachijoshima in the Izu Ridge. In addition, the equipment were installed for the first time at three points in Shikoku Basin (refer to the figure shown in the report of the *Keiten-maru* cruise). We also planned to take XBT readings, i.e., observations with dropsonde, if time permitted.

August 19, 10:00: Departure from Taniyama harbor. First three points in Shikoku Basin were to be arranged, followed by arrangements at points off Cape Shio, Hachijoshima, and Cape Toi. Immediately after leaving the harbor we were extremely busy in preparations such as hauling in the rope for mooring, preparing the buoy, starting the autorelease unit, starting the current meter, preparing anchor, etc. 22:40: Wire test for autorelease equipment starts. Test is over by 23:26. Rope, autorelease equipment, current meters, and buoy for all points in Shikoku Basin (SB) were ready. Now they only have to be launched. The autorelease equipment was tested aboard the ship and found to be in proper order. The piston and 0 ring of this equipment had been deformed, probably due to excess heat to which it was exposed during transportation from Tokyo in containers. The necessary parts were replaced.

August 20, 05:30: Standby. 05:38: lauching starts. As with other cruises, the first trial was always successful. We were worried as to whether the equipment could be installed without trouble (it was quite surprising to find no trouble at all); whether there is no damage done to rope; whether the rotor and shackle of the current meter are in order, or whether the chain of the autorelease equipment will become twisted or not (checking all these). "Go" signal for anchor was given at 06:19. The splash of water that arose as the anchor went down gave a feeling of satisfaction for the labor of so many months. Really, the current measurements using a mooring system is a great gamble. However, for all those who spent their time for so many weeks waiting for this moment, "it is not our subject", the job is over and we can look for another headache!

After anchoring, the position is found immediately using LORAN-C. This is a very convenient piece of equipment. In this oceanic region the error (in location) would probably be even less than 20 m. In addition (to position) the speed of the vessel, its direction, and the distance to be traveled to reach the target can also be found immediately. Even if there is no particular target of identification (for the user), the equipment can still be useful to know one's own position. With the availability of such equipment, the recovery of mooring system should be extremely smooth.

10:05-13:18: The other two points in SB were located respectively and we headed toward Cape Shio. Preparations for each off Cape Shio (CS) points begin. Again the rope is hauled in, the velocity meter starts, the weight preparation, etc. is undertaken. All the mooring systems of CS points have been recovered since the early morning of the following day and the new systems installed. It looked like a busy day.

August 21, 04:30: Standby. The vessel is on point CS 4. The skies are not clear yet. 05:00 Release order. We could see a sound wave propagating toward the sea bottom (or we felt like that). This is the deciding moment between success and failure. Will it meet a peak or a valley? To be frank it was the most anxious period. It is like waiting at home while the wife is in the labor room. After a few seconds, a response of release activity was detected from the seabed 4,600 m deep. All the crew members who were intently watching the monitor almost instantaneously shouted "Released. Released." However, how frequently we can have a successful recovery of a mooring system? Can anybody develop a recovery system with 100% guarantee? When shall we enter an era where one failure may not affect the project academically as well as financially.

05:25: Main buoy is seen. It lies 200 m away from the stern. In general since we started using the LORAN-C there was never an occasion when we could not see the main buoy with the naked eye. Last October, at the same CS 4 point, the buoy was floating a few meters left of the vessel and recovery was also easy. (When designing a mooring system, a large amount of buoyancy is provided for the main buoy and hence we cannot expect the rope to be cut by the screw.)

From now on we would be very busy. On the front side of the vessel, members are busy handling the mooring system, whereas, on the rear side, others are busy preparing the equipment. It was at this juncture that we noticed another power in the recovery of equipment from the mooring systems of the *Keiten-maru.* The rear side was always used for equipment preparations whereas the front side was used to recover buoy, current meters, rope, etc. with the hoist and line hauler. Rope recovered with the hauler, is coiled either on a doubler or cabler and there is no twisting or crimping at all. The buoy just recovered is examined at the rear and the damaged parts are replaced. One of the most important equipment to be recovered is the current meter. Immediately after its recovery the veins are removed, it is washed with water and taken to the research laboratory in order to study the last record.

06:50: Installation begins. So far three SB points have been installed and the crew members as well as research personnel are familiar with the job. The operation, therefore, proceeds smoothly. The ship moves at 4-6 knots and the rope tension is also enough. 07:25: "Go" signal for anchor. It took about 35 minutes to install a 200 m mooring system. While watching the beacon, the members on the back side start preparations for next installation. 08:10:XBT observation. At 200 m, the water temperature is 10°C. Probably the Kuroshio current is drifting too far southward. We traveled toward CS3. 08:35: Another XBT observation. The microcomputer installed on the ship immediately prepares a profile and output T-D, D-T tables. Credit, of course, goes to Mr. Kitagawa of the Institute of Oceanography, University of Tokyo. This excellent technician (he understands our intentions, makes the gadgets and assembles things) is the most important person, at least for a research group of our kind. His presence brings a sense of security to our group of Oceanographers. Without him we could not have thought of various measuring techniques or equipment to develop those concepts even on the continent. (I, the observer. You the brain behind).

10:15 CS3 recovered. 11:05 CS3 installed. 13:42 CS2 recovered. 14:21 CS2 installed. While traveling from CS3 to CS2, we observed a number of shiome (current trips) in northwest and southeast directions. The water temperature at 200 m, according to XBT measurements, remains unchanged at 10°C. The rope recovered from CS2 had an OAR vibrator (oscillator), (this was attached only to CS2 of all the CS points) but it did not work. Somehow the OAR was not working properly. The pressure switch could easily break and water leak in easily. We had repaired that equipment and also handled it with great care but even so the probability of its working satisfactorily after mooring could not exceed 70%. This equipment costs around $ 200,000. But we have no alternative but to buy it. We would like to develop this ourselves, if time permits. The OAR transmitter is another piece of equipment for the recovery of the mooring system but, since it is not necessary to measure the flow itself, it always gets low priority in development.

The current meters recovered at the CS points were used to record the data on paper tape after confirmation of the last record. According to the checks on board, all except one showed the right sequence. Since one out of six indicated some trouble, we thought it may be allowed. The last time, there were problems in three out of seven readings. Probably there might be some error during the manufacture of the Andera current meter. Particularly the batch between serial No. 3500-4500 was bad. The poor reliability of the current meters is a very painful thing for researchers. Actually, at three SB points, two current meters are installed at the same level. (Thus two Aanderaa are connected. If a current meter with better reliability of data collection is developed in the future, it can also be tried.) Simultaneously, a technique to measure current with the mooring system should be developed. Using the same techniques, Schmitz and his colleagues at the famous Woods Hole Oceanographic Institution installed equipment in the Kuroshio region in summer, 1980 and recovered them at the beginning of summer, 1981. Of 38 current meters (VACM) recovered, 28 were working properly but there was either major or minor trouble in the remaining ten. At our institute about 200 moorings have been tried with a recovery ratio of about 98%. So far the percentage of perfectly normally working current meters is quite high, at 80%. Recently, the development of ultrasonic current meters is in progress, an order to replace Aanderaa and a prototype has already been made.

15:00: We leave CS and move toward Hachijo. West (IR) scientists on board are busy resetting the automatic release equipment recovered from CS, checking current meters, repairing, and testing them. Bicycle operation. It is quite sad to know that if we are not successful this time, our cruise will be considered the most costly. Then who would make CTD diagram of measuring points? When the preparations are complete on board, the fact that we have to carry the replaced equipment from all points really reduces the efficiency rate in addition to the fact that we do not know (until CTD measurements are made) whether the equipment is good or bad. A solution to this problem may be only a fond hope for the next generation of mooring system. (There are so many aspects still unknown, whereas the equipment and manpower available is so limited.)

19:40: Wire test of autorelease unit. A signal is passed from the ship to the autorelease unit and a response received. This time a tape recorder was also suspended from the wire along with an autorelease unit in order to conduct some basic tests involved in the development of IES (Inverted Echo Sounder). This was to listen to the reflection of pinger sounds from the ocean surface.

21:06: Wire test over. Immediately tape was played back. First there was a sound like "Chiri, chiri, chiri, . . ." and then the signal call from the ship 3000 m above. This was followed by "Picoan" and a response from the autorelease equipment. Immediately afterward we hear another "Picoan" sound and a reflection from the seabed. Again after few seconds a weak "Picoan."

This second, after traversing to and from about 6,000 m, was quite stable. Looking at the results we feel quite sure that IES can be possible by simple means, at least for the reflections from the seabed.

The typhoon which was advancing westward turned in our direction. We wished that it would advance as slow as possible.

June 22, 05:30: Standby. 06:11 Autorelease. At IR, M type equipment made by Nippon Sekiyu was used for release. Since the pinger output of the release equipment is too small, it is really difficult to analyze the response from a depth of 1,800 m at IR, however, this time we could use LORAN-C and therefore a good response can be heard. The distance to the autorelease equipment was 1,782 m. Right below. The mooring history at IR point is quite long. They say it is the first wife of Prof. Taira (Institute of Oceanography). Data of 1,000 days is available. That means a number of days must have been spent at this point. For example, we have so far collected data of about 400 days off Cape Shio. The man days necessary to collect this much data is around 40. One cannot say that after the first installation of flow meters, the subsequent measurement will be just easy. It is a dry conclusion.

6:47: IR Located. 07:12: IR installed. Now next is Cape Toi. It is about one day and one night's journey. If hauling in rope, testing and starting the flow meter or anchor, etc. are ready, then we may rest for a day. All the crew members have been continuously working day and night and must be tired.

It is about two days since the typhoon changed its course to northeast. It appears to continue its course in this direction. If it continues this way, Tanegashima will come under (its epicenter) by midnight of the 24th. If all CT points are to be recovered and reinstalled, it will take the full day of 24th. If we take into consideration the possibility of the vessel escaping, then by gaining about 12 hours, we can complete our task. But the typhoon speed is not likely to decrease. What does the captain have in mind?

August 23: After the recovery and installation at CT 4 (the most distant point), the captain declared a critical condition. We were late by just half a day. The remaining points will be serviced after the typhoon dies down when the ship can be safely cruised to those points. But after how many days?

17:13: Reached CT 4 and immediately the order for launching autorelease equipment is given. No response is received the first time, and we had to repeat three times. This is really the first time that first trial was not successful and we had a feeling of uneasiness. By 18:30, recovery of mooring system was over and the installation of the new mooring began at 18:45. The sky was still clear but the undulations were increasing. 19:29: "Go" signal for anchor. After watching for 20 minutes, the *Keiten-maru* began its preparations to avoid damage from the typhoon. If only we get another half a day, we can finish all points and proceed with a sense of satisfaction. But . . . .

August 31: Again we headed toward Cape Toi. The typhoon has passed over Cape Miya and turned northward. This year's typhoon has properly landed on the continent. The winds were flowing at a speed of 5-6 m. Generally at the end of August the temperature drops. Now the three mooring points to be installed are only CT - 1, 2, 3 of Cape Toi.

15:33: Reached CT 1, release, 16:10: Recovery complete. The flow meters at CT 1 belong to Kagoshima University and are made by TSK. However, water has clearly leaked in one of them. This is probably the failure of the locking nut on the bolts closing the top cover. Fortunately, the circuit is unaffected and we had to service the belt alone. 17:31: Reinstallation complete. We check the autorelease equipment recovered on August 23 and set the batteries or fuel again. 20:00: The last wire test of this cruise begins.

September 1: To save time, we decided to start recovery and installation of equipment from CT 3 point. 5:00: Stand by. 5:30: Release order. The very first trial is successful. 6:26: Recovery complete. The operational sequence carried out at the previous five points has been well assimilated and this time the operation proceeds very smoothly.

06:54: Installation begins. 07:15: Installation complete. Now another point and all planned items are over. We ultimately made it. 09:00: CT 2 released. This time, there was no trouble of any kind in release. We would like such trouble-free operations in the future or, say, next time let us also have the same sequence. We now begin the last installation. Is there any abnormality in the links of rope? Is the rope extension tool O.K? Did we launch the current meter properly? After checking everything, and confirming that there is no problem, the last "Go" is signalled.

I am just a beginner in oceanographic observations using mooring systems. It may be strange to call myself a beginner but maybe I do not know much about it. Why do we turn to this method? It is because the mooring technique has been studied by earlier researchers and has been found to yield reliable results. We express our gratitude to all our seniors who have established this technique. The crew members who always met our demands also deserve our sincere thanks. Now the current measurements using moorings appear to be reaching a turning point. This technique necessitates flow measurements to a shallow thermocline layer or improvements in current meters, etc. As far as the results are concerned, it may be a proper time to establish a standard method for analysis or setting the points of observation, etc.

Now one last word. Probably a flying boat would be most ideal for observations by physical oceanographers, including the mooring operations carried out this time. The recent flying boats have much better water landing properties and have a much higher pay load also. We may soon be planning, "Today I have to take CTD readings and moorings in the Marianas and I may be late returning tomorrow."

# ADMINISTRATIVE REPORTS

## 12th Executive Meeting

The twelfth meeting of the Executive Committee was held on September 25, 1982 (Saturday) at the Institute of Oceanography, University of Tokyo.

*The agenda for the fifth meeting of the Managing Committee was discussed.

## 5th Executive Meeting

Fifth meeting of the Managing Committee was held on October 5, 1982 (Tuesday), at the Institute of Oceanography, University of Tokyo.

*The future administrative schedule of the Special Research Project was placed before the meeting.

## Fiscal Year 1982

| | |
|---|---|
| End January, 1983: | All research group leaders will submit a report on their studies and results obtained so far (the manuscript papers will be provided) for fiscal year 1982 |
| End February, 1983: | Managing Committee meeting: Reports on activity and results of each committee to be discussed, (Oral presentations.) |
| March 1983: | The report on the results so far collected will be submitted to Mombusho (Ministry of Education), separately for each group. |

## Fiscal Year 1983

| | |
|---|---|
| June to first half July: | Research Conference Managing Committee meeting: The guidelines for preparing final report to be discussed. The group leaders shall submit a report of their activity for fiscal year 1982 and proposals for fiscal year 1983. (Manuscript paper shall be provided.) |
| October: | Steering Committee meeting: Outline of the final draft will be decided. Future plans for carrying out studies will be discussed. |

| | |
|---|---|
| End January, 1984: | All group leaders will be asked to submit a report on their activities and results (Manuscript papers will be provided) for fiscal 1983. |
| February: | Managing Committee meeting. |
| March: | Reports regarding results so far collected will be submitted to Mombusho (separately for each group). |

### Fiscal Year 1984

Preparation and printing of final report.

*After completion of the research project, we enter the phase of the preparation of the final research report. However, we would like to discuss whether the explanatory notes regarding the research background should be mentioned or not.

*During the third year of this project, the proceedings of general research will not be published. As such, among various topics handled by this body, the planned research and other items closely connected to it will be put together in planned research groups. These will be further detailed as

| | |
|---|---|
| Research Group 1 (Physics): | Kanari—Aota group |
| | Masuda, and others — Teramoto group |
| Research Group 2 (Chemistry): | Shigematsu—Kanamori group |
| | Matsunaga — Kanamori group |
| | Umezawa — Horibe group |
| Research Group 4 (Measurements): | Nakamura — Yamazaki group |
| | Kawatate—Mochizuki group |

The plan proposals to be submitted with applications for the receipt of research grants during the next fiscal year will be prepared along the outlines of the initial plans submitted by each group. In case of the coordination group, instead of the CTD related equipment planned initially, four each of the acoustic release equipment and current meters as well as an amplifier for the acoustic output will be included. To finalize the proposals of each study group, discussions will be held after the main meeting of the managing committee.

*The acoustic release equipement (4 Nos.) mentioned above will generally be handled by the Mooring Committee (Group Leader: Prof. Teramoto) but since they can be loaned to other groups, the groups should send their requests with schedules (Period required and No. of units) to the group leader before the end of October.

*A small committee was proposed to be formed for IES (Inverted Echo Sounder). This group is supposed to develop IES and its applications.

Group leader: Prof. Teramoto (Institute of Oceanography, University of Tokyo)

Members: Taira and Fukazawa (Institute of Oceanography)
Mochizuki and Takeuchi (Dentsu University)
Okujima and Otsuki (Tokyo Institute of Technology)

*Active discussions were also conducted on the future plans of the Institute of Oceanography after this special project is over. It was decided to discuss the detailed steps to be taken to develop these plans in the Executive Committee meeting.

* * *

## *TANSEI-MARU-II* IS READY

The research vessel *Tansei-maru-II* shown on the cover page of this issue was handed over to the Institute of Oceanography on October 15 and was launched in the blue seas for two days on October 21-22. After carrying out necessary tests, it will be used for joint research from next year. The details of the vessel will be discussed in the next issue for want of space. Here we will just reproduce the "Introduction" by Nasu, Director of the Institute which is printed on the brochure regarding the *Tansei-maru.*

### Introduction

*Tansei-maru-II* has been manufactured with a new body.

The Institute of Oceanography was established in 1962, and the *Tansei-maru-I* was built in the following year, 1963. Since then it has been used for about 20 years on ocean expeditions. During this period the *Tansei-maru* has been extensively used for basic studies of oceans as the first Japanese Oceanographic study vessel. Oceanographers from all over Japan were able to use it in their studies.

*Tansei-maru-I*'s performance can be summerized as follows:

Total number of days on cruises: 3,457 (Average 187 days/year)
Total number of cruises: 394 (Average 21/year)
Total distance traveled: 223,649 Ri (414,198 km) (Corresponding to 10:34 per meters of earth)
Number of researchers who traveled: 3,783 (yearly average 200)
From the Institute of Oceanography: 1,652
Outside the Institute of Oceanography: 2,131

The *Tansei-maru-II* made this time, may not be actually double the size, but is considerably larger than its predecessor.

The research area for this ship will also be mainly the Japan sea but it is also possible to go a little further. New guidelines will be prepared for its use which can accelerate oceanographic research.

The name of the vessel remains unchanged, *Tansei-maru* like its predecessor.

We request all research members to make maximum use of this opportunity.

October 1982

Nasu, Noriyuki
Institute of Oceanography
University of Tokyo

## Guideline for Authors

Newsletter welcomes enterprising articles. The present activities under this project, results obtained, etc. can be published in this Newsletter. So that the connections between various study groups can be closer and help the entire project to go smoothly toward completion.

The contents are generally classified as:

| | |
|---|---|
| Highlights: | Quick reports of results of this Special Research Project |
| Bulletin: | Domestic or international news reports closely related to this Special Research Project |
| Reports: | Reports regarding the activities of each section or working group |
| Announcements: | Announcements regarding the proposed meetings or conferences of each group |
| Bibliography: | Listing of papers published by the members on this Special Research Project or other reports |

Manuscripts should be written on standard 400 character size paper (handwritten material also accepted). The figures should be drawn in such a way that they can be reproduced as they are.

It is proposed to publish Newsletter every six weeks.

**Note:**

When reproducing or citing the material published in this Newsletter, due acknowledgment should be given. Authors' names, wherever mentioned, should also be quoted. A copy of the publication containing such a reproduction, may be sent to the Editorial Section of the Administrative Office.

Published by: Administrative Office, Special Research Project "The Ocean Characteristics and Their Changes"

Published at: Administrative Office,
Physical Oceanography Department,
Institute of Oceanography,
Tokyo University,
1-15-1 Minami-dai, Nakano-ku,
Tokyo 164
Tel.: 03-376-1251, Ext. 251

Editorial staff: Taeko Kitano and Noriyo Kimura.

Special Research Project

# The Ocean Characteristics and Their Changes

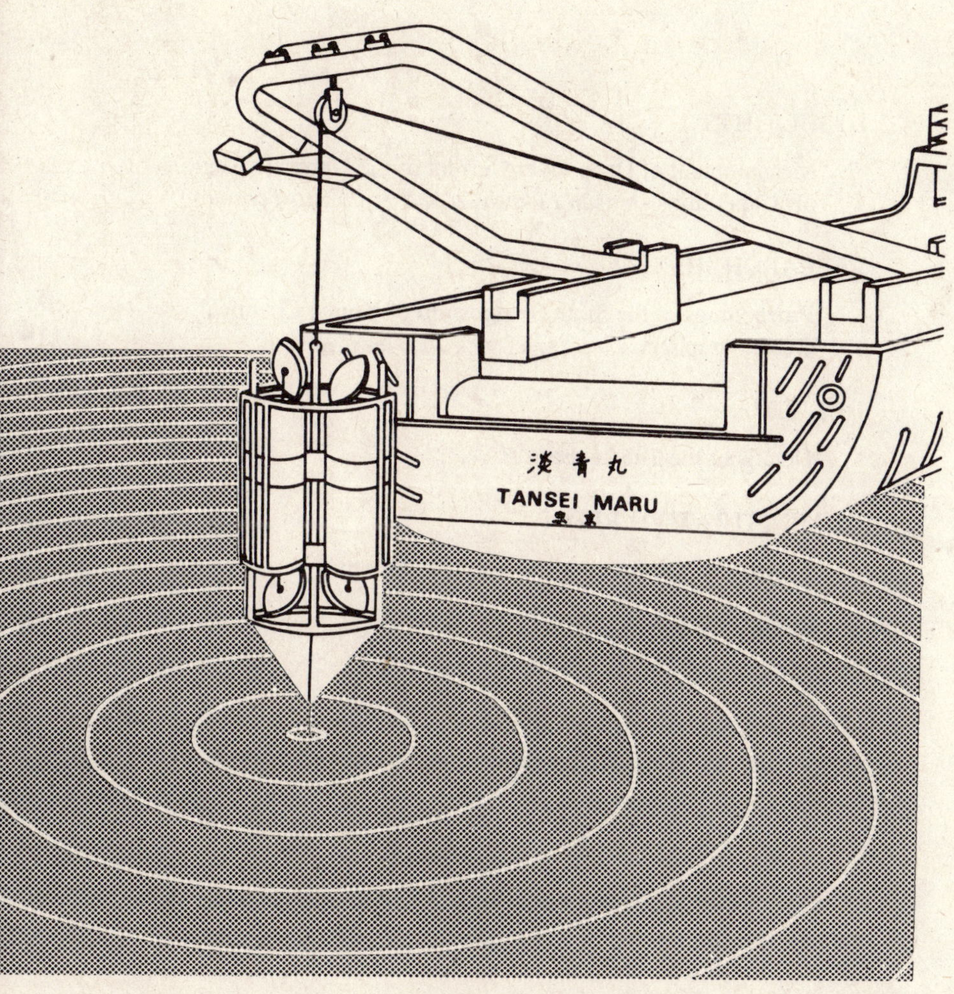


NEWSLETTER NO. 10, JANUARY, 1983

# Contents



Researchers engaged in this Special Research Project, who wish to submit their papers in English and to acknowledge receipt of a research grant for this special study, should indicate the above title of the project.

## Cover Page

This month's cover shows a large quantity water sampler (two 250 l containers) passing over the gantry. Sketch by Kitagawa (Institute of Oceanography, Tokyo University).

HIGHLIGHTS

# Measurement of Deep-layer Current of the Kuroshio Region off Cape Shio

*Masao Fukasawa and Prof. Toshihiko Teramoto*

Institute of Oceanography, Tokyo University
(Received: January 7, 1983)

## 1. Introduction

Monitoring the Kuroshio current is essential to analyze the characteristics of the Kuroshio current, particularly its dynamic properties and also for its comparison with gulf current. Many oceanographers have desired to monitor it. Earlier, the surface currents have been estimated by a number of people using the GEK method, but subsurface currents have rarely been monitored. The reports by Taft (1978), Taira and Teramoto (1981), Nishida (1982), Schmitz, et al. (1982), etc. are generally concerned with the measurements conducted to record current variations. The working group "Variation of the Kuroshio current and the effect of a topographic profile," set up as a part of this Special Research Project (Group Leader, Prof. T. Teramoto), began long-term, continuous measurements of the deep layer currents off Cape Shio from April, 1981 and off Cape Toi since October, 1981. So far, the group has collected data to August, 1982. However, the material presented here is based on the data collected during April-October, 1981. This period includes a period known as the "straight flow period." The later data generally corresponds to a long wave period and is also quite important to us.

The following part of this paper deals with the records of the basic parameters of the Kuroshio current observed during those 180 days, such as the basic flow statistics, periods of variation in space scale, etc. Comparing the results obtained for gulf currents, we find that this is a good monitoring method to study the effects of the Kuroshio.

## 2. Method of Measurement and Data

Measurements were made in seven layers at four points south of Cape Shio. The locations of these four points are given in Fig. 1. A cross sectional view of these points at seven layers is shown in Fig. 2. A buoyancy of 250–

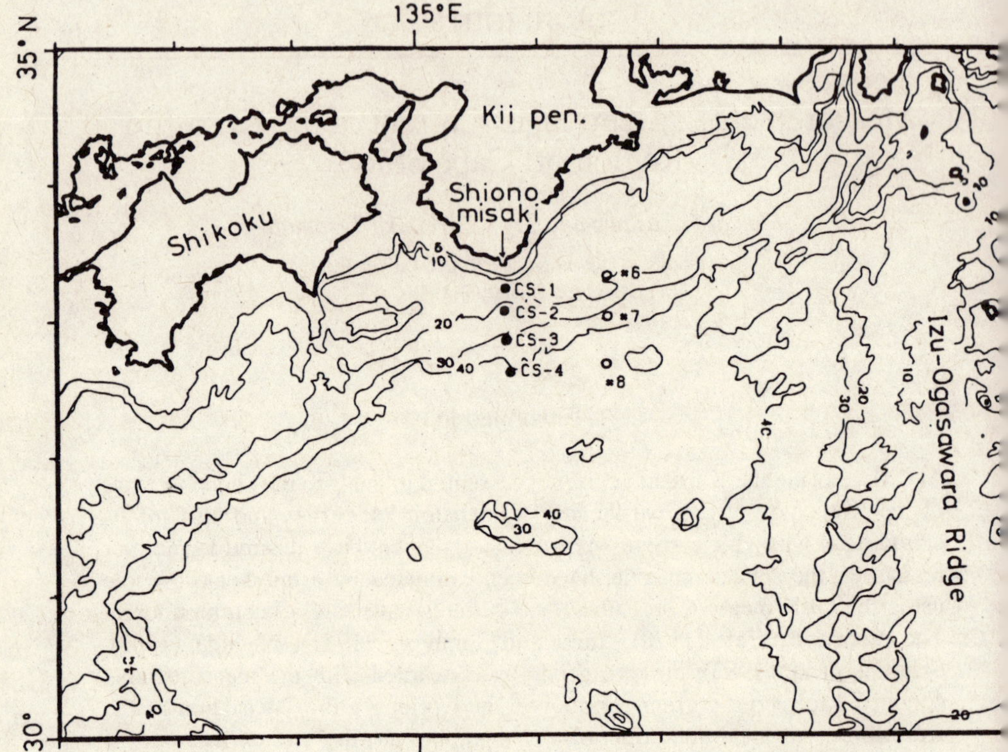


Fig. 1. Location map of measuring points [#6, #7, #8 correspond to Taft's (1978) locations].

350 kg has been imparted to the main buoy of the shallowest section of the measuring system. The current meter and automatic release equipment are respectively given a buoyancy of 50 kg and 75 kg so that the system is generally balanced at its center of gravity. At every point except CS 1, the measurements are made at two layers. In each case, the lower layer is selected as 400 m above the seabed. Fig. 2 also shows the water temperature at the time of installation of the CS moors and according to Kawai (1969), the axis of the Kuroshio current at the time of installation was between CS 2 and CS 3.

To measure current velocity we used the RCM5 model of Aanderaa Co. Details of each current meter are shown in Table 1. Among the different current meters, the data obtained from CS 11, CS 21, and CS 32 was not good enough to be subjected to statistical analysis. The remaining four points viz., CS 22, CS 31, CS 41 and CS 42, however yielded data for over 180 days. This data was processed using a 25-hour running mean and was considered

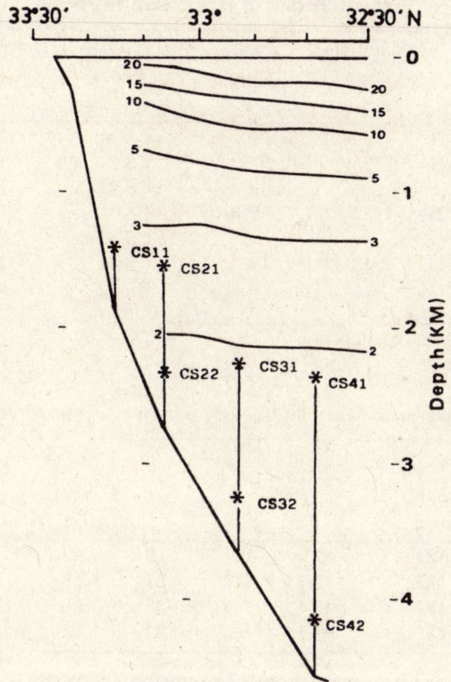


Fig. 2. Cross sectional view of measuring points [water temperature (°C) at the time of installation].

the data base for subsequent processing. Now Taft (1978) has used a running Gaussian of 25-hour half-power level as the low pass filter. When our data is used, there is no material difference due to the two different processing methods as far as basic statistical quantities are concerned.

## 3. Basic Statistical Quantities

As the basic statistical quantities, we calculated the time average ( $\bar{u}$, $\bar{v}$ ) of the east-west component ($u$) and north-south component ($v$) of the current velocity, their variance($\overline{u'^2}/\overline{v'^2}$). Excess energy of momentum ($K_e$) as defined by Taft, the mean energy of momentum ($K_m$), and their ratio ($K_e/K_m$) are all shown in Table 2.

We notice that at all points $\bar{u}$ is a digit larger than $\bar{v}$, whereas $\overline{u'^2}$ is larger by few times than $\overline{v'^2}$. If we take into consideration the fact that the seabed profile in this region is quite uniform in the east-west direction, then it becomes quite clear that the deep layer currents in this area are subjected to good bottom control. Now, $K_e$, at all points, is considerably larger than $K_m$.

Table 1. Details of each measuring point

| Station | Lat. | Long. | Water Depth | Current Meter | Nominal level | Length of data (day) |
|---|---|---|---|---|---|---|
| CS 1 | 33°15.3'N | 135°45.0'E | 1800 m | CS 11 | 1410 m | 263* |
| CS 2 | 33°05.6'N | 135°43.7'E | 2730 | CS 21 | 1530 | — |
| | | | | CS 22 | 2330 | 183 |
| CS 3 | 32°52.6'N | 135°45.4'E | 3650 | CS 31 | 2250 | 183 |
| | | | | CS 32 | 3250 | — |
| CS 4 | 32°39.1'N | 134°47.9'E | 4560 | CS 41 | 2360 | 182 |
| | | | | CS 42 | 4160 | 182 |

*Erroneous records are included.

Table 2. Basic statistical quantities of current as observed off Cape Shio

| Current meter | Record length (days) | $\bar{u}$ (cm/sec) | $\bar{v}$ | $\overline{u'^2}$ (cm²/sec²) | $\overline{v'^2}$ | $K_m$ | $K_e$ | $K_e/K_m$ |
|---|---|---|---|---|---|---|---|---|
| CS 22 | 183 | −2.5 | 0.1 | 13 | 7.5 | 3.1 | 10 | 3.2 |
| CS 31 | 183 | 2.7 | 0.2 | 27 | 8.8 | 3.7 | 18 | 4.9 |
| CS 41 | 182 | 2.4 | 0.3 | 35 | 6.6 | 2.9 | 21 | 7.2 |
| CS 42 | 182 | −5.3 | 0.3 | 93 | 12 | 14 | 53 | 3.8 |

We cannot easily explain any physical meaning from this observation and it is necessary to more carefully observe the points, particularly where the upper layer (i.e., still the layer below the main thermocline) is flowing eastward and the lower westward.

Table 3 shows the results obtained by Taft (1978) arranged in the same order as Table 2. Taft's observation points are all east of the CS system (ref. Fig. 1). His lower layer is 50 m from the seabed and is thus different from the CS system. However, if we try to find a correspondence in terms of water depth, then we find similarities between CS 22 and No. 6; also CS 42 and No. 8. Taft's period of observation is quite short and it is not possible to compare his results with that of the CS system directly. However, we can mention that, compared with the CS system, the value of $K_e$ observed by Taft is smaller and as a result $K_e/K_m$ at points 7 and 8 is also small. In Table 4 and Table 5, we have shown the results obtained by Luyten (1977) off Cape Hatteras in the same order as in Table 2. Table 4 is a set of observations for a level 1,200 m above seabed and Table 5 is for a level 200 m above seabed. In both these tables, the point separated by the dotted line can be considered to be below the gulf current, considering the temperature distribution, etc. If we compare the results of Table 4 with CS 31, CS 41 and those of Table 5 with CS 22, CS 42 we find that points CS 31, CS 41, and CS 42 match with the points directly below the gulf current and point CS 22 match is a point north of the gulf current as far as $K_e$, $K_m$, and $K_e/K_m$ are considered. If we assume that during observations of CS points, the Kuroshio current was above CS 2–CS 4, then the magnitude of the fluctuations of the deep layer currents remains more or less the same in the Kuroshio region as well as the gulf region. This is contrary to the Taft's conclusion (1978), viz., "the fluctuation of deep layer currents on the west side of Izu ridge is less than that in the gulf region."

**Table 3. Basic statistical quantities from Taft (1978)**

| Current meter | Record length (days) | $\bar{u}$ (cm/sec) | $\bar{v}$ (cm/sec) | $\overline{T'^2}$ ($cm^2/sec^2$) | $\overline{N'^2}$ ($cm^2/sec^2$) | $K_m$ ($cm^2/sec^2$) | $K_e$ ($cm^2/sec^2$) | $K_e/K_m$ |
|---|---|---|---|---|---|---|---|---|
| 6 | 30 | -1.3 | 0.5 | 5.7 | 6.1 | 1.0 | 5.9 | 5.9 |
| 7 | 64 | 2.8 | 0.6 | 7.8 | 0.8 | 4.2 | 4.3 | 1.0 |
| 8 | 30 | 4.2 | 2.3 | 18 | 17 | 11 | 17 | 1.5 |

**Table 4. Basic statistics for deep layer currents (1000 m above seabed) off Cape Hatteras (from Luyten, 1977)**

| Current meter | Record length (days) | $\bar{u}$ (cm/sec) | $\bar{v}$ (cm/sec) | $\overline{u'^2}$ ($cm^2/sec^2$) | $\overline{v'^2}$ ($cm^2/sec^2$) | $K_m$ ($cm^2/sec^2$) | $K_e$ ($cm^2/sec^2$) | $K_e/K_m$ |
|---|---|---|---|---|---|---|---|---|
| 5232 | 237 | -4.1 | -0.1 | 16 | 5.6 | 3.4 | 11 | 1.3 |
| 5246 | 235 | -3.0 | -0.4 | 12 | 9.5 | 4.6 | 11 | 2.4 |
| 5261 | 234 | -4.0 | -0.3 | 34 | 9.2 | 8.0 | 21 | 2.7 |
| 5291 | 234 | -3.5 | -0.2 | 51 | 14 | 6.1 | 33 | 5.3 |
| 5312 | 240 | -0.4 | 0.0 | 41 | 29 | 0.1 | 35 | 350 |
| 5331 | 240 | +0.7 | 2.4 | 74 | 43 | 3.1 | 59 | 19 |
| 5341 | 241 | 1.9 | 3.7 | 71 | 68 | 8.7 | 69 | 79 |

**Table 5. Basic statistics for deep layer currents (200 m above seabed) off Cape Hatteras (from Luyten, 1977)**

| Current meter | Record length (days) | $\bar{u}$ (cm/sec) | $\bar{v}$ (cm/sec) | $\overline{u'^2}$ ($cm^2/sec^2$) | $\overline{v'^2}$ ($cm^2/sec^2$) | $K_m$ ($cm^2/sec^2$) | $K_e$ ($cm^2/sec^2$) | $K_e/K_m$ |
|---|---|---|---|---|---|---|---|---|
| 5233 | 237 | -2.6 | -0.2 | 16 | 4.6 | 3.4 | 10 | 2.9 |
| 5247 | 235 | -2.5 | -0.6 | 15 | 22 | 3.3 | 19 | 5.7 |
| 5262 | 235 | -3.7 | -1.0 | 59 | 22 | 7.3 | 40 | 5.5 |
| 5292 | 234 | -5.0 | -1.1 | 75 | 29 | 13 | 52 | 4.0 |
| 5313 | 239 | -2.5 | 0.9 | 61 | 49 | 3.5 | 55 | 16 |
| 5332 | 240 | -2.1 | 2.3 | 82 | 42 | 4.9 | 62 | 13 |
| 5342 | 241 | 1.0 | 4.2 | 91 | 85 | 9.3 | 88 | 9.4 |
| 5373 | 239 | -0.4 | 4.4 | 60 | 55 | 9.8 | 57 | 5.8 |

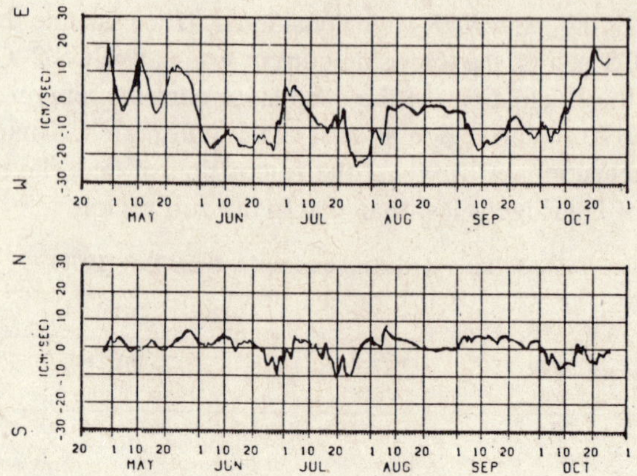


Fig. 3. Velocity fluctuations at CS 42.

As far as $K_e$ is concerned, the values obtained by Taira and Teramoto (1982) west of Hachijoshima (1800 m, 800 days) and those obtained by Schmitz, et al. (1982) in the Kuroshio region (4,000 m, 300 days) are in good agreement with the data at the CS points. (It is our subsequent topic of investigation to find the meaning of the smaller value of $K_e$ as reported by Taft but our attention is always drawn to the fact that under the sea on the eastern side of Taft's point No. 8, there is a 1,000 m high ridge with a base diameter of about 50 km.)

## 4. Time Scale of Fluctuations

One example of observed values of current velocities is shown in Fig. 3 in terms of $u$ and $v$ at point CS 42. It can be easily seen from this graph that in the total data of 180 days, there is a significant spectrum observed between the 10th and 20th as far as $v'$ is concerned. The fluctuations of $u'$ are more vigorous but instead of showing a clear peak, we observe a red shift. We calculated the autocorrelation of $u'$ at each point to find information on the time scale of these fluctuations. The first zero crossing is determined and shown in Table 6. In this table (excluding the figures in brackets) the values at a layer 1,000–1,500 m above the seabed tend to increase for points CS 31 and CS 41 but the data at both these points show a trend and if the trend is subtracted from the mean line, we get the values shown in brackets. Thus at all points a lag in the range of 14–17 days can be observed which means a period of about 60 days. Taira and Teramoto (1982) have reported a sharp spectral peak of 30 days in the velocity fluctuations along the current axis at points west of Hachijoshima. This differs from the results at CS points. One reason for this difference may be the different current layers (selected in these two cases) but a real cause is not yet known. However, it is necessary to mention that the observation period of first report corresponds to the serpentine tide/current period of Kuroshio. We plan to continue the observations at CS as well as west of Hachijoshima and we hope that the above point can be resolved in the near future.

**Table 6. The leg at which the autocorrelation function of $u'$ becoms 0 for the first time**

| Current meter | 1st zero-crossing lag (days) |
|---|---|
| CS 22 | 14 |
| CS 31 | 52 (17)* |
| CS 41 | 21 (16)* |
| CS 42 | 15 |

*without linear trend of $u'$.

## 5. Correlation between Data

In the CS system, the current at a depth of about 2,300 m is being measured at CS 22, CS 31, and CS 41. However, the current for two layers at the same point was measured only at CS 41 and CS 42 (ref. Fig. 2). Although not shown in the figure, studies on the correlation function for $v'$ between each point indicates that a high degree of correlation (0.6–0.7) exists between them for a lag 0, except for a combination of CS 4 and CS 22. For this combination of two points of CS 4 and CS 22 the peak value drops by about 0.2 but still the peak exists in lag 0. Accordingly as far as this observation is concerned, at all points, $v'$ occurs "in phase" quite frequently at the same time. On the other hand, as mentioned in the previous section, the period of fluctuation is large and CS 31 and CS 41 maintain a comparatively larger trend which makes it difficult to express the characteristic of the correlation func-

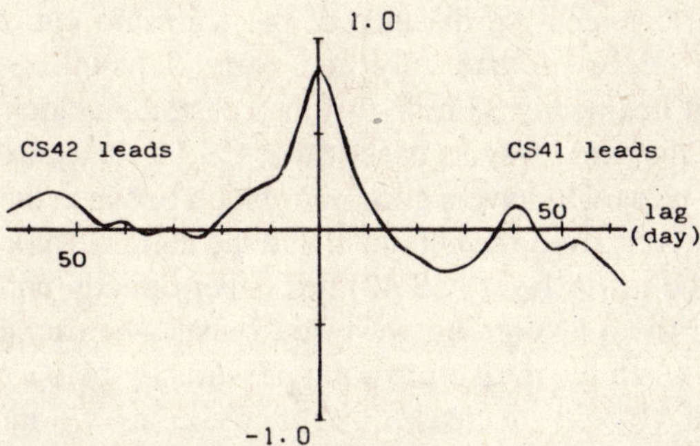


Fig. 4. Mutual correlation of $u'$ between CS 41 and CS 42.

tion of $u'$. For all combinations except that of CS 42 and CS 41, the correlation peak is not in lag 0 but its position was scattered. The correlation function between CS 41 and CS 42 is particularly shown in Fig. 4.

As seen during the studies of basic statistical quantities, the sign of $\bar{u}$ in CS 41 and CS 42 was reversed. Fig. 4 shows that the correlation of $u'$ between the two has a high value of 0.8 at lag 0 (CS 41 leads CS 42 by few hours). Thus, the fluctuation over about a 60-day period mentioned in the previous section is in the same phase for points CS 41 and CS 42 but fluctuations with longer periods (or deep layer recycling) are not necessarily in phase for CS 41 and CS 42. In other words, the reversal of sign in the $\bar{u}$ of CS 41 and CS 42 may have been caused by the fluctuations which have a period much longer than our observation period.

## 6. Relation between Kuroshio and the Obtained Data

During October, when our observation period is almost over, Kuroshio has started its serpentine movements. This can be expected from the Oceanographic Bulletins issued by the Maritime Safety Agency. The previous studies have never surmised whether the changes in position of the Kuroshio current axis can be seen in the velocity field of deep layer currents. Only Nishida (1982) has touched that aspect while discussing his results obtained off the Tokai region.

This time, we measured the distance of the Kuroshio current axis from Cape Shio from Oceanographic Bulletins (200 m, 16°C) and this is shown in the lowest section of Fig. 5. The upper four parts indicate the stick diagram for CS 22, CS 31, CS 41, and CS 42 respectively. This is obtained by subsampling the velocities at these points at every 0 hour and then separating the eastward component. If we look at the readings from CS 41, we notice that current velocity reduces whenever the Kuroshio current axis lies under 50 miles from Cape Shio. During this period, a large westward burst occurs at CS 42. At CS 22, which is closest to the continent, no particular correlation can be observed between the distance of the Kuroshio current axis and the current velocity. Between late July and early September, i.e., when the Kuroshio current lies within 35 miles off the coast, the velocity at CS 31 also becomes less in the same way as observed at CS 41. Thus, except for CS 22, other points can be said to have a good correlation between the distance of the Kuroshio current axis from the continent and the current velocity. To be more specific, when the lower layer (CS 42) lies either directly under or alongside the current axis, there is a strong westward burst; whereas, around 2,000 m (CS 31, CS 41), when the measuring point lies directly below or alongside the axis, the current velocity decreases and increases as we move toward the shore. If we look at this result of a 2,000 m depth, from the assumption of a

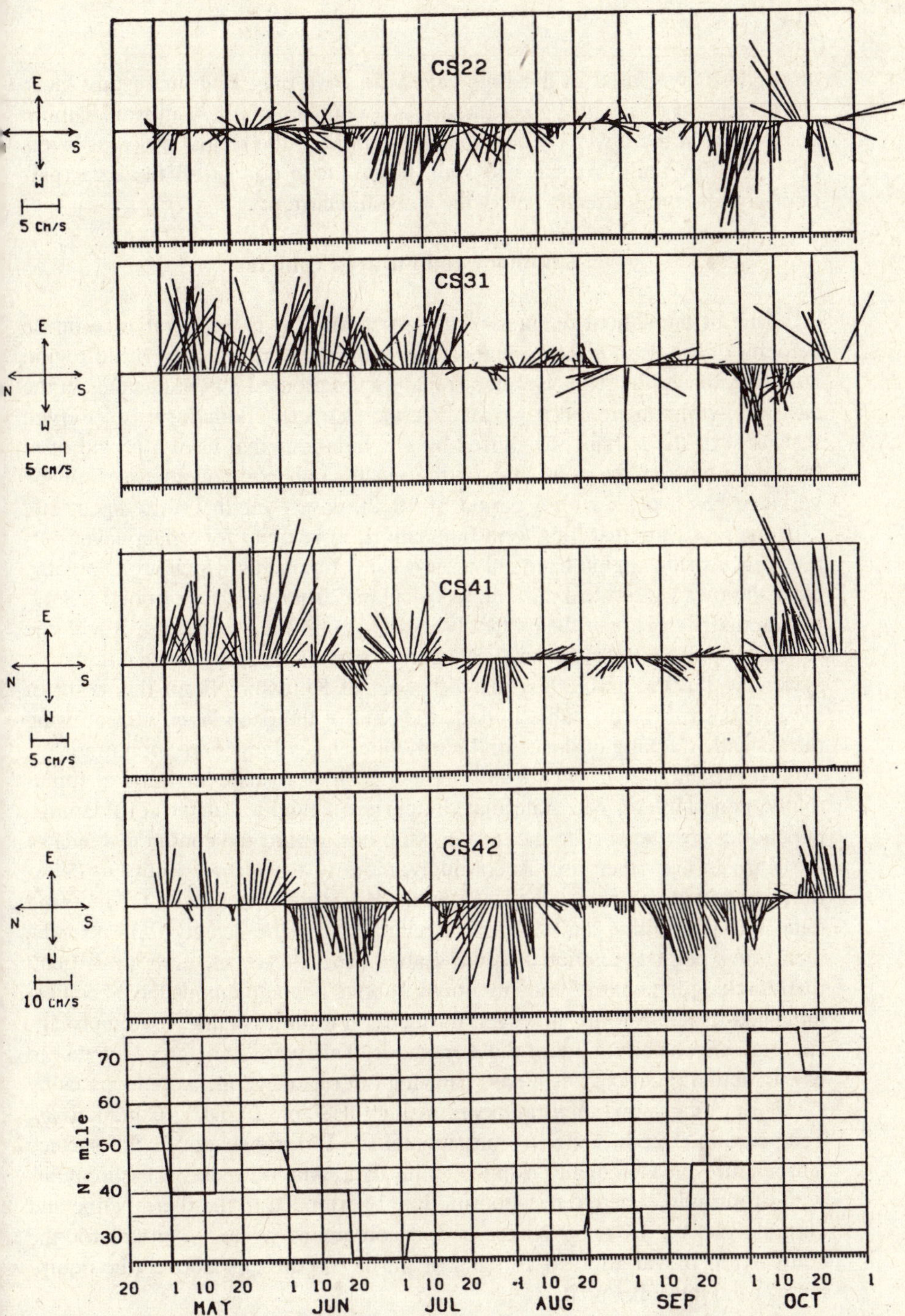


Fig. 5. Stick diagram of current velocity sampled every 0 hour and the distance of the Kuroshio current axis from Cape Shio.

motionless layer used in dynamic calculation, we may find interesting facts. Another point of equally great interest is a comparison of southward current component in the lower layer of the gulf current to the results of lower layers observed this time. We have to wait, however, to collect all the data (particularly for layers directly below the Kuroshio current).

## 7. Summary and Future Problems

One of the salient features of the observation is that the variance in the velocity of the deep layer currents is quite larger in both the Kuroshio region and the gulf region. This variance is mainly contributed by the changes in the east-west component and if we analyze the frequency domain, it is observed that most of the variance is caused by the variations that have a period of at least few tens of days. In case of the north-south component, we find the variance has a peak with a period of 10–20 days over the entire spectrum. This may indicate that long term fluctuations in velocity for a deep layer current in Kuroshio region have only east-west components which are energetic. Now the mean westward current, as calculated from the data of point CS 42, may actually exist over the entire observation period if we compare it with the results at point CS 41, and may indicate a part of the deep layer circulation system in the Philippine Sea that lies west of Kuroshio. From this point of view, it is strongly desirable to trace the path of the deep layer current using the acoustic tracking of the central buoy which has a longer duration time.

At present though we are still not in a position to state anything quantitatively, generally we can state that the current velocities differ considerably, depending on whether the measuring point lies toward the continental side of the Kuroshio or otherwise. Accordingly, there is a good possibility to determine the structure of the deep layers in the Kuroshio current with the mooring data. The possibility can be further strengthened if the density field near the measuring point is restricted to the seabed. This, however, is more difficult than tracking the control buoy mentioned above (some methods have been established to achieve this, however, for the field team it is either extremely difficult or impossible to adopt those methods. This point is of great interest to us). In actual practice, this means carrying out repetitive measurements using CTD or long-term measurements using IES or CTDC (Conductivity, Temperature, Depth, Current) measurements with dropsonde; or the surface temperature measurements using satellite data. Among these, measurements with dropsonde, wherein the mooring line is extended to the thermocline and currents observed over a longer period continuously, are important for the comparison of variation in currents of the layers directly above (the equipment?) and those below it.

So far we have broadly discussed topics of future studies which arose from the analysis of data obtained this time and the general approach to these problems. Academically the most effective methods of monitoring Kuroshio could be IES, satellite data, or mooring line extensions. (As long as it is used for a short period, the dropsonde proves to be good equipment for a case study but if used for a period specified by government agencies, the most effective techniques for ascertaining the dynamic phenomena of the Kuroshio are quite clear from the example of the gulf current.) Of the three methods mentioned above, some attempts have been made to obtain thermal image processing from satellite data, at the Physical Oceanography Department of the Institute of Oceanography of Tokyo University. Practical tests using IES will be conducted this year when the computer simulations are carried out regarding extensions of mooring lines. Some improvements were achieved over the existing mooring system but there is a good possibility that the objective will be met with and it is proposed to study this technique further during the next year.

## Acknowledgments

During these measurements we received extensive help from Prof. Takahashi and Prof. Maeda of Kagoshima University, Lecturer Sakurai and Schikawa, Captain Watami and all crew members of the *Keiten-maru,* Captain Tadama and crew members of the *Hakucho-maru* to whom we are extremely grateful.

Also Prof. Taira, Kawabe, and Kitagawa of the Institute of Oceanography guided us continuously in the field for data analysis and helped develop new methods of measurement and we are extremely thankful.

## References

1. Kawai, H. 1969. Statistical estimation of isotherms indicative of the Kuroshio Axis. *D.S.R.* 16, 109–115.
2. Nishida, H. and S. Kuramoto. 1982. Deep current of Kuroshio around Izu ridge. *Rep. of Hydro. Res.,* 17, 241–255.
3. Luyten, J.K. 1977. Scales of motion in the deep gulf stream and across the continental rise. *J.M.R.,* 35, 49–74.
4. Schmitz, Jr., J. William, P.P. Niiler, R.L. Bernstein, and W.R. Holland. 1982. Recent long-term moored instrument observations in the western north Pacific (submitted to J.G.R.).
5. Taft, B.A. 1978. Structure of the Kuroshio, South of Japan, *J.M.R.,* 36, 77–117.
6. Taira, K. and T. Teramoto. Fluctuations of the Kuroshio near the Izu Ridge and their relationship to the current path.

# Proceedings of the Sixth Symposium of Young Chemical Oceanographers

1. Objective of the symposium

The aim of this symposium was to provide a platform for introspection for young researchers through discussions on recent topics in chemical oceanography or geochemistry. The discussions were further extended to geophysics and geology in order to find their links with chemistry. The discussions were lively.

2. Place and time

University Seminar Hall (4th Seminar Room). October 22, 1982 3:00 p.m. or 23 October 1:00 p.m.

3. Participants

The names of 25 participants who attended with their parent organizations are listed at the end of this report.

4. Program

October 22

3:00–3:30 p.m. Program explanation and self introduction
3:30–5:00 p.m. Keynote address I

Chairman for addresses I and II: Osumi Takashi.

"Experiments with very high temperature and pressures and the materials in the deep layers of the earth".

Speaker: Prof. Akimoto, Physical Research Laboratory, Tokyo University.

The speaker used the latest experimental results to introduce methods to obtain very high pressures corresponding to the center or lower mantle of the earth, discussed whether the structural materials of the mantle or other nucleus could be found from the very high pressure experiments and the role of $H_2O$ in the model of deep layers of the earth.

5:00–6:00 p.m. Keynote address II. "The warm-water circulation system in the central ridges of ocean—A review".

Speaker: Prof. Kawagata, D3; Oceanographic Laboratory, Tokyo

University.

The speaker discussed the warm-water circulation system in the deep ocean beds of the East Pacific Rise or Galapagos ridge, particularly the properties of the ocean floor profile and the factors responsible for the warm-water circulation system.

Chaiman for Keynote Address I and II: Prof. Osumi.

8:00–9:30 p.m. Keynote address III. "Variation in biomass during the reversal of the earth's magnetic field".

Speaker: Prof. Niizuma, Department of Physics, Shizuoka University.

The speaker indicated a method to measure the degree of isotope formation for carbon and oxygen in the microorganisms that exist in minute quantities in sediments at the bottom of the sea. He also explained how this method could be used to study the variation (in biomass) during the past corresponding to the reversal of the earth's magnetic field. (Chairman: Prof. Kato).

9:30–11:00 p.m. Group discussions. Some basic aspects of oceanography (Data provided by Mr. Endo) were actively discussed. Thereafter the participants were divided into smaller groups and open discussions continued.

October 23

Chairman for Keynote Address IV and V: Mr. Suzuki.

9:00–10:30 a.m. Keynote address IV "General Circulation of Ocean".

Speaker: Mr. Endo, Meteorological Research Laboratory.

The speaker introduced the numerical analysis results and the GEOSECS observations regarding the circulation of ocean waters in general. He also indicated the future plan of research.

10:30–11:30 a.m. Keynote address V "Circulation of the deep-layer waters in the north Pacific as observed from the material trace".

Speaker: Kaneko, D3; Institute of Oceanography, Tokyo University.

The speaker explained how physical and chemical oceanographic techniques could be used. In order to make the patterns of countercurrents in the deep layers of the north Pacific around Izu Ridge and the Philippine Sea quite clear, with half the available results, he discussed the problems involved.

11:30–12:00 a.m. Concluding session. It was decided to hold the next symposium at University Seminar Hall after the annual convention of geochemists scheduled in October, 1983.

Lunch and valedictory function.

**Names of Participants** (*indicates speakers)

| *Name* | *Organization* |
|---|---|
| *Akimoto | Biological Research Laboratory, Tokyo University |
| *Niizuma | Faculty of Science, Shizuoka University |
| *Endo | Meteorological Research Laboratory |
| Hirose | Meteorological Research Laboratory |
| Fushimi | Meteorological Research Laboratory |
| Suzuki | Meteorological Research Laboratory |
| Sato | Department of Oceanography, Tokai University |
| Kato | Department of Oceanography, Tokai University |
| Kurami | Department of Oceanography, Tokai University |
| Tsuruoka | Department of Oceanography, Tokai University |
| Otana | Department of Oceanography, Tokai University |
| Kofugata | Department of Oceanography, Tokai University |
| Nutani | Department of Oceanography, Tokai University |
| *Gamo | Institute of Oceanography, Tokyo University |
| *Kanegoko | Institute of Oceanography, Tokyo University |
| *Kawabata | Institute of Oceanography, Tokyo University |
| Korean | Institute of Oceanography, Tokyo University |
| Fujiwara | Faculty of Science, Tokyo University |
| Akagi | Faculty of Science, Tokyo University |
| Hashimoto | Faculty of Science, Tokyo University |
| Toyota | Faculty of Science, Tokyo University |
| Nakayama | Faculty of Science, Kyoto University |
| Hitoiro | Faculty of Science, Kyoto University |
| Osumi | Faculty of Science, Tokyo Institute of Technology |
| Yamada | Department of Fisheries, Hokkaido University |

(Convenor and Rapportein: Toshitaka Gamo.)

# Details of the *Tansei-maru II*

The outline of the *Tansei-maru II* is taken from the pamphlet on the *Tansei-maru* (issued by Institute of Oceanography, Tokyo University) and arranged separately.

**Observation Equipment:**

Survey Winch:

No. 1 winch (Tsurumi Equipment)
Drive: Hydraulic/Electric
Specified load: 3 t × 43 m/min (1 layer)
Maximum velocity: 1 t × 120 m/min (31 layers)
Wire used: 9.14 mm ø × 5000 m.

This winch is used for water sampling of deep seas and the collection of sediments or biomass.

No. 2 winch (Tsurumi Equipment)
Drive: Electric/Hydraulic
Specified load: 1 t × 76 m/min (1 layer)
Maximum velocity: 0.5 t × 150 m/min (30 layers)
Cable used: 6.37 mm ø × 4000 m.

A station for CTD measurement is established underwater using armored coaxial cable.

No. 3 winch (Tsurumi Equipment)
Drive: Electric/Hydraulic
Specified load: 1 t × 75 m/min (1 layer)
Maximum velocity: 0.5 t × 150 m/min (39 layers)
Wire used: 4.76 ø × 7000 m (3 strands)

This winch is used for measurements, water sampling, sediment sampling, and the collection of biomass.

No. 4 winch (Tsurumi Equipment)
Drive: Electric
Specified load: 150 kg × 115 m/min
Electric motor: 3.7 kW × 440 V×1680 rpm
Wire used: 3 mm ø × 1500 m (stainless)

This is used for different types of BT and optical measuring equipment, as well as for the collection of samples of water, sediments, and biomass.

Line Hauler: (Izumi Steel Industries)
Drive: Electric/Hydraulic
Specified load: 150 kg
Specified velocity: 268 m/min
Electric motor: 11 kW × 440 V×1160 rpm
Pump discharge: 60 1/min
Designed pressure: 95 kg/cm$^2$

GEK Reel: (Motogigo)
Winding: Manual
Cap tire cord: 11 mm ø × 310 m with 2 poles.

Front Deck (Davit):
Drive: Manual
Specified load: 500 kg
Maximum radius: 1.0 m
Height: About 2.5 m above deck
This lies on the right side of the front deck and is to be used with No. 4 winch.

Right Radial Davit (Tsuji Industries)
Drive: Electric
Specified load: 1 ton × 0.3 rpm
Electric motor: 1.5 kW × 440 V × 1170 rpm
Maximum radius: 2.1 m
Height: 3.0 m above deck
This is for use with winches 2 and 3 and it is possible to move the equipment under suspended condition from the side, through a bulwark opening.

Davit for Piston Corer: (Sonoda Pulleys)
Drive: (turning): Manual
(winding): Electric hoist
Load during discharge: 0.9 t
Maximum radius: 1.9 m
Height: 2 m above deck
This lies on the right back and uses an electric hoist. It can be mainly used to move the piston cover.

Collapsible Gantry (Tsuji Industries)
Drive: Electric/Hydraulic
Load when fully stretched: 5.0 ton
Load when collapsed: 1.0 ton
Width (effective) × Height (during vibration): 4.0 m × 4.4. m
Auto reach: 1.5 m

This is used for winches 1 and 3 and for moorings and is operated with a fixed stand. It can be supported by a stopper in such a way that the operation is possible even when the angle at the time of wire pay out is 90° or 45°.

HIAB Crane (HIAB Co.)

Drive: Electric

Specified load: 650 kg/4.0 m (2.6$^{t-m}$)

Winding speed: 30 m/min (5 layers)

Turning speed: 2.5 rpm

Maximum radius: 6.5 m

It is installed on the left back side of the top deck and is used to move heavy objects.

Work Boat (Mitsubishi Heavy Industries Ltd.)

Length × width × depth: 6.0 m × 2.0 m × 0.9 m

Planned speed: About 6 knots

Specified personnel: 6

Structural material: Aluminium alloy

Main Engine: Diesel, water cooled, 4 cycle. Normal output 22 ps.

It is stored on the left side of the top deck and is used in shallow waters or estuaries for minor tasks and for transportation.

In addition to the above, various other special equipment is provided at proper places such as winding boom, electric hoist, detachable platform, platforms for piston cover at the opening of the blue walk, platforms for each winch, mooring rope reel, etc.

**Research Equipments**

This vessel is mainly designed to study the structure of the ocean, the ocean floor, the status of various marine organisms, etc. Various pieces of research equipments necessary to carry out this study are provided.

Research Laboratory

The research laboratory is directly connected to the rear working area and occupies an area of 53 m$^2$ in the central part. This consists of dry, semi-dry, and wet sections.

The research laboratory of this vessel uses various measuring equipment necessary for a variety of oceanographic studies; it is possible to transport the necessary equipment in and out.

Dry Research Laboratory (Hull about 21 m$^2$)

Various Indicating Instruments:

Wind speed and direction indicators, gyro, electromagnetic log, Doppler current meter, hybrid distance meter, winch line meter. In addition to these, there are gadgets like a crystal clock, telephone, teletalk equipment, ventilat-

# GENERAL ARRANGEMENT

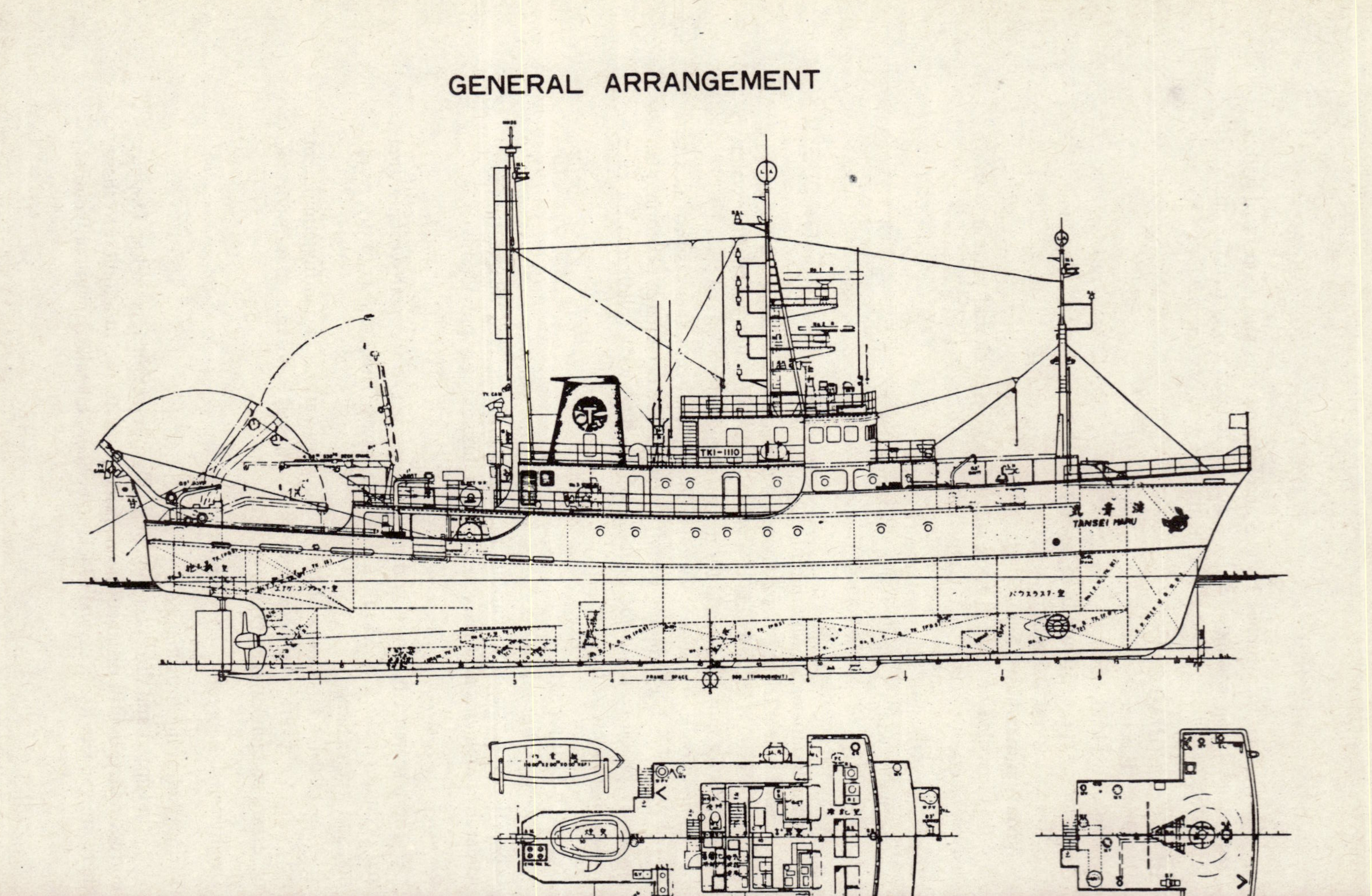

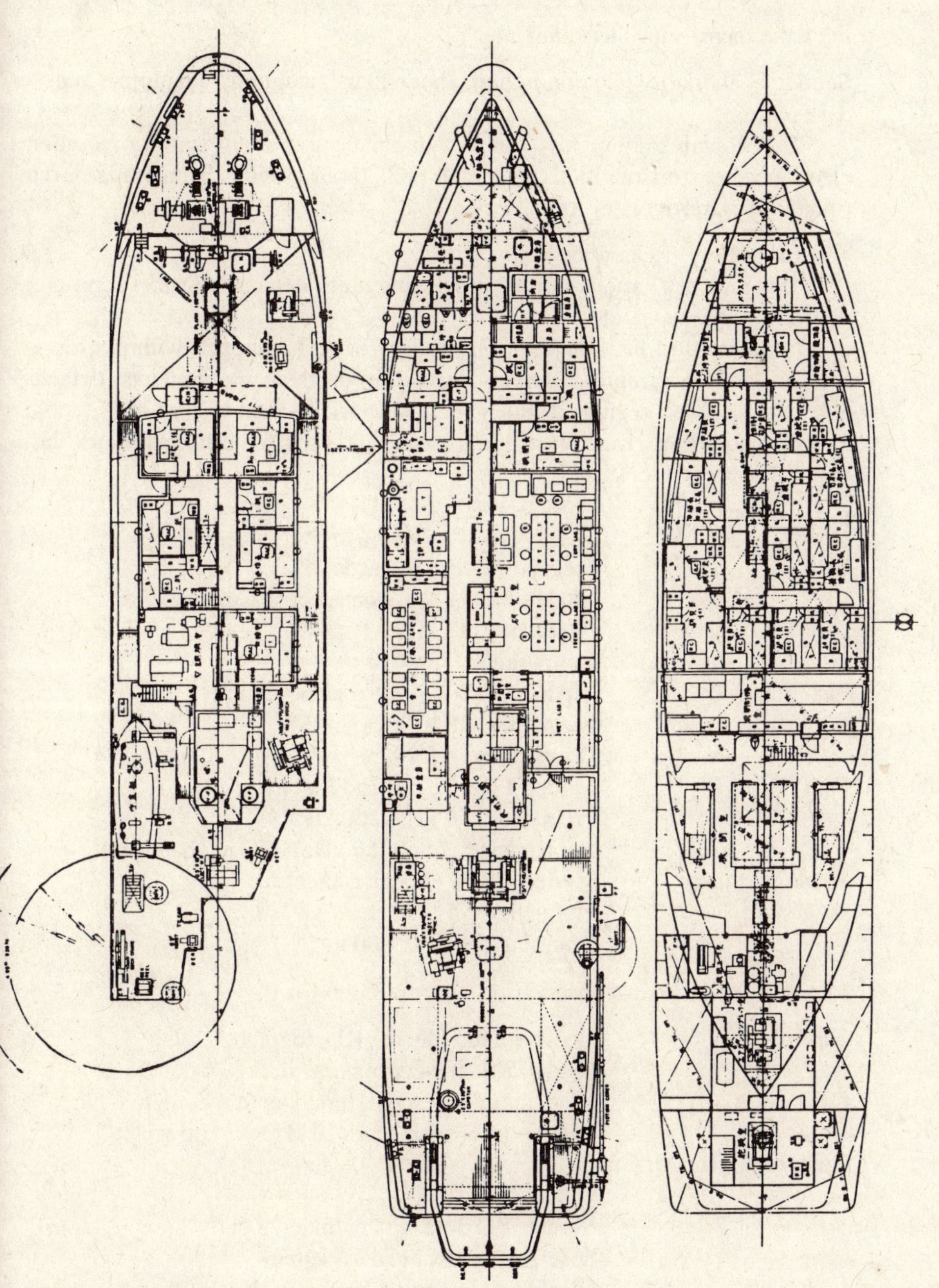

ing equipment, winch terminal, etc.

Semidry Laboratory (Central region, about 20 $m^2$, including quarantine and dark rooms).

This contains equipment such as automatic water temperature recorder, clean booster (quarantine), chemical rack, book shelf, disinfecting torch (quarantine), refrigerator, etc.

Wet Laboratory (Rear, about 12 $m^2$)

This includes various sample processing units, detachable water sampling platform, tensometer, etc.

In addition to the above, the laboratory is also equipped with platforms for various measuring equipments depending on their requirements, detachable work tables, regulated and general electricity supply terminals, cable hanger, cable hole, Rushing metals, basin, and taps for drinking water, hot water, sea water, etc.

Research Equipment

| | |
|---|---|
| PDR: | Precision Depth Recorder<br>Model: NS-74 (Nippon Electric Co.)<br>Frequency: 12 kHz<br>Depth range: 8,000 m |
| Fish Finder I | Model: FE-D 814 F (Yoshino Electric)<br>Frequency: 28 kHz<br>Depth range: 5,200 m |
| Fish Finder II | Model: FQ-50 (Yoshino Electric)<br>Frequency: 50 kHz-200 kHz<br>Depth range: 1,860 (50 kHz)-630 m (200 kHz) |
| Scanning Sonar | Model: CS-70 (Yoshino Electric)<br>Frequency: 75 kHz<br>Scanning depth: 0-800 m |

Marine meteorological observation system (Taiyo Co.)

| | |
|---|---|
| GEK: | Geomagnetic Electro-Kinetograph. 1 No.<br>Model: Riken electromagnetic flow meter. |
| CTD: | Conductivity, Temperature, Depth recorder. 1 No.<br>Type: Niel Brown Mark III B (Niel Brown Co.) |

Compressor for Air Gun.

Gun: 1 No.
Model: YQ 3-45 Y. 4 cylinder, 3-stage compression.
Discharge pressure: 12 ø $kg/cm^2$
Discharge amount: 1.8 $m^3$/min (Kachi Steels).

## ADMINISTRATIVE REPORT

The Thirteenth Meeting of the Executive Committee was held on November 8, 1982 (Monday) at A wing, Institute of Oceanography, Tokyo University.

The plans for this special project for the fiscal year 1983 (general plan) were submitted to the Ministry of Education.

The steps to be taken for the promotion of oceanographic studies after completion of this project were discussed.

The Fourteenth Meeting of the Executive Committee was held on December 27, 1982 (Monday) at A wing, Institute of Oceanography, Tokyo University.

The application of funds available for this year (Administrative Group) was considered.

It was decided to hold the Sixth Meeting of the Managing Committee on February 28 (Monday) at the Institute of Oceanography, Tokyo University. The agenda points to be included were discussed.

It was decided to invite proposals from each group for their future research plans after the special study project is over.

## ANNOUNCEMENT

At the end of January, a convention on "Considerations of Tomorrow's Geoscience" was held in the University Seminar Hall at Hachiego. During the convention, about 60 people, mostly young scientists from various fields such as geology, geophysics, geochemistry, etc., took an active part in the discussions. Another such convention is planned in the manner indicated below. This convention has been arranged in four sessions, one each on geology, paleontology, geophysics, and geochemistry. A common theme for discussions in each session has been decided to be "Earth Science of the Active Margin." One scientist in each field will be asked to take responsibility for the session. As a result, although the theme is common, each session may be conducted in a different manner and the peculiarities (gaps in the knowledge) of each field may be highlighted.

Persons desirous of participating are requested to contact the following address. The registration fee for students will be around 5,000 yen and for lecturers and other teaching staff, will be about 10,000 yen.

### Circular

| | | |
|---|---|---|
| Date and time: | February 26, 1983 (Saturday) 1:00 p.m. to February 27 (Sunday)<br>4:00 p.m., i.e., 2 days and 1 night. | |
| Place: | University Seminar Hall, Hachiego, Tokyo. | |
| Programme: | | |
| | February 26 (Saturday) | |
| | 13:00 | Registration |
| | 14:00 | Welcome |
| | 14:15 | Introduction |
| | 15:30–17:00 | Session I |
| | 17:30–19:00 | Session II |
| | 21:00 | Discussions and party |
| | February 27 (Sunday) | |
| | 10:00–11:30 | Session III |
| | | Lunch break |
| | 13:00–14:30 | Session IV |
| | 16:00 | Conclusion and Summary |
| | 16:00 | Valedictory |

For information and registration contact:

Fujioka and Tanaka, Institute of Oceanography, Tokyo University, 1-15-1 Minami-dai, Nakano-ku, Tokyo 164, Tel.: 03-376-1251, Ext. 265.

Coordinators: Oda, Minoura, Fujioka, Nishimura, Sugi, Yamashina, Gamo and Osumi.

## Guidelines for Authors

Newsletter welcomes enterprising articles. Current activities under this project, results obtained, etc., can be published in this Newsletter so that the connections between various study groups will be closer and the entire project will move smoothly toward completion.

The contents are generally classified as:

| | |
|---|---|
| Highlights: | Quick results of this Special Research Project |
| Bulletin: | Domestic or international news reports closely related to this Special Research Project |
| Reports: | Reports regarding the activities of each section or working group |
| Announcements: | Announcements regarding the proposed meetings or conferences of each group |
| Bibliography: | Listing of papers published by the members on this Special Research Project or other reports. |

Manuscripts should be written on standard 400 character size paper (handwritten material also accepted). The figures should be drawn in such a way that they can be reproduced as they are.

**Note:**

When reproducing or citing the material published in this Newsletter, due acknowledgment should be given. Authors' names, wherever mentioned, should also be quoted. A copy of the publication containing such a reproduction should be sent to the Editorial Section of the Administrative Office.

| | |
|---|---|
| Published by: | Administrative Office, Special Research Project "The Ocean Characteristics and Their Changes" |
| Published at: | Administrative Office,<br>Physical Oceanography Department,<br>Institute of Oceanography,<br>Tokyo University,<br>1-15-1 Minami-dai, Nakano-ku,<br>Tokyo 164<br>Tel.: 03-376-1251, Ext. 251. |
| Editorial staff: | Taeko Kitano and Noriyo Kimura. |

Special Research Project

# The Ocean Characteristics and Their Changes

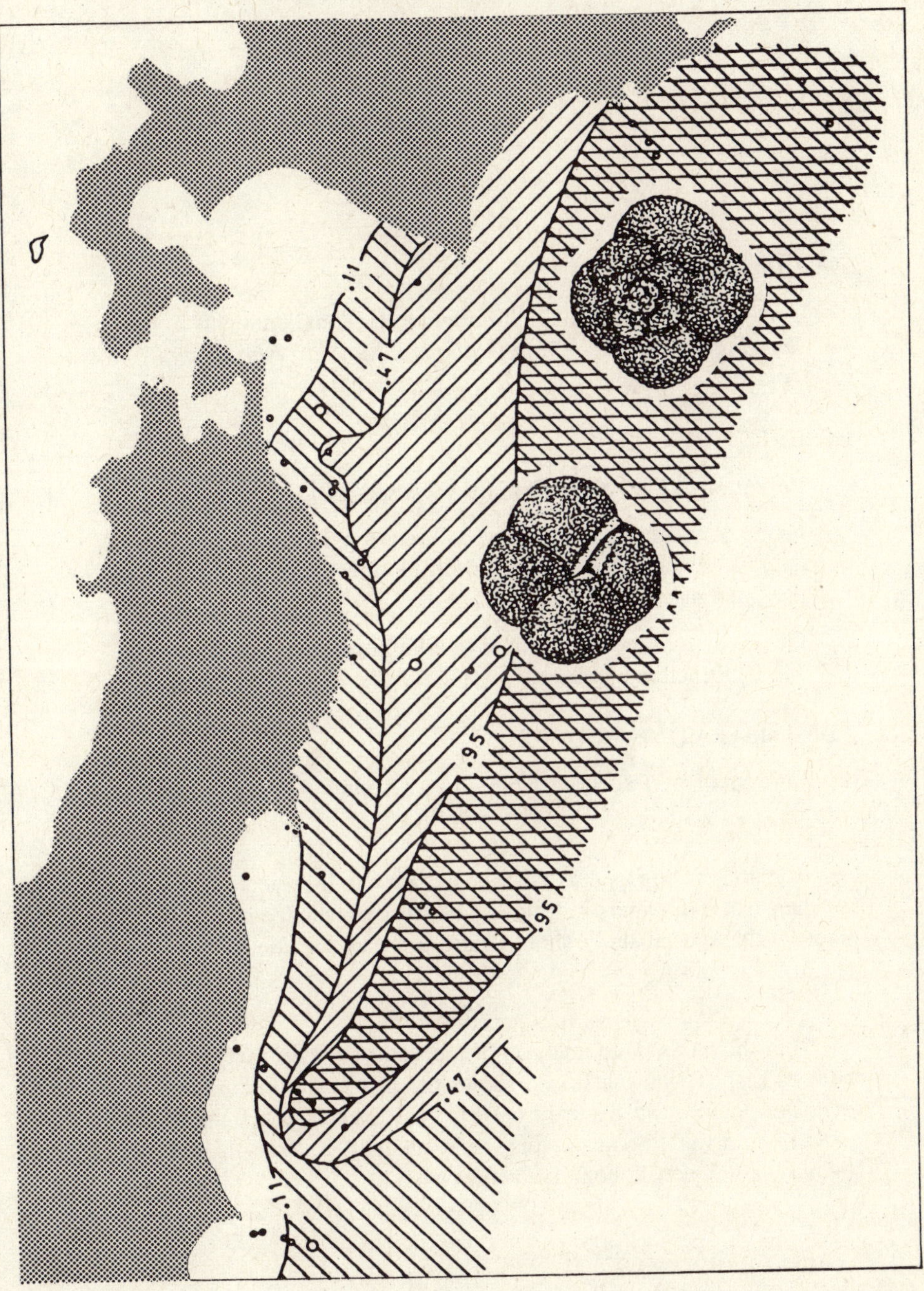


NEWSLETTER NO. 11, MARCH, 1983

# Contents



Researchers engaged in this Special Research Project, who wish to submit their research papers in English and to acknowledge receipt of a research grant for this special study, should indicate the above title of the project.

Cover Page

This shows the load analysis of a second barimax factor as observed in the seas east of Honshu. It also shows the planktonic foraminifer *Neogloboquadrina pachyferma* depicted on this background. The figure indicates the effect of cold water bodies in this sea. Courtesy Oda, Ishisaki and Takayanagi (Faculty of Science, Tohoku University).

HIGHLIGHTS

# Analysis of Planktonic Foraminifera in Bottom Sediments in the Seas East of Honshu

*Motoyoshi Oda, Kunihiro Ishisaki and Yokichi Takayanagi*

(Faculty of Science, Tohoku University)

(Received: January 10, 1983)

## 1. Introduction

While the plans for exploring the paleoclimate of the Quaternary period were in progress in the United States, a new technique was proposed whereby the main elements of paleoenvironments are expressed numerically. The transfer function in this case is a recurrent expression between the regeneration frequency of the bio-organisms concerned (explanatory variable) and the fluctuations of environmental elements (temperature, salinity, depth, etc.). Earlier it was also assumed that such a recurrent relationship exists (between these two) and based on this, paleoenvironmental elements such as temperature, salinity, etc., were estimated for the sediment core samples obtained in different waters of the world.

As a result, questions were asked as to whether the same transfer function could be applied to all seas and many professed that it is necessary to use a transfer function that takes into account regional properties. The living creatures adjust themselves to their respective environments. In other words, the zonal properties are also reflected in the distribution of bio-groups, which should be an obvious conclusion. The authors studied planktonic foraminifera in the seas east of Honshu and postulated a preparatory transfer function based on the regeneration data of foraminifera, in order to estimate the paleoenvironments of this region.

## 2. Sample

During this study, about 180 sediment samples were collected from the region that extends off Erimo to Choshi on the east and Tsugaru Strait on the west to study the planktonic foraminifera. Of these, we used 42 samples in which more than 100 planktonic foraminifera had been reproduced. (For depths between 38-3, 260 m, two or three samples were put together if the reproduction was less in individual cases and they were then treated as one sample.)

### 3. Significance of Study

Planktonic foraminifera are single cell creatures and have a high reproduction rate in surface waters (0-200 m). A geographical differentiation of these biogroups, depending on the weather belts, is quite clear. After their death, they submerge toward the ocean floor and are precipitated. The group structure of these planktonic foraminifera in the surface sediments of the seabed varies with period. This had long been noticed as a step to revive the environmental elements of the ocean in the Quaternary period at other locations, such as the North Atlantic. Recently, Imbrie and Kipp (1971) determined a transfer function from the recurrent analysis using the relative regeneration frequency of planktonic foraminifera, which is the main component of surface sediments at the seabed, to explain fluctuations in specific elements (temperature or salinity) of the ocean. This method was subsequently applied to the Pacific Ocean, as well as using Radiolaria, calcareous deposits, or foraminifera, etc. Particularly, Thompson (1981) has confirmed six groups by subjecting the planktonic foraminifera of the west Pacific to factorial analysis and has plotted the distribution of each group. According to him, the seas east of Honshu mainly have a subpolar assemblage. However, the sampling points in the Japan seas were very few in his studies and as such the details of distribution are not really clear. We, therefore, feel that it is necessary to ascertain the main (bio) groups in the surface-layer sediments of the seabed by analyzing the planktonic foraminifera. Also a transfer function should be developed, which will clearly explain the fluctuations of specific ocean elements, based on the geological (geographical) variation in regeneration frequency. This would directly help progress during the project "Paleoenvironments in the fourth period at the ocean floor."

### 4. Analysis of Data

As a first step in data analysis, Takayanagi and Ota (1983) conducted a sample survey in the region 140°-147°E, 35°-43°N. The area was divided by each latitude and longitude into different pockets and each pocket (quadrant) was sampled in such a way that it included one or many samples containing more than 100 planktonic foraminiferas. There were 42 samples which satisfied this condition. (However, of these, in six cases, two or more samples were clubbed together from the same quadrant since the reproduction number of individual sample was quite low.)

The existence of 31 "Takusa" (30 types, two forms) was ascertained in these samples. Based on this, a 31×42 row column matrix was prepared and the principal factor analysis was carried out with a Q-mode. As a result we found four groups (varimax groups) of planktonic foraminifera in this region.

As ocean environmental elements, we also determined for each quadrant the mean temperature and salinity of the surface and 100 m layers in summer and winter from the Ocean Environmental Chart (Oceanographic Data Center Ed. 1978). In addition, the depth of water at each sampling point was also recorded.

It was assumed that the geographical fluctuations of the ocean environmental elements can be explained in terms of the variation in relative reproduction frequency of the varimax group and, accordingly, a recurrent analysis was done from these. The recurrent expression obtained in this manner is called end transfer function (Imbrie and Kipp, 1971).

i) *Principal Factor Analysis*

Recently a technique called Factor Analysis or Principal Factor Analysis has become widely known. The technique used by Imbrie and Kipp (1971) during the volume analysis of microfossils (which he calls simply the Factor Analysis) is basically the same as that used by Imbrie and Purdy (1962) or Imbrie and van Andel (1964) during the analysis of carbonate sediments in the Bahama Sea or for the analysis of heavy metals. The same technique is also used for the analysis of various microfossils. However, this method does not take into account the effects of special factors included in the Factor Analysis or the possible errors and hence it can be classified as the Principal Factor Analysis.

In this Principal Factor Analysis method, a new matrix is determined for the structural resemblance coefficient from the data matrix using percentages. (This is the cos θ matrix of Imbrie and Purdy and not the relative coefficient matrix.) Subsequently, an eigenvalue and eigenvector are calculated. Using these values, the factor load calculated directly is subjected to a similar analysis. It is different from the basic Principal Factor Analysis method since it does not use the principal factor grading at all. Now, the coordinates are transposed in such a way that the dispersion of the sum of squares of factor loads reaches a maximum in each row, and then the varimax factor load is calculated. The factor grading calculated from the factor loads of this time is positively used in the group composition. In this respect this partially resembles the Factor Analysis Method.

One of the salient features of the Factor Analysis Method that deserves mention here is that a very detailed analysis becomes possible based on the structural ratio (percentage) without assuming any modification in basic data.

As can be seen from Table 1, the eigenvalues calculated for actual data explain more than 97% of the total fluctuations using the first four factors.

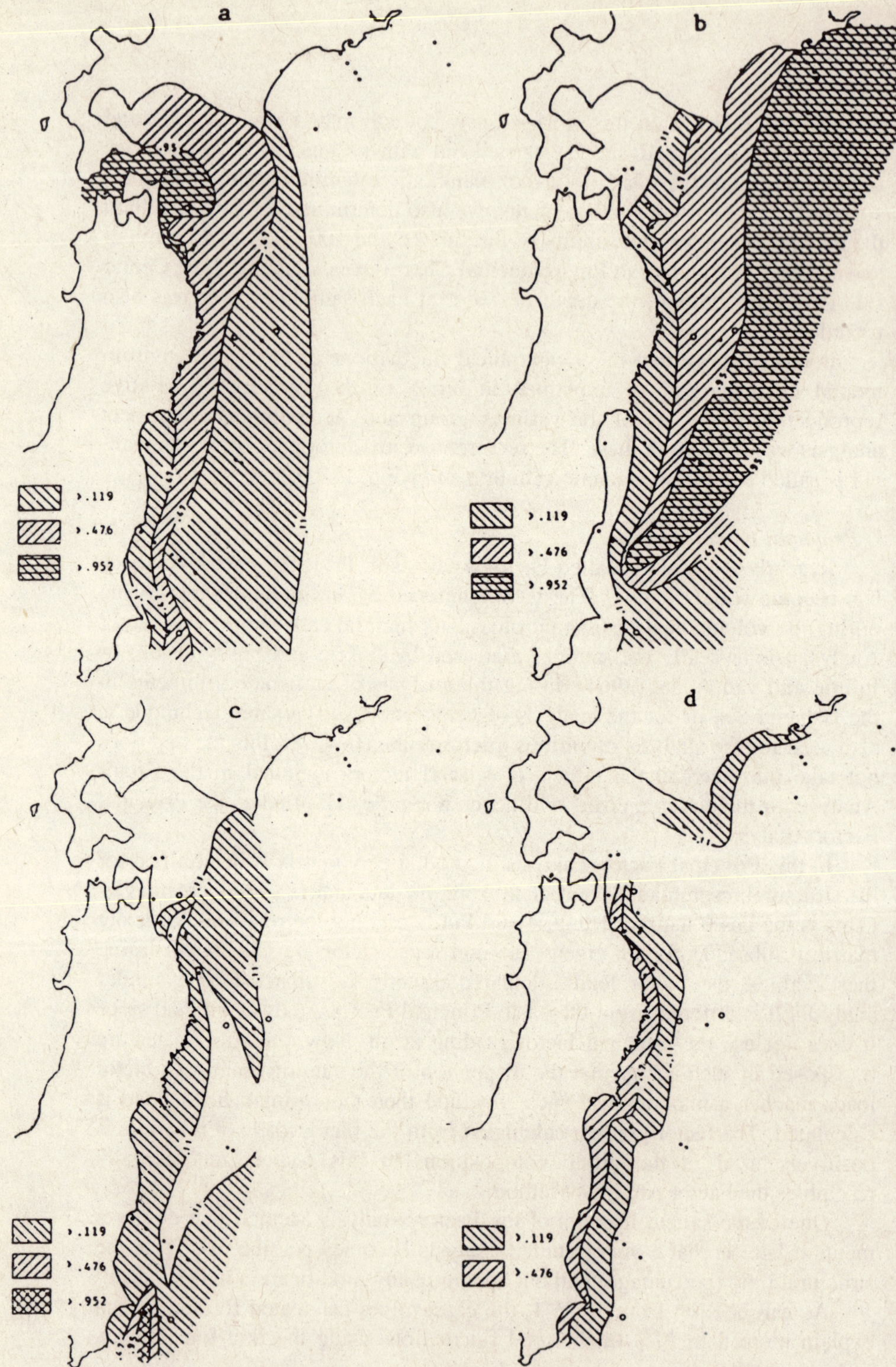


Fig. 1. Distribution of varimax factor loads. a—Factor 1, b—Factor 2, c—Factor 3, d—Factor 4.

**Table 1. Outline of the Principal Factor Analysis results (the eigenvalue and cumulative percentage to the 10th factor)**

| Factor | Eigenvalue | Cumulative % |
|---|---|---|
| 1 | 25.3519096 | 60.36 |
| 2 | 10.7602399 | 85.98 |
| 3 | 3.1186289 | 93.41 |
| 4 | 1.8955382 | 97.41 |
| 5 | 0.3981914 | 98.65 |
| 6 | 0.2317999 | 99.21 |
| 7 | 0.1115936 | 99.47 |
| 8 | 0.0741828 | 99.65 |
| 9 | 0.0540403 | 99.78 |
| 10 | 0.0339420 | 99.86 |

This percentage is much more than required during the usual microfossil data analysis. Accordingly, we have decided to restrict our discussions only to those four factors hereafter. The factor load readings under the transposed condition are shown in Fig. 1a–d.

Factor 1 explains more than 60% of the total fluctuations and is the most important factor within the limits of survey (Table 1).

Although not very accurate, the significance of the factor load can be subjected to $t$ test based on the standard error. Thus the load factor $a$ satisfying the condition

$$1.96 < \frac{a}{\sqrt{a\frac{(1-a)}{n}}},$$

(where $a$ is the load factor, $n$ is constant) is considered to be significant at the 95% level. The calculations resulted in a value of 0.119 (will satisfy the above condition). The figure divides factors in three categories: 1) with $a$ value of 0.119, 2) with $a$ value four times higher, i.e., above 0.476 and 3) with $a$ value eight times or higher, i.e., above 0.952, and shows their distribution. As is clear from Fig. 1a, the first factor has maximum load in Tsugaru Strait and in the eastern seas whereas it drops to 0.476–0.952 near Cape Erimo, Ojika Peninsula off Choshi. On its outskirts and south of Sendai Bay it is 0.119–0.476. We can conclude from this distribution that the first factor strongly exhibits the effect of the Tsugaru warm current.

Looking at Factor 2, we notice that the factor load between Muroran and Kashima is above 0.952 and its west and south boundary has a value of 0.476–0.952. A factor load of 0.119–0.476 is observed south of Cape Erimo (Fig. 1b). Such a distribution shows that Factor 2 reflects the effect of the parent tide.

In case Factor 3, a higher load distribution can be noticed in the narrow region off Choshi, whereas south of Cape Erimo and up to Choshi the load varies from 0.119–0.476. However, in the crescent-shaped region between Miyako and Kashima, no significant load of Factor 3 could be detected (Fig. 1c). Such a factor load distribution can be considered to reflect the effect of Kuroshio. We hope that when our area of observation is further extended southward, the third factor will also assume importance.

As far as Factor 4 is concerned, a significant positive load is found nowhere. Generally, this factor seems to be distributed over a limited region from the south coast of Hokkaido to the east coast of Honshu and is increasing along the east coast of Honshu (0.476–0.952) (Fig. 1d). Such a load distribution may represent factors corresponding to other elements of the ocean (such as the inshore current).

The factor ranking, calculated on the basis of factor load, is shown in Table 2. As is very clear from this table, only *Neogloboquadrina incompta* supports Factor 1 in exceptionally great proportions. This also corroborates the relationship between Factor 1 and the Tsugaru warm current established from the distribution of load factor.

Factor 2 is exceptionally strongly supported by *Neogloboquadrina pachyderma.*

Factor 3 is rather complicated and is corroborated by five species, viz., *Globigerina bulloides, G. quinqueloba, Globigerinoides ruber, Pulleniatina obliquiloculata, Neogloboquadrina eggeri.* From its load distribution, Factor 3 was assumed to represent the effect of the Kuroshio, but when we look at the above information, our present knowledge indicates that *Globigerinoides ruber, Pulleniatina obliquiloculata* and *Neogloboquadrina eggeri,* etc., do exist in the Kuroshio system. However, Factor 3 has further extended to *Globigerina bulloides.* It therefore corroborates the analysis of the distribution of previous factor loads. Now, according to the ranking of Factor 3, the classified groups also include *Globigerina quinqueloba* which represent a boreal group. However, under the present form the species which can be further classified in 2–3 types are grouped under one class and we would like to study this further in the near future.

Factor 4 mainly constitutes *Neogloboquadrina eggeri.* Looking at the distribution of the factor load, this generally appears to represent the effects of outer ocean elements. This type is mainly observed in the Kuroshio main current region. Thus the load of Factor 4 may represent the degree of Kuroshio agency (Fig. 1d shows a negative distribution of factor load and this reflects the degree of estrangement from the Kuroshio agency). However, since each factor bears no mutual correlation with the other, this type of analysis for Factor 3 needs to be studied in more detail.

The groups thus classified according to factor ranking as discussed above are generally called varimax groups.

**Table 2. Factor ranking up to Factor 4**

| Sp. | Factor 1 | 2 | 3 | 4 |
|---|---|---|---|---|
| 1 | 0.087 | 0.317 | 1.077 | -0.466 |
| 2 | -0.015 | -0.018 | 0.560 | 0.162 |
| 3 | 0.362 | -0.256 | 1.339 | -5.320 |
| 4 | 0.025 | -0.010 | 0.260 | 0.096 |
| 5 | 0.010 | 0.006 | 0.060 | 0.004 |
| 6 | -0.010 | -0.003 | 0.037 | -0.028 |
| 7 | -0.016 | -0.010 | 0.265 | 0.078 |
| 8 | -0.050 | -0.038 | 0.785 | -0.077 |
| 9 | -0.030 | 0.027 | -0.029 | -0.283 |
| 10 | 0.057 | -0.014 | 0.095 | 0.090 |
| 11 | 0.153 | -0.107 | 1.519 | 0.448 |
| 12 | -0.043 | -0.025 | 0.670 | 0.137 |
| 13 | 0.056 | -0.014 | 0.045 | 0.032 |
| 14 | 0.006 | 0.001 | 0.054 | 0.047 |
| 15 | -0.001 | -0.002 | 0.040 | 0.007 |
| 16 | 0.074 | 0.327 | 0.882 | 0.157 |
| 17 | 0.001 | -0.003 | 0.016 | -0.001 |
| 18 | 0.013 | 0.006 | 0.212 | 0.009 |
| 19 | -0.010 | 0.025 | -0.018 | -0.022 |
| 20 | -0.009 | -0.002 | 0.079 | 0.006 |
| 21 | -0.015 | -0.027 | 1.191 | 0.425 |
| 22 | -0.001 | 0.005 | 0.000 | 0.001 |
| 23 | -0.007 | -0.012 | -0.025 | -0.173 |
| 24 | 0.007 | 0.010 | 0.025 | 0.037 |
| 25 | 0.059 | -0.007 | 0.076 | 0.063 |
| 26 | 0.036 | 0.062 | 4.648 | 1.306 |
| 27 | 0.011 | 0.001 | -0.020 | -0.032 |
| 28 | 0.146 | 5.544 | -0.065 | -0.226 |
| 29 | 0.193 | 0.130 | 0.069 | 0.121 |
| 30 | 5.554 | -0.140 | -0.205 | 0.357 |
| 31 | 0.086 | 0.011 | 0.005 | -0.050 |

ii) *Regression Analysis*

In order to explain the structural ratio of the above four varimax groups classified as discussed in the above section, we carried out a multiple regression analysis between the summer and winter temperature, salinity and the depth of water. There are different methods available to select the important variable from among the various variables. They can be broadly classified as: 1) Based on the significance of the coefficient of regression, the number of variables are first reduced to a minimum possible and then the fluctuations are explained effectively. 2) Without taking into consideration the efficiency aspect, the variables are selected so that the sum of residual square becomes minimum. The latter method was used based on a progressive selection method. Here four varimax groups were formed and their square or cube values were treated as variables. The result of analysis is shown in Table 3.

Looking at the water temperature in this table, we notice that the service ratio at a depth of 100 m is higher (0.8215) than that at the surface (0.7960) during summer whereas during winter, the service ratio at the surface (0.7865) is considerably larger than at the 100 m layer (0.7685) (Table 3). We shall call these two regression expressions with a larger service ratio transfer functions NWPTs 82 and NWPTw 82. As far as salinity is concerned, the service ratio at the surface (0.7099, 0.7407) is smaller than the service ratio at the 100 m layer (0.8268, 0.7605) during both summer and winter. These two regression expressions shall be called NWPSs 82 and NWPSw 82.

**Table 3. Outline of the regression analysis results. $R^2$ is service ratio. $T_s$, $T_w$, $S_s$, $S_w$ and D represent, respectively, temperature of water at a depth of 100 m during summer and winter, salinity of water at a depth of 100 m during summer and winter, and the water depth. $F_1$–$F_4$ is the structural composition of the 1–4 varimax groups**

TEMPERATURE

Summer: 100 m layer $R^2=0.8215$; surface $R^2=0.7960$

$$T_s=13.103+53.634F_1-129.873F_1^2+84.696F_1^3$$
$$-30.528F_2+45.599F_2^2-25.456F_2^3$$
$$-10.218F_3+7.747F_3^3$$
$$-27.870F_4+382.230F_4^2-719.611F_4^3$$

Winter: surface $R^2=0.7865$; 100 m layer $R^2=0.7685$

$$T_w=21.210+69.158F_1-227.508F_1^2+158.147F_1^3$$
$$-59.639F_2+73.696F_2^2-34.293F_2^3$$
$$-94.140F_3^2+101.021F_3^3$$
$$-50.927F_4+473.892F_4^2-896.856F_4^3$$

SALINITY

Summer: 100 m layer $R^2=0.8268$; surface $R^2=0.7099$

$$S_s=34.203+4.381F_1-11.276F_1^2+7.679F_1^3$$
$$-3.551F_2+6.305F_2^2-3.668F_2^3$$
$$-1.081F_3+0.949F_3^3$$
$$+1.694F_4$$

winter: 100 m layer $R^2=0.7605$; surface $R^2=0.7407$

$$S_w=34.9732+5.893F_1-19.466F_1^2+13.954F_1^3$$
$$-5.448F_2+7.097F_2^2-3.471F_2^3$$
$$-8.491F_3^2+9.273F_3^3$$
$$-3.322F_4+35.809F_4^2-69.150F_4^3$$

DEPTH

$R^2=0.6626$

$$D=2.1967-1.660F_1^2$$
$$-11.070F_3+21.370F_3^2-13.621F_3^3$$

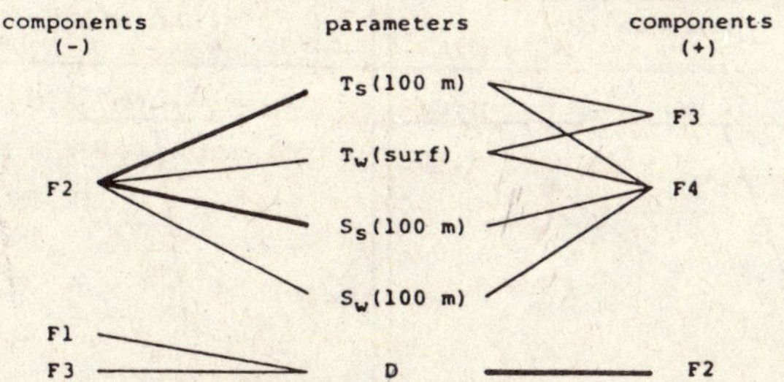


Fig. 2. Correlation between the ocean environmental elements and varimax groups F1–F4 indicate the frequency of varimax groups 1-4. $T_s$ (100 m), $T_w$ (surf), $S_s$ (100 m), $S_w$ (100 m) represent the water temperature at a depth of 100 m during summer, at surface during winter; salinity at a depth of 100 m during summer and winter. (+) or (-) represent positive or negative significant correlation. Bold lines represent a high correlation above 0.7.

As regards the water depth, the service ratio is smaller at 0.6626 which means that this regression expression can explain only 66% of the variations (of this parameter). Accordingly we will not discuss this regression expression beyond this point.

While carrying out the regression analysis as discussed above, we found a significant, simple correlation coefficient between the parameters. This is shown in Fig. 2. We notice from the figure that a very high correlation coefficient above 0.7 exists between the parameters connected by the bold line. Also a high positive correlation exists between the 2nd varimax group and water depth whereas a high negative correlation exists between this group and the water temperature and salinity of summer at the 100 m layer.

Thus within the limits of this survey, we can state that the fluctuations of the main ocean environmental element is extensively reflected in the frequency distribution of the left winding *Neogloboquadrina pachyderma.*

## 5. Applications and Discussions

The sediment samples in this sea were obtained off Kashima using the prototype samples SK-1. We shall conclude this paper by trying to determine the water temperature and salinity over a geological period from the relative frequency of planktonic foraminifera actually present in the sediment sample using the four transfer functions indicated above.

Here we considered three samples at a depth of 195 m, 180 m, and 130 m. The age of all these samples is in the renovation period.

As a result of our analyses, we obtained the values shown in Table 4 and Fig. 3. Thus looking at the water temperature of the 100 m layer in summer we find that, it drops slightly at a depth of 180 m. The winter temperature of the surface layer also similarly drops to a depth of 180 m but it rises suddenly

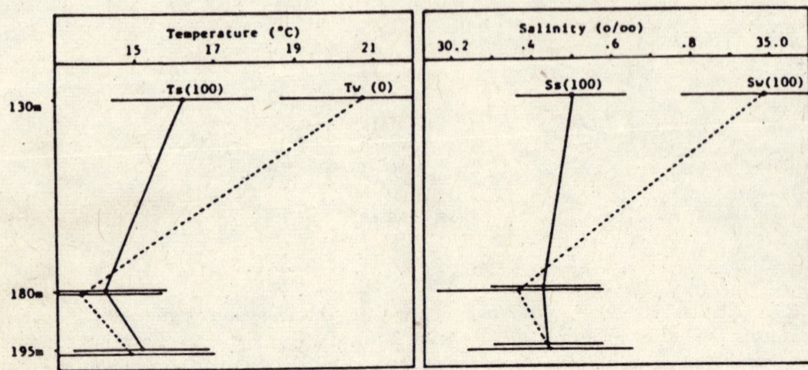


Fig. 3. Results of analysis of sample SK-1 (off Kashima). This shows fluctuations in summer temperature $T_s$ (100) and salinity $S_s$ (100) at a depth of 100 m, winter temperature at surface $T_w$ (0), and winter salinity at a depth of 100 m $S_w$ (100).

to above 20°C for a depth of 130 m. While preparing the transfer function, the surface layer sediment at the ocean floor was in the temperature range of 1°–16°C, thus exceeding the range as defined by the transfer function. This temperature under the present ocean conditions is 12.5° both for a 100 m layer and surface layer. As such the winter temperature calculated from the transfer function appears to be rather abnormal.

Salinity shows almost similar fluctuations. Referring to the structural composition of the varimax groups, the significance of the third group changed considerably from 56% at 195 m to 48% at 180 m but the frequency drops to half or 28% at 130 m. For this reason, there is considerable fluctuation in groups at depths 180–195 m and 130 m.

**Table 4. Results of analysis of sample SK-1 (off Kashima). The notation used is the same as explained in Fig. 3.**

| | Depth from the core top | | | |
|---|---|---|---|---|
| | 130 m | 180 m | 195 m | SE |
| $T_s$ (100) | 16.361 | 14.204 | 15.129 | 1.810 |
| $T_w$ (0) | 20.739 | 13.548 | 14.872 | 2.111 |
| $S_s$ (100) | 34.501 | 34.297 | 34.367 | 0.142 |
| $S_w$ (100) | 34.984 | 34.364 | 34.435 | 0.211 |

One salient point to be mentioned here is that the sample SK-1 off Kashima does not show any presence of *Neogloboquadrina pachyderma.* As mentioned earlier, while preparing a transfer function this foraminifera had contributed to fluctuations of most of the parameters. As such, in this kind of sample where it is not reproducing at all, it may not be proper to use such a transfer function. The Kashima range lies to the southern tip of the area surveyed for the seabed surface layer sediments and lies in shallow waters

(38 m) very close to the seashore. Also looking at the Oceanographic Environmental Chart (Ocean Data Center, Ed. 1978), we can clearly notice that isotherms or isohalines for this region are extremely complicated indicating a balanced pattern in the north-south direction. This indicates a possibility of a lot of noise in the data. Thus from any consideration, the Kashima offshore is under marginal conditions.

During this study, we divided the subpolar assemblage of Thompson (1981) into four types of varimax groups to form the basic material for future detailed analysis. However, in order to continue the study of the SK-1 sample off Kashima it is necessary to extend the sampling range of seabed surface samples further southward and a more generalized transfer function also needs to be formed.

Also, to explore the fluctuations of past ocean environments, it may be necessary to select samples from those regions where noise is comparatively less.

## Bibliography

Imbrie, J. and N.G. Kipp. 1971. A new micropaleontological method for quantitative paleoclimatology: Application to a Late Pleistocene Caribbean Core. In: K.K. Turekian (Editor)—The Late Cenozoic Glacial Ages. Yale Univ. Press, New Haven Conn., pp. 71-181.

Imbrie, J. and E.G. Purdy. 1962. Classification of modern Bahamian carbonate sediments. In: W.E. Ham (Editor), Classification of Carbonate Rocks. Amer. Assoc. Petrol. Geol. Mem., 1: 253-272.

Imbrie, J. and T.H. van Andel. 1964. Vector analysis of heavy mineral data. *Geol. Soc. Amer. Bull.,* 75: 1131-1156.

Ocean Data Center, Ed. 1978. Ocean environmental chart. North-West Pacific Ocean-II. Nihon Surio Kyokai.

Takayanagi and Ota. 1983. Distribution of Planktonic foraminifera in surface layer sediments in seas east of Honshu. *Monthly Oceanography,* 1983—2 (In press).

Thompson, P.R. 1981. Planktonic foraminifera in the western north Pacific during the past 150,000 years. Comparison of modern and fossil assemblages. *Paleogeogr., Paleoclimatol., Paleoecology.,* 35: 241-279.

RESEARCH REPORTS

# *Hokusei-maru* Observations

*Jiro Fukuoka*

(Department of Fisheries, Hokkaido University)

(Received: January 24, 1983)

It is almost two years since the "Special Research Project on the Ocean Characteristics and Their Changes" was launched. Our group studying the "Ocean structure around the Kuroshio" consists of five persons and is divided into members who work mainly on observation and data analysis around the Kuroshio current and members who are carrying out numerical experiments. Although it is necessary to illustrate the activities of both these subgroups while preparing the report of our group, we would like to take this opportunity to discuss the positions and observations of the mooring buoys that have been installed in the Kuroshio current region since August, 1982.

The members of this group from Hokkaido University had planned direct observations on current velocity in the Kuroshio region in 1978. However, these plans did not materialize immediately and they had to wait until 1980. Then, when this special project was launched in 1981, our proposals were included as a part of that study and the full year was spent completing the preparations. The second year of the project, i.e., fiscal 1982, was the time when a long-term mooring buoy could be installed. For the installation of mooring buoys and subsequent observations we used the Hokkaido University Department of Fisheries' training vessel *Hokusei-maru* (No. 892-92). So far there are only two or three reports critical of the activity of training vessels from universities engaged in studies of fisheries. We would like to make it clear, however, that such a vessel is always ready to be used for research cruises in addition to its standard training trips. This fact is amply attested by the use of *Hokusei-maru* cruise or the activities of *Keiten-maru* of Kagoshima University or *Shinkei-maru* of Tosui University.

*Hokusei-maru* left Hakodate at 14:55 hrs on August 20, 1982. There were indications of typhoon somewhere in the south Pacific but the weather was so far good. The observation group of 8 persons consisted of Prof. Fukuoka, Norigi, Mitaku, Murakami; 2 postgraduate students and 2 undergraduate students from Tokyo University.

The main task of this cruise was to install 2 mooring systems, install and

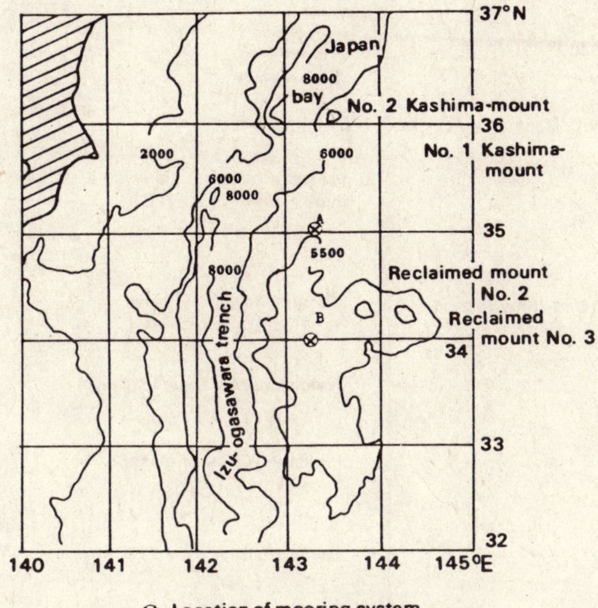


Fig. 1. Location map of mooring systems and area under observation.

recover sediment trap and observe the ocean conditions.

The area to be covered is broadly shown in Fig. 1. The initial task was to install sediment trap.

On August 22, we reached the desired location of 7°00′N, 144° 30′E. All members, including Mr. Norigi, were busy preparing for sediment trap.

The equipment was launched into the ocean at 06:45 a.m. The water depth is about 5,400 m. The equipment is attached to the release unit and if we trace the depth at proper time interval, we can know the time of reaching the bottom. The signal from the release unit was received at 07:55 a.m. indicating a depth of 5,400 m. This sediment trap consists of eight sampling tubes and, due to its automatic movement, it is possible to know the time variation in sediment collections.

August 23, the weather is good. Our mooring system is shown in Fig. 2. Total length is 4,350 m. The surface layer current meter is 4,210 m from the seabed whereas the lower layer current meter is 590 m from the seabed. With the help of these two meters, we propose to monitor the fluctuations of the Kuroshio current in the surface layer and the deep layer over a long period.

We got up early in the morning and inspected different sections of the mooring system. The preparatory operation began at 08:00 a.m. using the slipway. Since the vessels used in fisheries are generally equipped with a trawl winch, a long rope such as we used can be wound on the winch drum which makes the work easy.

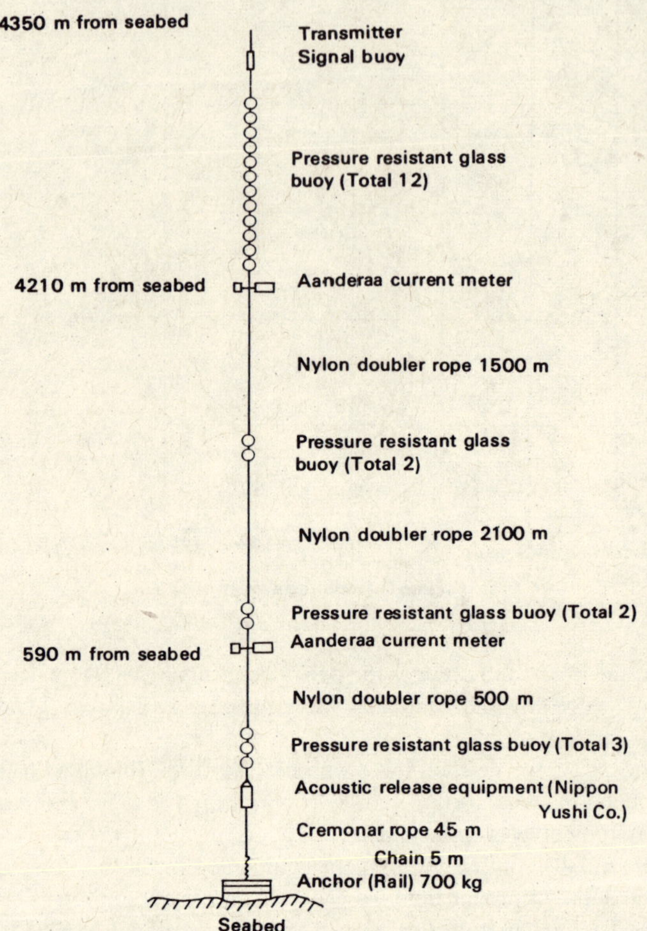


Fig. 2. Mooring system.

At 08:26 a.m. the front transmitter was installed, followed by the surface-layer current meter at 08:30 a.m. and deep-layer current meter at 09:18 a.m. The last 700 kg weight was dropped into sea at 09:24 a.m.

We feel it (weight) must have reached the seabed at 10:56 a.m. This point corresponds to point A in Fig. 1. It is at 35°00′N, 143°20′E and the water depth here was about 5,700 m.

Thus the weight reached the seabed $2^1/2$ hours after the front transmitter was launched into the water or 1 hour and 32 minutes after the weight was dropped.

Thereafter we moved further south toward point B (34°00′N, 143°20′E).

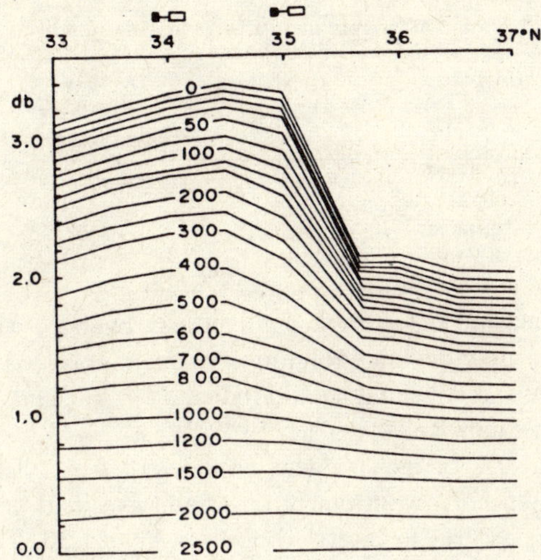


Fig. 3. Distribution of geopotential height along 143°20′E (2,500 db standard level). ▬◻ indicates the position of the current meter.

August 24, the weather was good. The transmitter was launched at 06:19 a.m., the surface-layer current meter at 06:21 a.m., the deep-layer current meter at 07:00 a.m. and the weight was dropped at 07:09 a.m. Thus the installation of the two mooring system is over.

The water depth at this point is about 5,600 m. Since the ultrasonic method of measuring distance using a release unit indicated 5,579 m at 08:41 a.m., we assumed the weight had reached seabed at that time.

The oceanic observations began at 12:14 p.m. In the area surrounding these two mooring points in the north-south and east-west directions, we would observe 20 hydrocast points (around 3,000 m) and 21 XBT points.

The data from these observations and that obtained from the mooring system will be compared later. Here we shall indicate only one of the Hydrocast results.

A north-south cross section along 143°20′E is shown in Fig. 3. This graphically represents the slope of sea level. In order to simplify things, the figure shows the geopotential height of each plane, assuming 2,500 db as the standard.

We notice from the figure that point A lies on the southern edge of the main Kuroshio current. Similarly point B lies in the area that indicates current fluctuations of low temperature vortex with a low atmospheric pressure cycle.

The location and depth of current meters can be summarized as follows:

**Location of current meters in a mooring system**

| Location | Coordinates | Depth, m | Depth of surface-layer current meter, m | Depth of deep-layer current meter, m |
|---|---|---|---|---|
| Point A | 35°00′N 143°20′E | 5,756 | 1,546 | 5,166 |
| Point B | 34°00′N 143°20′E | 5,560 | 1,350 | 4,970 |

We plan to measure the current continuously over a period of one year and ten months (during which a wrong measurement occurred in first month). We feel that the data so collected should indicate the substance of recent fluctuations in the ocean currents.

When almost 90% of its task was completed, the collision prevention radar of the *Hokusei-maru* went out of order and as it had to be fixed under any circumstances, we had to return to port on August 30. The repairs were over but by then Typhoon No. 15 was at 27°N, 147°E and moving north-west. However, only a little job was left and therefore, in consultation with the captain, we left the port again on August 31, at 09:40 a.m. to recover the sediment trap.

Early morning of September 1, we reached the desired location of 37°N, 144°30′E at 08:54 a.m. The release unit was commanded and a signal from the transmitter was received at 10:06 a.m. The state of the sea was around 2–3 and few white caps were seen. However, the captain noticed a yellow bonden at 10:43 a.m. and recovered it safely by 11:10 a.m. Thus, since all the tasks were completed, we took the route to Hakodate and reached there safely on September 3 at 09:30 a.m.

Now we would like to just mention how the ocean structure within the area of the mooring buoys was at this time.

During the April observations on *Toshitaka-maru* a low temperature vortex was observed with a diameter of about 100 km and its center at 34°30′N, 143°30′E; whereas the Kuroshio current along 143°E was observed at 36°N.

A similar low temperature vortex and the position of the Kuroshio was observed during July, when scientists sailed in the *Wakataka-maru* of Tohoku Fisheries Research Laboratory.

During October 23–30 a team from Tokyo Fisheries University conducted a detailed survey of this area in association with Prof. Otsuka of Tosui University, Prof. Yoon of Tokyo University and four graduate members of Tosui and Tokyo universities. They observed along three lines passing through 142°E, 143°20′E and 144°30′E in the north-south direction while focusing their attention on the mooring system. The results of their observations will be published soon. We would, however, like to collect as much data as possible during the mooring system period and then compare the fluctuations of the current velocity and the ocean structure.

So far we have mainly discussed the locations of the mooring system. It needs no mention that the ocean expeditions become possible only through the cooperation of a large number of people. We take this opportunity to thank them all, particularly the crew members of the *Hokusei-maru* and other research vessels who supported our studies and the researchers, technicians and students who assisted during the studies.

# Symposium on the Precision Surveys Regarding the Paleoenvironments at the Ocean Floor in the Quaternary Period

In continuation with last year, the above symposium was organized mainly for the researchers of the planned project "Paleoenvironments of Quaternary period at the ocean floor" and "Precision survey near the center of gravity area in the ocean floor." The members of general research projects and concerned scientists also participated.

Place and time of symposium: Assembly Hall of the Institute of Oceanography, Tokyo University, November 18, 1982, 10:00-18:00, November 19, 1982, 10:00-17:00.

Participants: Total 55.

Program:

Welcome: Prof. Saito (Faculty of Science, Yamagata University)

1. Introduction to the paleoenvironments of the Quaternary period of the Japanese islands as analyzed from the data from the ocean floor—Prof. Takayanagi (Faculty of Science, Tohoku University).

2. Paleoenvironments of the Quaternary period in Japan seas as indicated by the analysis of microfossils (Foraminifera nanoplankton)—Prof. Takayama (Faculty of Education, Kanazawa University).

3. Fluctuations in paleoenvironments of the Quaternary period in Japan seas by Piston-core analysis—particularly by the analysis of benthonic foraminifera—Kanoya Hiroshi (Department of Physics, Ryukyu University).

4. Paleoenvironments of the Quaternary period in Japan seas as seen from diatomic fossils—Koisumi (Osaka University).

5. Paleoenvironments of the Quaternary period in Japan seas based on measurements of the isotope ratio—Tadamichi Oba (Faculty of Science, Kanazawa University).

6. Modern paleoenvironmental fluctuations of the Ryukyu Islands as seen from the coral reef; particularly coral stones (Presentation)—Konishi, Futakuchi, Iguchi, Iwahara, Sato, Ohara and Takahashi (Faculty of Science, Kanazawa University).

7. The foraminifera fossils in the later Quaternary period and paleoclimatic changes in the Bering Sea—Yoneya and Inoue (Petroleum Resource K.K.)

8. Environmental changes observed in benthonic foraminifera of the early Quaternary period in the northwest Pacific Ocean—Tekiba (Department of Mining, Akita University).

9. Foraminifera during the Quaternary period in the area off Kashima at point SK-1 and isotope ratio fluctuations—Saito (Faculty of Science, Yamagata University) and Oda, Hasegawa and Takayanagi (Faculty of Science, Tohoku University).

10. Analysis of planktonic foraminifera observed in surface sediments of seas east of Honshu—Ishizaki, Oda and Takayanagi (Faculty of Science, Tohoku University).

11. The climate of Japan at the end of the Quaternary period and data on the ocean floor—Nakagawa (Faculty of Science, Tohoku University).

12. Gravitational abnormalities in trenches—Tomoda and Fujimoto (Ocean Research Institute, Tokyo University).

13. Precision survey of the South Sea trough off Ashizuri—Kagami and Takayama (Institute of Oceanography, Tokyo University); Sakurai and Sato (Hydrographic Department of Marine Safety Board).

14. Studies conducted with digital data recording equipment for an on-board acoustic survey—Kinoshita and Amamiya (Faculty of Science, Chiba University).

15. Recent measurements of earth crust heat flow in the seas surroundings Japan—Yamano, Honda, Ueda and Fujizawa (Earthquake Research Institute, Tokyo University); Kinoshita (Faculty of Science, Chiba University).

16. Tectonics of the Japan trench off Sanriku as understood from seismographs installed at the seabed—Kasahara (Earthquake Research Institute, Tokyo University).

17. The survey system for exploring the structure of the earth crust at the ocean-floor using an air gun-OBS-present and future—Suehiro and Nishizawa (Faculty of Science, Tohoku University); Yamada (Faculty of Science, Tokyo University); Shimamura (Faculty of Science, Hokkaido University).

18. Studies regarding earth's magnetism in the Daito ridge—Iguchi, Isezaki, Matsuda and Yasukawa (Faculty of Science, Kobe University).

19. Sedimentation environments in the Kishu Yonmanju belt—Matsuyama and Nakasawa (Faculty of Science, Kyoto University).

20. History of the formation of Hidaka River sediments—Kimibumi and Nakasawa (Faculty of Science, Kyoto University).

## General Summary of Symposium and Results

There were 11 presentations on the first day on the paleoenvironments of the Quaternary period, followed by lively discussions. The discussions can be divided according to region: 1) Sea of Japan, 2) From the northeastern Japan side of the Pacific Ocean to the northwest Pacific, 3) Ryukyu seas, 4) Antarctic Ocean, 5) Japanese islands and ocean floor.

The history of paleoenvironmental fluctuations is very clear in the Sea of Japan. The present form of the Sea of Japan was formed about 8,000 years ago. Due to the drop in sea level during the ice period 20,000 years ago, the waters of the Yellow River from China flowed into the Sea of Japan. This has been confirmed by the oxygen isotope ratio and age determination of Teflor $^{14}C$ (Oba). Also as a result of a drop in sea level, the surface layer of the third or earlier period was available for the formation of shore plains which occurred at a later date. The microfossils which were washed away from that crust (during such shore formation) precipitated widely in the Japan basin (Takayama and Kaizumi). Shika tried to corroborate the change in benthonic foraminifera with the time changes in depth of C.C.D.

In order to understand the fluctuations in the paleotemperature of the Pacific Ocean off northeastern Japan during the Quaternary period, Ishizaki used the transfer function to planktonic foraminifera in the bottom material and was successful in estimating the water temperature and salinity during medieval and modern periods. His estimated values are in good agreement with Saito's results who arrived at his values from measuring the oxygen isotope ratio in foraminifera. Tekiba selected the case of benthonic foraminifera and explained that between the Bering Sea and northeastern Japan, their breathing depth varied with time and went deeper in the southern part.

If we study the core material of coral with a longer growth line we can obviously notice the effects of the drop in temperature. There are also periods when normal growth is hindered by stresses and this can be seen at different places in coral reefs. This should lead to another comparison method (Konishi). Komeya and Inoue tried to study how the fluctuations of paleoenvironments in the Quaternary period could be reconstructed in the higher latitudes in the southern hemisphere. Here the foraminifera is in excellent condition and calcareous structures could reach a stage of perfection during a warm period. Nakagawa compared the knowledge gained on the Quaternary period environments distributed in the Japanese islands with the paleoenvironments determined from the data at the ocean floor. During the first glacier period around 120,000 years ago a fan-shaped structure was developed in the Japanese islands sometime around 65,000 years ago; whereas the glaciers developed in two periods: 20,000 and 40,000 years ago.

When analyzing the paleoenvironments in the Early Ice Age in the Japanese islands, not just this temperature variation but also the changes in humidity are equally important. The paleoenvironment of the Sea of Japan affect the humidity considerably and its analysis therefore is closely connected with the reconstruction of the paleoenvironment of the continent.

On the second day, there were nine presentations on the precision gravitational survey of the ocean floor. These can be divided into the regions of 1) Japan trench, 2) South Sea trough and Yonmanju belt, and 3) Daito ridge. In addition to the above, 4) there was also a discussion on future studies and the development of new equipment.

1) Papers were presented on gravitational force, warm currents from the earth's crust and earthquake studies on the Japan trench. The reason for a positive abnonnal gravitational force can be assumed as follows: When it sinks the heavy mass appears divided on both sides of the trench as an effect of topography and a positive abnormality occurs in the axial part of the trench Lisosphere which is quite elastic, can change shape in the trench and sinks. Hence this was supposed to form a sufficient mass on both sides of the trench as suggested in the above proposition. According to Tomoda, since the later structures could not be detected in the seismological studies, the thickness of the Lisosphere, which is 0.1 time heavier than the Asenosphere, can be assumed to increase in the trench and, as a result, the mass increases. This is considered to be the prime reason for the abnormal gravitational force. If we look at the gravitational force from geoid size as obtained by Stokes integration, then the amplitude of the gravitation force of 4,000 km wave length in the New Guinea-Australia belt is much higher than the Japan seas presently under active consideration. A number of reasons can be offered for this phenomenon, such as a large mass might have sunk there in the past, etc.

According to the submerged heat model of northeastern Japan, a (V) wedge-shaped mantle can be considered to be present on the seaside, as pointed by a seismic front. It was noticed that in this region no heat transfer takes place and the same behavior as observed in earth's crust can be seen (Honda). It is also assumed that a frictional heat under the pressure of 500-1,000 bars is generated on the upper surface of a sinking plate or an endothermic reaction occurs during dehydration. In the vicinity of a volcanic front, when a mayma tries to rise through the earth's crust, heat transfer is assumed to take place. Such model calculations reveal that the temperature profile around Japan can be explained on the basis of the mantle countercurrent that develops on the continental plate.

As a result of seismological measurements at the ocean floor, various interesting facts came to light. Thus it was noticed that two microseismic belts run parallel to the trench axis, over 30-100 km on the outskirts of the Japan trench along 39°N-41°N off Sanriku (Kasahara). The center of this earthquake was quite shallow between 0-10 km and lay on the periphery of a linear gura vein that is seen in the profile of the seabed. The microseismic zone near the trench corresponds to the epicenter of the earthquake off Sanriku as recorded by Kanamori (1971) and is reported to have a layered structure. Earthquakes are comparatively less in the 30 km wide region from trench axis towards continent around the water depth of 4,000 m. On the inner side between 4,000 and 3,000 m, active microseismic movements were noticed. As it is in phase T, it must be occurring at a shallow depth of 0-5 km. Studies on the mechanism of such occurrences showed on inverse layer structure.

2) As regards the studies on the South Sea trough and the Yonjuman belt, the results on acoustic survey, heat flow from the earth's crust, and the geological survey were presented during the symposium. The appendages on the continental side of the trench in the South Sea trough were studied in detail (Kagami). The appendages were broadly classified as the proto thrust and principal thrust belts. The proto thrust belt is characterized by a 60° inverse layer. The existence of this 60° layer is considered to be the reason behind a minor slope in this belt on the seaside. Another characteristic is the existence of a capsized tile structure and this showed a peak value according to the deflection of the vertical motion of the appendages, contrary to expectations. The principal thrust is characterized by the existence of a 30° inverse layer and a plain layered mound formed as a result, which is known as a plateau. The following characteristics were noticed from the geographical distribution. a) The length of the continental slope at Cape Tosa, where the seashore of the Shikoku basin has sunk, is quite short and the slopes are extremely steep. Here the appendage is simply seen as a plateau. b) At Cape Muroto, the width of the proto thrust belt is wider and large upwarped structure has developed there. c) At Cape Ashizuri, a plateau has developed whose width may be as much as 5 km. d) As we reach Cape Hyuga, the plate shows a capsized tile structure due to a number of inverse layers. If we consider the fact that such a capsized tile inverse layer structure is quite common in the appendages, plateau development or upwarping in this region, in a characteristic manner, shows that this is a shallow region and with sunken slope and (its characteristics) depend on the thickness or properties of sedimentations.

Regarding the thermal current distribution in the South Sea trough, new data was added this year. Along the trough axis, particularly at the center of Cape Tosa and Cape Shiono, a distribution of 3.0 HFU was ascertained (Yamano). The factor responsible for hot currents at the bottom of the trough, is the temperature of the Philippine Sea plateau which, for some reason, is much higher than what we should expect from its age and the heat is transferred with the movement of water around the sediments.

The studies on the sediment conditions and the history of the Yonjuman belt formation advanced significantly. The Yonjuman layer (Hidakagawa layers) in the Kii Peninsula, can be divided from bottom into the Yugawa, Misan and Ryujin layers. The Yugawa layer had long pebbles and was like a sand table due to the shallow seas. The Misan layer consists of Turbidite rich in acidic or neutral volcanic rocks. It also includes chart, greenish stones. It might have been deposited in the straits. There were also signs of horizontal pressure and hence it was thought that the Misan layer is made from an appendage formation. The Ryujin layer consists of quartz particles but there is not much deformation.

The chart period in the Misan layer is about 145 Ma-130 Ma and is different from the sedimentation period (about 80 Ma) of this layer (Matsyama). Also the period turns out to be more recent as one travels from north to south along the Misan layer and naturally the chart periods included there are also modern on the southern side. The standard layers and periodicity of this layer strongly indicates the formation of appendages, and is very important in the development of appendage theory.

3) The surveys also progressed along the Daito seashore. Here the shape of volcanic clusters and the formation of coral reefs was ascertained from gravity and abnormalities in the earth's magnetism (Iguchi). When calculated from the rock model and abnormalities in the earth's magnetism, it was found that Daito Island lay around 17°N. Since presently this island is at around 27°N, it means a shift of more than 1,000 km. The thickness of the coral reef was 500-800 m and from the results of site 445, the coral reef period on the volcanic cluster comes out to be minimum 30 Ma.

4) Two papers were presented on the corresponding research steps and development of equipment necessary for this project. A four-channel recorder was made. It converts the sound records from the seismological observations to digital records (Kinoshita). It consists of a converter section, control terminal, and MT console. It includes a JJY's time correction mechanism, high frequency wave generator mechanism, etc. The sampling period may be varied from 0.1-10 mSec according to frequency. Also an error message is fed to the terminal if there is any error during the writing operation of the MT.

The current status of survey systems on the structure of the ocean floor using an air-gun OBS and the development trends there were summarized (Matsuhiro). The submarine seismometer OBS became a reality in the 70s. Compared to other seismic receivers, this has a better S/N ratio. Since it is a fixed point apparatus, the positions are accurate and it can even record S waves. Due to its merits this apparatus has assumed an important role in surveying the structure of the ocean floor. There is a restriction in that it can only be used on-line but because of improvements in the equipment, both installation and recovery have become easy. The depth control of the air-gun affects the improvements in recording. Advances in digital processing were also illustrated.

The symposium, this time, discussed various themes in much more detail as compared with last year and hence was a great success.

# Proceedings of the Symposium on Monitoring the Kuroshio and Its Structure

## Symposium on Monitoring the Kuroshio and Its Structure

The Institute of Oceanography Symposium was held in its assembly hall. A large number of scientists engaged in this special project participated in the symposium and lively discussions were held.

### Program

January 25, 1983 (Tuesday)

Inauguration: Toshihiko Teramoto (Institute of Oceanography, Tokyo University), 10:00-10:10 a.m.

Chairman: Takeuchi (Faculty of Science, Hokkaido University)

Fluctuations in the Kuroshio current after tidal waves subsided in 1980.

Discharge results of the satellite tracking buoy—Nishida (Hydrographic Department), 10:10-10:55 a.m.

Fluctuations of the Kuroshio around Cape Shio—Takeshita (Kagoshima Hydroresearch Lab.), 10:55-11:30 a.m.

Ocean structure around the Izu ridge—Tomosada (Tokai Regional Hydroresearch Lab.), 11:30-12:05 a.m.

Chairman: Kubota (Department of Oceanography, Tokai University)

Flow around a round mass which is itself in a rotating body.

Effect of island or peninsula on the ocean current—Matsuura (Applied Research Laboratory, Kyushu University), 13:05-13:35.

Time variation observed in an inflow and outflow at a central latitude—Sakai (Department of Culture, Kyoto University), 13:35-14:00.

Inertia inflow and outflow currents in a bay—Masuda (Applied Research Laboratory, Kyushu University), 14:00-14:30.

Kuroshio serpentine wave and its dependence on flow mass—Yoon (Faculty of Science, Tokyo University), 14:30-15:10.

Kuroshio current region as manifested by a numerical model—Ishizaki (Meteorological Research Laboratory), 15:20-15:50.

Drift circulation according to one and two-layer models offshore current on the west coast border—Suginohara (Faculty of Science, Tokyo University), 15:50-16:30.

Formation mechanism of a subtropical front—Takeuchi (Faculty of Science, Hokkaido University), 16:30-17:00.

Social gathering, 17:15-18:15.

January 26 (Wednesday)

Chairman: Matsuyama (Tosui University)

The Kuroshio current around the east China Sea and the flow conditions around the Kuroshio—Maeda (Faculty of Engineering, Kagoshima University), 09:30-10:05.

Structure of the Kuroshio south of Kyushu and its flow mass—from the results of direct flow measurements—Takematsu, Mizuno and Kita (Applied Research Laboratory, Kyushu University), 10:05-10:40.

Deep layer currents in the Kuroshio region off Cape Toi—Ichikawa (Department of Fisheries, Kagoshima University), 10:50-11:25.

Deep layer currents in the Kuroshio region off Cape Shio—Fukazawa (Institute of Oceanography, Tokyo University), 11:25-12:00.

Chairman: Sakurai (Faculty of Engineering, Kagoshima University)

Currents around the seabed on the continental slope off Cape Omae—Ishizaki (Meteorological Research Laboratory), 13:00-13:35.

Deep layer currents in the Kuroshio on the west side of the Izu ridge—Nishida (Hydrographic Department), 13:35-14:10.

Characteristics of deep layer current fluctuations on the east and west sides of the Izu ridge—Taira (Institute of Oceanography, Tokyo University), 14:10-14:45.

Deep-sea current measurements by the Meteorological Department—off Shikoku, 137°E south of Kuroshio, Point B—Ishikawa (Meteorological Department), 15:00-15:35.

Medium fluctuations in the sea east of Torishima—Imawaki (Faculty of Science, Kyoto University), 15:35-16:10.

Discussions, 16:10-17:00.

## ADMINISTRATIVE REPORT

Minutes of the Fifteenth Executive Committee Meeting held on January 22, 1983 (Saturday) at the Institute of Oceanography, Tokyo University.

The agenda of the sixth Managing Committee meeting were discussed.

The proposals received from each committee on future study plans were discussed.

A special committee will be formed to work out future study plans.

## Guidelines for Authors

Newsletter welcomes enterprising articles. Current activities under this project, results obtained, etc., can be published in this Newsletter so that the connections between various study groups will be closer and the entire project will move smoothly toward completion.

The contents are generally classified as:

| | |
|---|---|
| Highlights: | Quick results of this Special Research Project |
| Bulletin: | Domestic or international news reports closely related to this Special Research Project |
| Reports: | Reports regarding the activities of each section or working group |
| Announcements: | Announcements regarding the proposed meetings or conferences of each group |
| Bibliography: | Listing of papers published by the members on this Special Research Project or other reports. |

Manuscripts should be written on standard 400 character size paper (handwritten material also accepted). The figures should be drawn in such a way that they can be reproduced as they are.

**Note:**

When reproducing or citing the material published in this Newsletter, due acknowledgments should be given. Authors' names, wherever mentioned, should also be quoted. A copy of the publication containing such a reproduction should be sent to the Editorial Section of the Administrative Office.


| | |
|---|---|
| Published by: | Administrative Office, Special Research Project "The Ocean Characteristics and Their Changes" |
| Published at: | Administrative Office,<br>Physical Oceanography Department,<br>Institute of Oceanography,<br>Tokyo University,<br>1-15-1 Minami-dai, Nakano-ku,<br>Tokyo 164<br>Tel.: 03-376-1251, Ext. 251. |
| Editorial staff: | Taeko Kitano and Noriyo Kimura. |

Special Research Project

# The Ocean Characteristics and Their Changes

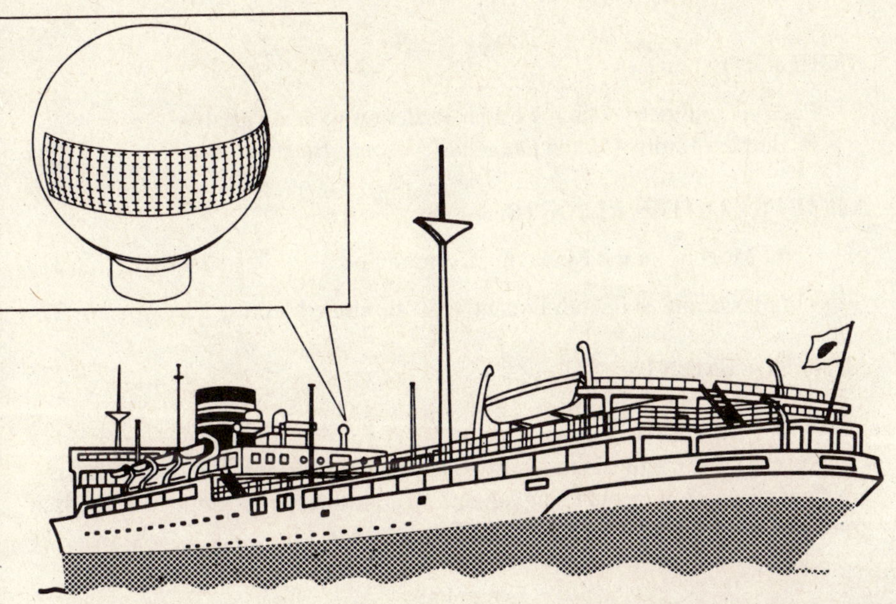

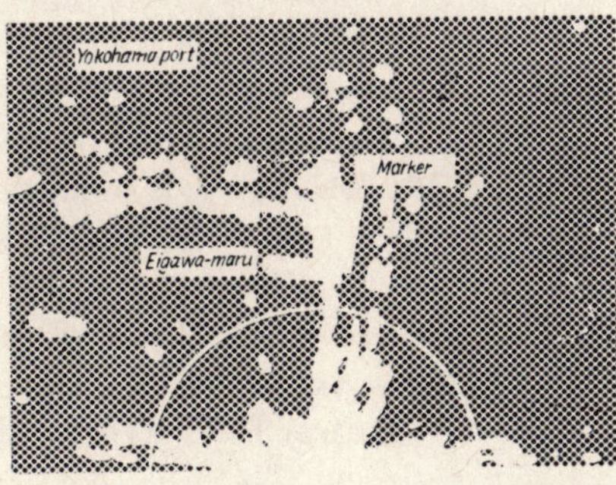


**NEWSLETTER NO. 12, JUNE, 1983**

# Contents



Researchers engaged in this Special Research Project, who wish to submit their research papers in English and to acknowledge receipt of a research grant, should indicate the above title of the project.

Cover Page

Radar reflector with variable reflection ratio.

This reflector uses the resonance effect of slots. By feeding the modulating signal to switching diodes mounted on its conductor plates, it is possible to vary the amplitude of incoming radar signals and retransmit them.

This type of reflector with a diameter of just 30 cm was installed in a large vessel as shown in the figure. The frequency echo was detected at a radar station and displayed on the PPI scope as a marker. It is also possible to transmit a certain kind of data.

Courtesy Hasebe (Faculty of Engineering Physics, Japan University).

HIGHLIGHTS

# Passive Telemetry Using a Radar Reflector with a Variable Reflection Ratio

*Mario Onoe*

(Production Technology Research Center, Tokyo University)

*Nozomu Hasebe*

(Faculty of Engineering Physics, Japan University)

(Received: April 22, 1983)

## 1. Introduction

To date, the collection of oceanographic data has been based on observations from continental areas or the data obtained by instruments installed on research vessels. As a result, the observations were restricted to a comparatively narrow range and steady-state measurements were difficult. Recently, however, with the advance of telemetry techniques, the data collected by mooring systems installed at the sea could be obtained by ratio transmission. The data could also be collected via satellites. However, in this method, it is necessary to provide a transmitter at the measuring point. And it involves problems such as interference from other signal sources or higher power requirements demanded by equipment at the observation post. Also, this method calls for special equipment such as a Doppler signal processing unit or multiple direction sensing equipment to determine the position of the observation post, etc.

It is quite common for a special agency to collect the oceanographic data and pass it on to actual users after proper processing. It is, however, expected that as far as the localized collection of data is concerned, the user can directly obtain the data from the observation point so that it can be widely used.

With this background, the authors gave more thought to cruiser radar which is compulsory even for small vessels. A radar reflector is supposed to efficiently reflect the electrical signal from the radar in the desired direction and has been found extremely useful in detecting small vessels or identifying their course. However, the reflector itself does not return any special type of echo signals and therefore it is difficult to distinguish it from other objects. If

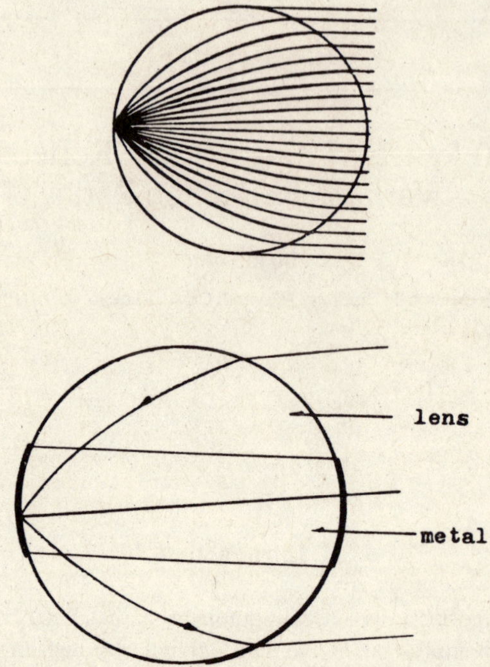


Fig. 1. a—Runeberg lens, b—Omnidirectional radar reflector.

only we can vary the reflection ratio of the reflector and attach a special meaning to this fluctuation, then it may be possible to distinguish the radar reflector from other objects whenever a meaningful reflected wave is detected. Thus it will also be possible to have a kind of information transmission.

The merit of this method is that, since the reflector itself does not generate the electrical signals, it will not cause problems to other communication channels. This is called passive telemetry. It can save the power requirements of the observation post. Also, if (another) radar is used for collecting data, the position of the observation post can also be decided simultaneously.

The following paper discusses the radar reflector with a variable reflection ratio developed with the above considerations, and explains the principles of passive telemetry. It also outlines the results of experiments conducted so far, followed by a discussion about the use of this system in the collection of oceanographic data.

## Radar Reflector with Variable Reflection

A radar reflector is used (in the radar system) to return the signals from

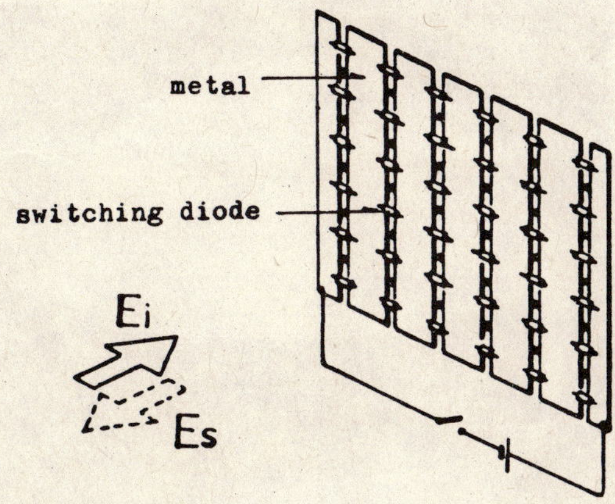


Fig. 2. Element with variable reflectivity.
Ei—Incident electric signal, Es—Incident electric field.

radar with efficient reflection. One such reflector used with the above objectives is the Runeberg lens. This is a spherical lens whose refractive index varies with $\sqrt{2 - r^2}$ ($r$ = specified radius). As shown in Fig. 1(a) the incident waves can all be collected on the reflecting spherical surface. Accordingly, if a conducting surface is provided on this surface, as shown in Fig. 1(b), then all incident waves will be reflected in the direction of incidence by conductive plates, thus making it an omnidirectional reflector [1].

Now if the conducting plates are removed when electromagnetic waves occur, it cannot return the echo signal to the radar station. Thus by attaching and detaching the conductive plates such an echo can be codified. This enables the transmission of information from the reflector station to the radar station. With an electrically variable reflection radar reflector [2], this operation is obtained electrically (hereafter this is called the variable reflector).

If a resonance slot of the same dimensions as the incident wavelength (in our case we used a slot of half the wavelength) is provided in the conductive plates, then the cross sectional area of radar (and the reflection ratio of electrical waves) is greatly suppressed. This is because the electrical signal phase reflected at the conductive part and that of the waves reflected at the resonant slot are 180° apart and hence cancel each other. Thus, if we provide the resonant slots at proper intervals to the conductive plates fixed at the focus, then the waves will not reflect even if a conductive plate is present. When the same part is to be used as a reflector, we can either short circuit the gap or make it nonresonant.

The element with variable reflectivity using the above principle is shown in Fig. 2. Here switching diodes operating in the microwave range are arranged at an interval of half wavelength between the rectangular conductive plates separated by a very fine slot. When an electrical source is connected at

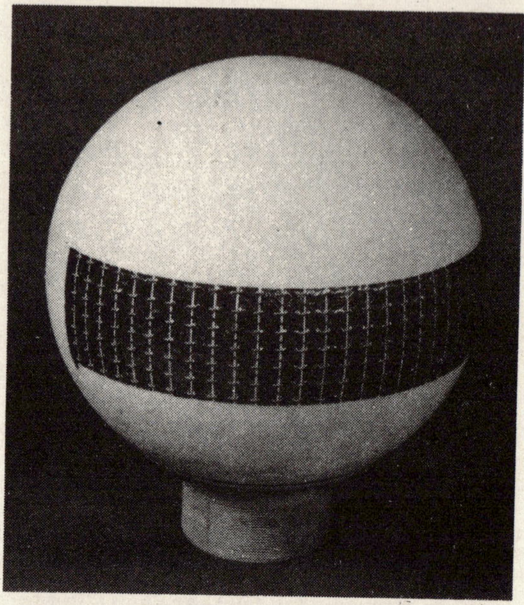

Fig. 3. Prototype radar reflector with a variable reflection ratio.

the end of these conducting plates, diodes are short circuited and form a resonant slot between each diode and the electrical waves can pass through as explained above. When a source is not connected, a discharge occurs between the conducting plates but, as the gap between two plates is much smaller than the wavelength, the incident waves are reflected to almost the same level as if there were a continuous conducting surface. Thus by switching the diodes to ON and OFF positions, it is possible to reflect the incident waves with variable echo amplitudes.

Figure 3 shows a photograph of such a prototype reflector made to operate around 9,000 MHz. Slot arrays are provided over 150° in the horizontal range (longitude) and 30° in the vertical range (latitude) and the echo can be varied for all waves incident from the opposite direction. With the prototype reflector, when the diodes are in the OFF position the radar cross sectional area is about 0.7 (-1.5 dB) times the area when the full conducting surfaces were available. Similarly, the difference (in echo level) is above 10:1 (10 dB) due to the ON and OFF positions of the diodes. For a 7×30 diode array, it is necessary to use the 20V, 10 mA bias source, which is much smaller than that required by the transmitter.

## 3. Application of Passive Telemetry

Figure 4 shows the block diagram of the passive telemetry system with a variable reflector. The operational waveforms of this system are shown in Fig.

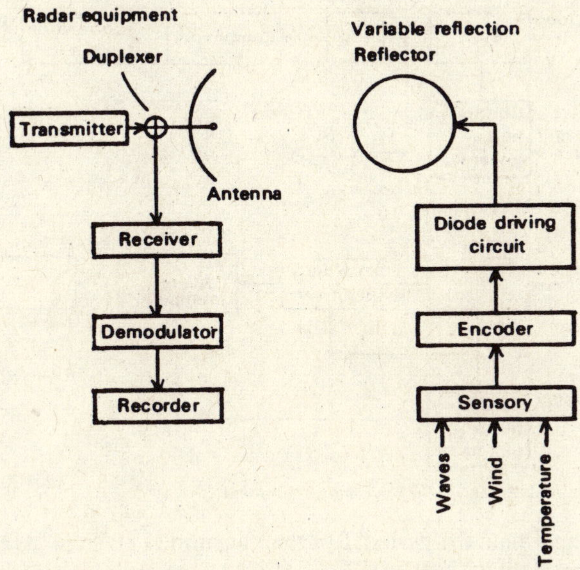


Fig. 4. Block diagram of passive telemetry with a variable reflection reflector.

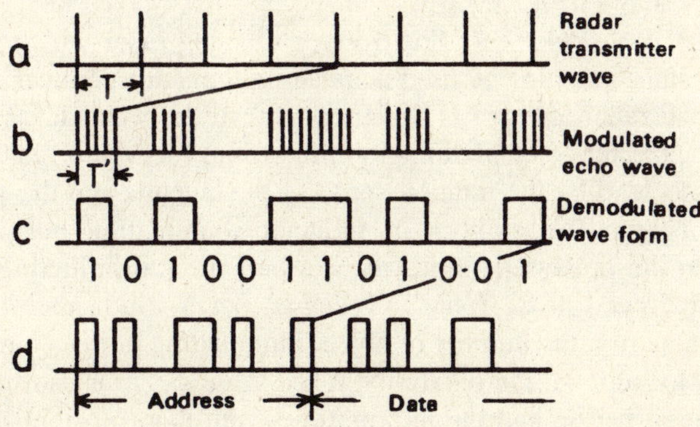


Fig. 5. Operational waveforms.

5. Thus a radar station consists of radar transmitter, receiver, and antenna as well as demodulator and recorder. Similarly the reflector side consists of a sensor, an encoding circuit that will codify the measured values, a diode driving circuit, and a variable reflector.

Radar station transmits a pulse with a period T (Fig. 5a). When this is incident on the reflector, it is modulated according to the encoding signals obtained from the measured values (Fig. 5b). The modulation period T′ is

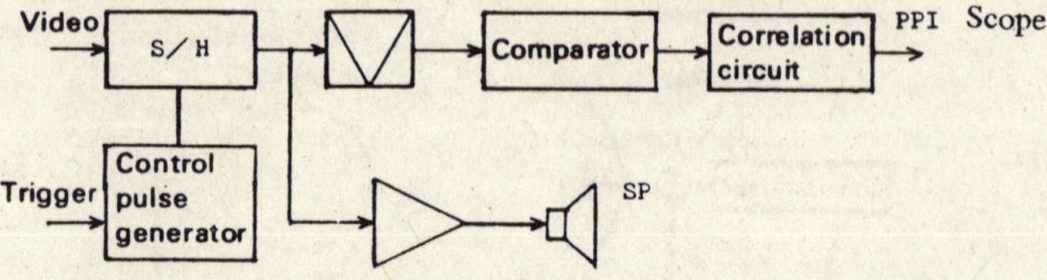


Fig. 6. Analog modulator circuit.

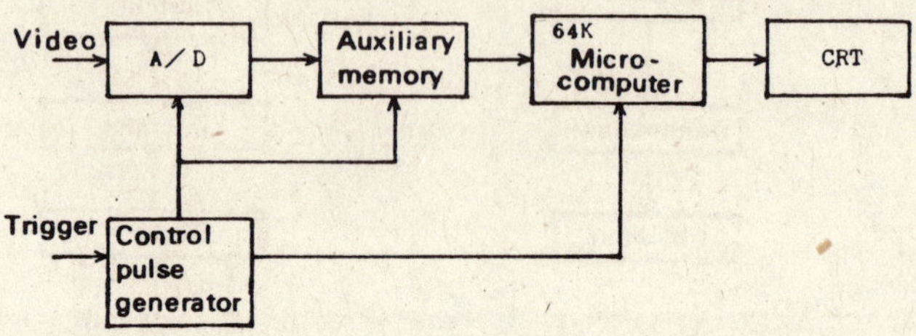


Fig. 7. Digital modulator and data-processing equipment.

generally longer than the period T of the transmitted pulse (4 T in figure). The echo modulated by the reflector is received by the radar antenna and forms a video signal through its receiver. This output is demodulated by the demodulator (Fig. 5c). Thus it is possible to retrieve information transmitted by a reflector station (Fig. 5d).

The usual PPI (Polar coordinates) type cruising radar which is generally installed on ships can also be used as radar equipment to collect data. In this case, since the radar antenna is designed to scan the water level by rotating around itself, the reflector illumination time (time for which radar waves are incident) is decided by the rotation speed of the antenna and the width of the underwater beam. The number of reflector illumination pulses varies in proportion to the pulse repetition frequency of the transmitted signal. Thus, for a rotation speed of 22 rpm, 1.07 degree beam width, and 4 kHz pulse repetition frequency, the number of object illuminating pulses comes out to be 32. In order to improve the reliability, if the same signal is sent twice within the illumination period and the information width is to be eight bits then the clock frequency after modulation by reflector would be two transmitted pulses, i.e., half the pulse repetition frequency of the transmitted signal.

When the location of the reflector equipment is known, the video signal can be isolated with filters and then processed. This enables the identification of signals received from multiple reflectors.

As far as information transmission is concerned, we can consider the analog method as well as the digital method. In the former, the voltage (or current) corresponding to the measured value is converted into a modulating signal, whereas in the latter the measured values are converted into a pulse code.

Figure 6 shows the analog block diagram modulator used in the experimental system. In this case, the echo duly modulated at the reflector station is obtained as video output and is held in Sample Hole (S/H) before the trigger pulse. The output which is then selected with a band filter is compared with the predesigned pulse in the comparator and the modulation signal is isolated by obtaining a correlation with the previous scan in the correlation circuit. This (isolated signal) is then displayed as a fan-shaped marker on the PPI scope.

Block diagrams of the digital system are shown in Fig. 7. The video signal obtained before the transmission pulse specifies the distance range. Analog data in this range is A/D converted into a digital signal and stored in the auxiliary memory. This data is then transmitted to the main memory of the microcomputer for data processing using the free time in antenna rotation. The processing results are displayed on the CRT.

## 4. Propagation Experiment

These experiments with variable reflectors were carried out on land as well as on sea using small and big ships [3, 7]. This confirmed the usefulness of the subject method and also made clear the problems involved in the system construction.

The land experiments confirmed that it is possible to distinguish between various modulating waves from the reflector using a radar receiver output. In another sea experiment using a small wooden ship, it was confirmed that modulated waves can still be detected at a distance where the echo from the ship is totally absent. Sea experiments conducted on a large vessel indicated that just a 30 cm diameter reflector can easily distinguish the modulated echoes, thus completing the development of the coding system. On the other hand, we also considered the use of cylindrical waves, which are not much affected by reflections from the sea surface, and developed a variable reflector for cylindrical waves [5].

The problems that surfaced during the experiments can be summarized as follows:

1) If an echo with the same base frequency is present during the analog modulation, it causes an image interference.

2) Since the dynamic range of the receiver in a general purpose radar is narrow, the receiver can become saturated if the level of the object echo is higher; then it is difficult to detect the modulating wave. Also the received signal level varies with time and it is not possible to continue detecting the modulated wave at an optimum level.

3) The phase relationship between the echoes emanating from the objects in the vicinity of the reflector and the modulated waves coming from the

Fig. 8. Onboard experiment using a large ship.

Fig. 9. Radar antenna and variable reflector setup.

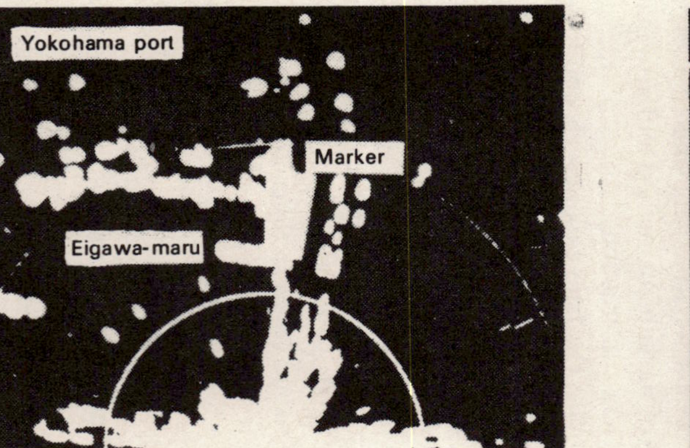

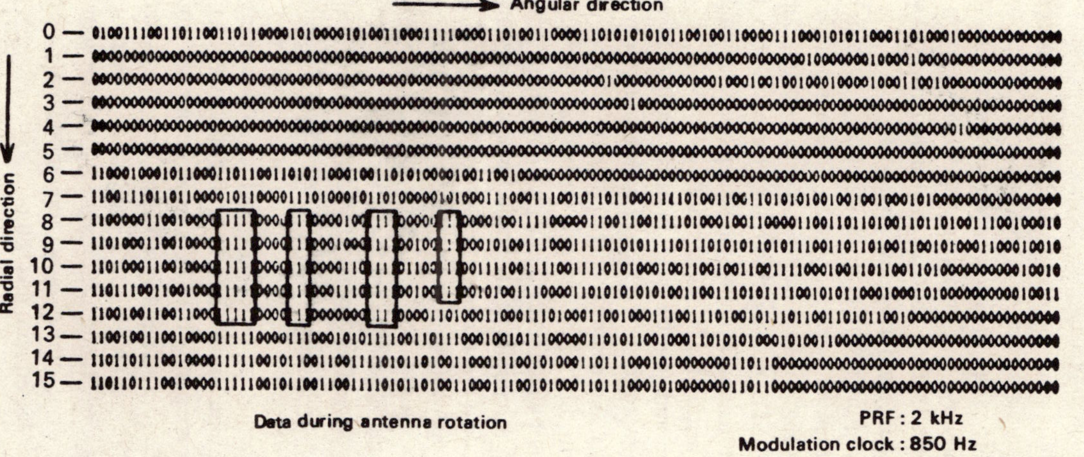


Fig. 12. Data obtained during antenna rotation.

reflector itself changes, depending on the route of sea surface propagation and other periodic fluctuations. As a result phase difference occurs and the coded signal may be reversed in phase.

The first problem could be solved by properly selecting the modulation frequency for correct distinction. The second problem can be attacked by expanding the dynamic range of the receiver and providing AGC (Automatic Gain Control) capability. The errors arising from the third problem can be minimized by using a coding structure that eliminates phase reversed codes and then taking proper care during data processing.

As an example of the above system, we carried out a transmission test on a large vessel the *Eigawa-maru* moored in Yokohama harbor. This is shown by a photograph in Fig. 8. A cruiser radar was installed 500 m away on the roof of a bar (Fig. 9). The propagation route connecting the radar antenna and the reflector lies on the sea, whereas the effectiveness of this propagation can be controlled by adjusting the land conditions.

There are radial and angular gates in radar equipment which identify the modulated waves coming from a reflector and display them on a PPI scope as a fanned out marker. An example of this marker display on a PPI scope is shown in Fig. 10. In this figure the echo that spreads below the center corresponds to the harbor wall and the left extension is the *Eigawa-maru*.

The output of a sample hold circuit, when a rectangular modulating waveform is applied to the reflector, is shown in Fig. 11. In this figure, the reflector echo is superimposed on the large echo signal coming from the *Eigawa-maru*. In this photo, the reflector echo is superimposed on the large echo coming from the *Eigawa-maru* but since it (the reflector echo) is modulated, distinction is easy. Thus, even when a reflector of just 30 cm dia is installed on large steel ships, the echo from the reflector can be identified from its modulation. In the same *Eigawa-maru* experiment we transmitted with a modulating clock frequency of 1.7 kHz and ten codes of eight bits each. The data was processed with a microcomputer and displayed on CRT.

Based on the experience from this experiment, we would like to plan a future system. In the meantime, a variable reflector propagation experimental setup was installed [8] on the roof of a production technology research center of Tokyo University, to obtain basic data to solve the problems involved in propagation. These days, studies on various modulating methods or diversity effects are being conducted with this setup.

Figure 12 shows an example of simple level detection of the data obtained from 16 radial points (corresponding to 120 m) and 128 angular points (corresponding to 5.8 degrees). We can observe here a repetition of the transmission code 11001000 in this data. In this the case pulse repetition frequency

(PRF) was 2 kHz and the modulating clock frequency was 850 Hz, which means that for one modulating bit there are two or three carrier frequency bits.

## 5. Collection of Oceanographic Data

From the results of the above experiments, it was possible to confirm the usefulness of passive telemetry using a variable reflector.

When general purpose cruiser radar is used in this system, the horizontal beam width of the antenna, the rotation frequency, and the carrier frequency (PRF) together restrict the velocity of information propagation. Accordingly, such a system is useful for comparatively slower and simpler data transmission. Particularly when installed on vessels, it not only acts as a means of telemetry but can also be used to prevent collision during the voyage.

When the system is to be used specifically for telemetry, the antenna can be sequentially directed to a number of predetermined reflector locations for data transmission, thereby increasing the transmission speed and minimizing the error. This method may be ideal for collecting large volumes of data from a number of observation points.

We would like to add that passive telemetry using a variable reflector can also be used with satellite mounted radar [4].

Similarly, by encoding the radar carrier wave, the variable reflector can also be used as an antenna. Thus with the proper receiver, we can get bidirectional communications with a single transmission station [6]. This aspect will be further discussed later, after necessary improvement.

## 6. Summary

In this paper we have proposed a passive telemetry system using a radar reflector with a variable reflection ratio and discussed its usefulness.

Merit of this method lies in the fact that the reflector sending information does not generate the electric energy itself, thereby causing no interference for other communications. This is quite a significant step over the previous methods. Also, compared with the telemetry system using a regular transmitter, the power consumption is many times smaller. Also it is possible to obtain data without interference from any number of radar equipment using the same carrier frequency and hence this system may be useful in future for general safety.

We would like to conclude this paper with our sincere thanks to all those who extended their cooperation, particularly Iyama, the Director of Tokyo

Measurement Research Laboratory, Mr. Ota, the Asst. Director, and Mr. Okada. The *Eigawa-maru* experiment was also recorded as a joint research program. Thanks are also due to Mr. Yamada and Nakamura of Tokyo University's Production Technology Research Center as well as to students of the engineering faculty of Japan University, who assisted during the experiment.

## References

1. Ota, Mizutani and others. 1972. Omnidirectional radar reflector with a Runeberg lens. *Proceedings of the 4th Conference of the Electrical Association,* Hokkaido Division, pp. 130–131.
2. Ogami, Hasebe and Zama. 1980. Radar reflector with variable reflection ratio, Shingakuron, 63-B, 3, 218–224.
3. Ogami, Hasebe and others. 1980. Ocean experiments using a radar reflector with variable reflection. *National Convention of Photo-Electrical Communication,* 71.
4. Ogami and Hasebe. 1980. Application of a radar reflector with variable reflection in remote sensing. *TV All Japan Conference,* September 12.
5. Hasebe and Ogami. 1981. Application of a radar relector with variable reflection for circularly polarized waves. *Proceedings of the 1981 Conference of Communication Engineering Association,* 649.
6. Ogami, Hasebe and others. 1982. Bi-directional communication system with a single transmitter station and variable reflection radar reflector. *Proceedings of the 1982 Conference of Communication Engineering Association,* 2383.
7. Ogami, Hasebe and others. 1982. Information transmission experiment using a variable reflection radar reflector. *Proceedings of the 1982 Conference of Photo-Electrical Communications,* 210.
8. Ogami, Hasebe and others. 1983. Experimental setup for a variable reflection propagation experiment. *Proceedings of the 1983 Conference of Communication Engineering Association,* 2337.

## ADMINISTRATIVE REPORTS

The sixth meeting of the Managing Committee was held on February 28, 1983 (Monday) at the Institute of Oceanography, Tokyo University.

Coordinators of various research groups presented the methodologies adopted in their studies and their results. Details of presentations:

The studies conducted after the last managing committee meeting (October 5, 1982).

1) The scheme for fiscal year 1983 of the Special Research Project was submitted to Mombusho (Ministry of Education).

2) It was decided to divert the unused funds from fiscal year 1982 to acquire the following equipment:

Underwater automatic release equipment (2)
Semiconductors for the detection of x-rays (1)
Electric furnace (1)
Digital volt meter (1)
Self-recording flow direction and velocity measuring equipment (2)

3) The executive committee meetings of November, December, and January mainly discussed future plans.

4) The Newsletter was published six times this year as planned, to No. 11.

5) Based on the reports received from each research group regarding the course of studies and results obtained, a book has been prepared on the progress of the research.

**Items discussed**

The research conference for 1983 will be held sometime in the beginning of July. This will be the third of its kind and this time the program will be printed and circulated to other people also. The conference will have two themes of general interest and two papers from each committee will be presented.

The executive committee will decide the date, time, place, and speakers, etc., as early as possible.

It was decided to form a separate committee to take care of the future course of this study. The concerned coordinator should take the opinions of members before deciding its scope, etc., and the members for this committee should be selected in the next executive committee meeting (March 26).

When preparing future plans, the results of surveys on various fields which are theoretically much advanced can be used as a good reference.

## Minutes of the Sixteenth Executive Committee Meeting held on March 26, 1983 (Saturday) at A Wing, Institute of Oceanography, Tokyo University

**Items announced**

It was decided to hold a symposium for oceanographers on July 13, on technical aspects proposed under the joint sponsorship of Kaiken renshuken and the Oceanographic Association.

Reports regarding the research experiments and the state of the art of the project, conducted under the 1982 Scientific Research Grant, were submitted to Mombusho (Ministry of Education). A copy of these reports was also distributed to members of the administrative committee.

A seminar on the Special Research Project was held on March 18.

**Items discussed**

The research conference will be held on July 11, 12 and 13 as a part of a general symposium (Ref. p. 276 for announcement).

Two members from each of the existing committee were selected to work on the "Committee on Future Plans".

| | | |
|---|---|---|
| Geology | Yokichi Takayanagi | Faculty of Science, Tohoku University. |
| | Hideo Kagami | Institute of Oceanography, Tokyo University. |
| Physics | Toshihiko Teramoto | Institute of Oceanography, Tokyo University. |
| | Seiichi Kanari | Faculty of Science, Hokkaido University. |
| Chemistry | Yoshio Horibe | Institute of Oceanography, Tokyo University. |
| | Nobuhiko Handa | Oceanographic Laboratory, Nagoya University. |
| Measurement | Hitoshi Mochizuki | Denstu University. |
| | Morio Onoe | Production Research Center, Tokyo University. |

The first meeting of this committee will be held on May 7, i.e., the next Executive Committee meeting day (Saturday), from 13:30 hrs onward in the Institute of Oceanography.

## Minutes of the Seventeenth Executive Committee Meeting held on May 7, 1983 (Saturday), Institute of Oceanography, Tokyo University

**Items discussed**

Progress regarding the general symposium was discussed. Chairman is not yet elected.

Currently, the program and posters are being printed and will soon be distributed to concerned persons.

The program will also be published in the Newsletter for general information.

The group leaders will be requested to consider bearing the traveling expenses of the desirous members so that members can attend in large number.

The traveling expenses of the members of the administrative group and those presenting papers will be borne by the administrative group.

After the symposium is over, the contents will be published as "Proceedings of the General Symposium." Contributors are requested to send their manuscripts beforehand.

After the amount of scientific research grant to be received for this project is decided, a publication entitled "Outline of Research Composition and Plans for Fiscal Year 1983" will be published. The group leaders are requested to send their manuscripts for this purpose.

With the advent of the "Committee on Future Plans," group leaders have sent information regarding current experience and future time schedules. These were passed on to the first meeting of the "Committee on Future Plans."

The next executive committee meeting will be held at 13:00 hrs on June 14, at the Institute of Oceanography, Tokyo University. This will be followed by the second meeting of the "Committee on Future Plans" from 13:30 hrs onward. Members of the Executive Committee will attend that meeting as observers.

## ANNOUNCEMENTS

The program of the general symposium under this Special Research Project is mentioned below. Persons interested in oceanographic studies are requested to attend the symposium.

Inquiries regarding the program and other aspects may be directed to Prof. Teramoto, Physical Oceanography Department, Institute of Oceanography, Tokyo University (1-15-1 Minami-dai, Nakano-ku, Tokyo 〒164. Tel.: 03-376-1251, Ext. 251).

Time: July 11, 1983 (Monday), 13:30–17:05
July 12, 1983 (Tuesday), 09:00–17:00
July 13, 1983 (Wednesday), 09:00–11:40.
Place: Aoi Kaikan, 4th Floor.
(2-2-6 Toranomon, Minato-ku, Tokyo
Tel.: 03-582-9721).

### 1st Day, July 11 (Monday)

Welcome/Inaugural address: Prof. Kajiura (Earthquake Research Laboratory, Tokyo University), 13:10–13:40.

Deep-layer currents of the Kuroshio as obtained from the mooring system for flow measurements and effect on the earth's profile—Fukazawa (Institute of Oceanography, Tokyo University) and Ichikawa (Department of Fisheries, Kagoshima University), 13:40–14:15.

Laboratory experiment of Kuroshio long waves—Masuda (Applied Research Laboratory, Kyushu University), 14:15–14:35.

Ocean structure around the Kuroshio—including a comparison with the gulf current—Fukuoka (Department of Fisheries, Hokkaido University, 14:35–15:10.

Coffee Break, 15:10–15:20.

Distribution and behavior of $^{14}$C as a Transient Tracer—Gamo (Institute of Oceanography, Tokyo University), 15:20–15:55.

Ocean flow characteristics at the continental shelf—around the Sea of Ohotsk—Aota (Low Temperature Laboratory, Hokkaido University) and Nagata (Faculty of Science, Tokyo University), 15:55–16:30.

Physical conditions and their changes at the front continental shelf of Seto Strait and the Hogo-Kii channels—Nishi (Faculty of Science, Kyoto University) and Yanagi (Faculty of Engineering, Eihime University), 16:30–17:05.

## 2nd Day, July 12 (Tuesday)

Geographical and vertical changes in the sedimentary particles in the Pacific and the process of material removal—Kakumi and Harada (Department of Fisheries, Hokkaido University), 09:00-09:35.

Studies on the mechanism supporting the vertical transportation of organic substances in the ocean—Handa (Oceanographic Laboratory, Nagoya University); Organic composition of fecal plate of zooplankton—Tagami (Department of Natural Sciences, Kobe University), 09:35-10:10.

Mutual examination of non-pulluting water samples—Murozumi (Muroran Institute of Technology), Tsubota (Department of General Science, Hiroshima University) and Kanamori (Oceanographic Laboratory, Nagoya University), 10:10-10:45.

Distribution of heavy metals in the North Pacific and its mechanism—Kanamori (Oceanographic Laboratory, Nagoya University), Tsubota (Department of General Science, Hiroshima University) and Murozumi (Muroran Institute of Technology), 10:10-10:45.

Molecular distribution of essential trace elements in natural water, particularly selenium—Shigematsu, Nishikawa, Tairagi and Goda (Kinki University), 10:45-11:05.

Behavior of heavy metals in red tides—Matsunaga (Department of Fisheries, Hokkaido University), 11:05-11:25.

Ocean water analysis using an ion selective electrode—Umezawa (Faculty of Science, Tokyo University), 11:25-11:45.

Lunch break, 11:45-13:00.

Paleoenvironmental changes in seas east of Honshu. Studies of a prototype SK-1 sampler off Kagoshima—Takayanagi (Faculty of Science, Tohoku University), 13:00-13:35.

Sedimentation operation in a trench and the structure of trench appendages—Kagami (Institute of Oceanography, Tokyo University), 13:35-14:10.

Tea break, 14:10-15:00.

**General lectures**

1) Circulation of ocean water and materials in deep layers of the North Pacific—Teramoto (Institute of Oceanography, Tokyo University), 15:00-16:00.

2) Dynamic condition of seabed around Japan—Nasu (Institute of Oceanography, Tokyo University), 16:00-17:00.

### 3rd Day, July 13 (Wednesday)

Free-fall type electromagnetic flow velocity profiler—Kanenai (Faculty of Science, Hokkaido University), 09:00-09:20.

Measurements of ocean currents using free-fall type ultrasonic flow velocity meter—Kawadate, Mizuno, Nagahama and Ishii (Applied Research Laboratory, Kyashu University), 09:20-09:40.

Studies regarding measurements in the Kuroshio region using acoustic waves—Okushima (Tokyo Institute of Technology), 09:40-10:15.

Measurement techniques for a large ocean area using the ultrasonic compound aperture—Yamazaki (Faculty of Engineering, Tokyo University), 10:15-10:50.

Studies of basic techniques involved in measurements of oceanographic data—Kumamoto (Dentsu University), 10:50-11:10.

Summary, 11:10-11:40.

## Guidelines for Authors

This Special Research Project has now entered the last year. Since this may be the time to summarize the studies conducted to-date, Newsletter welcomes enterprising articles on the state of the art of the research and results, which will go a long way to finalize this special research in general.

The contents are classified as:

| | |
|---|---|
| Highlights: | Quick reports of the results of this Special Research Project |
| Bulletin: | Domestic or international news reports closely related to this Special Research Project |
| Reports: | Reports on the activities of each section or working group |
| Announcements: | Announcements regarding the proposed meetings or conferences of each group |
| Bibliography: | Listing of papers published by the members on this Special Research Project or other reports. |

Manuscripts should be written on standard 400 character size paper (handwritten material also accepted). The figures should be drawn in such a way that they can be reproduced as they are.

It is proposed to publish this Newsletter 5 times more this year at two months intervals.

**Note:**

When reproducing or citing the material published in this Newsletter, due acknowledgment should be given. Authors' names, wherever mentioned, should also be quoted. A copy of the publication containing such a reproduction may be sent to the Editorial Section of the Administrative Office.


| | |
|---|---|
| Published by: | Administrative Office, Special Research Project "The Ocean Characteristics and Their Changes" |
| Published at: | Administrative Office,<br>Physical Oceanography Department,<br>Institute of Oceanography,<br>Tokyo University,<br>1-15-1 Minami-dai, Nakano-ku,<br>Tokyo 164<br>Tel.: 03-376-1251, Ext. 251. |
| Editorial staff: | Taeko Kitano and Noriyo Kimura. |

Special Research Project

# The Ocean Characteristics and Their Changes

NEWSLETTER NO. 13, AUGUST, 1983

# Contents



Researchers engaged in this Special Research Project, who wish to write their research papers in English and to acknowledge receipt of a research grant, should indicate the above title of the project.

## Cover Page

The cover shows a schematic diagram of a hydraulic hand-boring machine (lower right-hand corner) in operation which is used to excavate the scleractinia colony fossils. These are used to analyse the time-wise changes in ocean environments based on scleractinia skeletal chronology. Also shown schematically is a core sample divided in three sections after excavation. The sample is first divided in two equal parts. A plane sheet is sliced from one part for x-rays and from the width of the annual growth rings on these photographs, the rate of growth is measured, whereas the presence of a stress band is ascertained. After this, a part corresponding to one year's growth is separated from the above and used to measure yearly changes (or if the year is further divided in smaller parts, it will be seasonal changes) in oxygen and carbon isotopes. The remaining half is cut into pieces, corresponding to growth during every three years and then used to measure annual changes in the concentration of radioactive carbon. (The orginal diagram was prepared by Iwahara, Iguchi and Sato of the Faculty of Science, Kanazawa University.)

HIGHLIGHTS

# Analysis of Paleoenvironments of the Quaternary Period by Boring a Coral Reef (2)

*Kenji Konishi, Mahito Iguchi, Nohiro Sato and Toyoki Iwahara*

(Faculty of Science, Kanazawa University)
(Received: July 25, 1983)

The above studies were undertaken as part of the Special Research Project "Paleooceanography of the Seabed in the Quaternary Period" (Chief Coordinator: Prof. Takayanagi) which began in 1981. Results obtained during the first year were reported in the 7th issue of this Newsletter (Konishi and Niguchi, 1982).

Field operations during fiscal year 1982 were carried out again in the Holocene coral reefs of the Nakakuma region in the Kikai Islands (129°50′09″E, 29°20′08″N) just as last year. However, the details were divided in two sections depending on the objects and contents.

The first part which can also be considered an extension of studies conducted so far, involved excavation of a reef front layer from the seabed to the surface where the excavation ship was positioned. It is expected that the samples recovered may have the most continuous records of reef formation during the postglacial age. The actual boring was entrusted to a firm called Koken Shisui Kogyo Co. Ltd. Since 1977, three points—No. 17, 18 and 19—which lie on the northern extension of the measurement line, have been excavated to 11.70 m, 25.00 m and 25.00 m, respectively. Excavation record, depth of water, depth of excavation, depth of Saba surface of discontinuity (Holocene system/Pleistocene system surface of discontinuity), etc., for each hole are as shown in Table 1.

**Table 1. Record of excavation of coral reefs in the Chutaichi region of the Kikai Islands during fiscal 1982**

| Excavation hole No. | Depth | Depth of excavation (m) | Recovery length (m) | Depth up to Saba surface of discontinuity (m) |
|---|---|---|---|---|
| 17 | -7.37 | 11.70 | 9.92 | — |
| 18 | -9.42 | 25.00 | 10.41 | 19.70 |
| 19 | -1.80 | 25.00 | 15.72(+) | 23.50 |

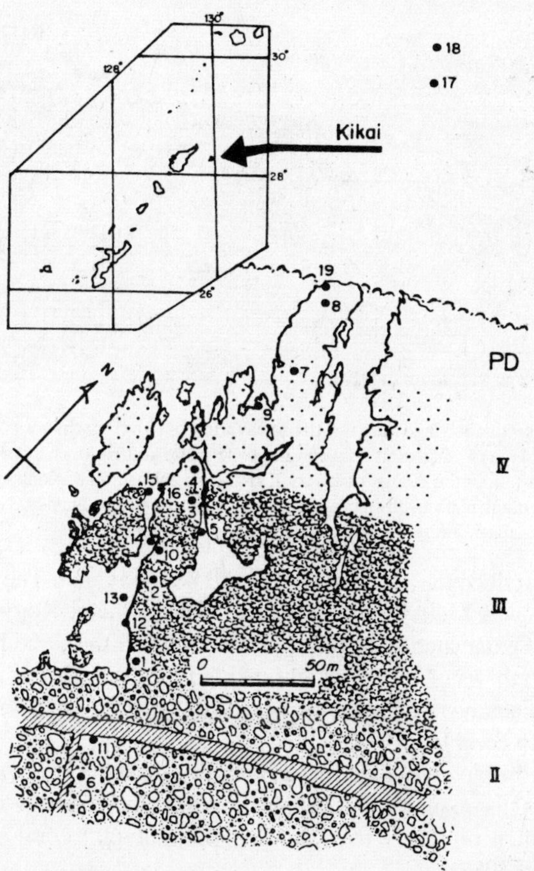


Fig. 1. The plane view showing the excavation site of a Holocene coral reef in the Nakakuma region of the Kikai Islands. Please note 4 regions including this coral reef (PD) and 3 other bench class (II, III and IV). The excavation point of the 1982 studies lies in the coral wall of this coral reef (No. 19) and the surface layers (Nos. 17 and 18).

Of the three holes, we did not reach the surface of discontinuity only in hole No. 17, for which the excavation was begun first. This may be because due to bad weather the ship had to be removed from the place during excavation and we could not avoid abandoning the hole. In hole No. 2 excavation vessel also met with bad weather while it was moving from hole No. 18 to hole No. 19 and broke down. Due to these operational troubles, a steel pipe structure was constructed from the land to hole No. 19. At the end of this structure a working platform was added and the excavation was carried out from there. As a result, this work far exceeded its estimates and called for both a longer period and more manpower. The cost naturally increased also

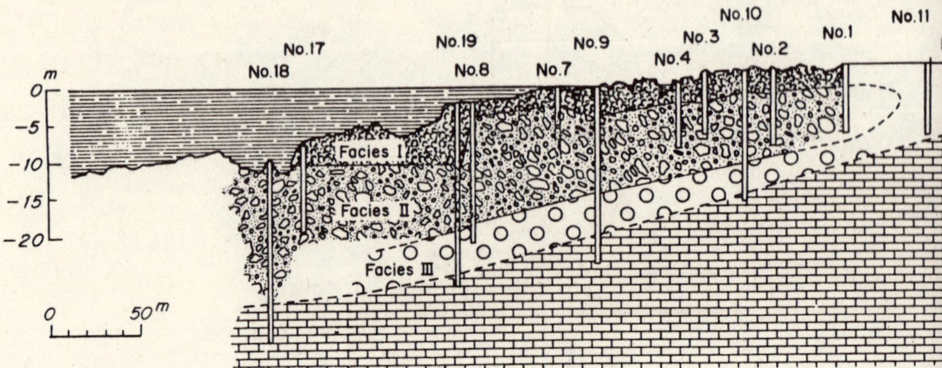


Fig. 2. The cross sectional view showing the excavation position, excavation depth, server discontinuity plane (plane of mismatch between the base plane and Holocene carbonaceous rocks), layer identification (about the details of layers I, II and III, please refer Konishi, et al., 1983, pp. 157) of Holocene coral reefs in the Nakakuma region of the Kikai Islands. Note the depths of water and those of excavation points tried in 1982 (Nos. 17, 18 and 19).

and we had to reduce the number of holes to be excavated. The machine used for excavation was Model OP-1 from the Koken Shisui Kogyo Co. and the diameter of the excavated rock pillar was 66 mm. In the three holes, the total excavation length was 61.7 m which enabled us to reach our desired target. The laboratory analysis of samples is still in progress and will be reported separately when complete.

The second objective added to the studies during 1982 was to carry out an accurate paleooceanographic analysis using the skeletal chronology of scleractinia which represent reef building (hermatypic) coral as indicated in Part 1 (Konishi, 1983).

One characteristic of sediments in a coral reef is their high periodic resolution. If we use the skeletal chronology of scleractinia, which have the highest growth rate ($10^{1}$ ~ 10 mm/y) of all aquatic organisms that form a hard shell, it is possible to clarify the periodic variations of seasons and years. We used a sampler (KURD-1), same as we used before, to recover rock samples either containing semi-spherical bodies in the center or regions very close to them. It is just a short time since the study was completed and the following discussion may include some points that may not be very clear, but the material is presented as an intermediate results in line with the policy of this Newsletter. The subject is interdisciplinary and we would therefore be happy to receive a word of advice, evaluation, and assistance from concerned researchers.

During the summer of 1982, we excavated coral stones (scleractinia, which corresponds to *Porites* sp. aff. *lutea,* a type of *Porites lutea*) from 16 holes in location 20 over a total distance of 19 m 73 cm. The number of each sample, position of excavation point (benchwise classification), excavation depth, scleractinia depth in this excavation, type of boring machine used, etc., are indicated in Table 2. In this table we have also mentioned the age of this sample as determined by x-rays of the pellets made from recovered samples in our laboratory. (The actual age of these samples is higher than the values indicated as the initial formation period will be added to it.) Positions of excavation points are also shown in Fig. 3.

Of these, sample 601, which was obtained from four adjacent cores, was a part of a semi-spherical colony with a radius of 2 m (Fig. 4) and exceeded 170 years of age. The following record is based mainly on the studies of this sample No. 601.

**Table 2. Record of excavating samples of coral beach on a Holocene coral reef in the Nakakuma region of the Kikai Islands, used for skeletal chronological studies (fiscal 1982)**

| Excavation point | Bench class | Total excavation length of scleractinia (total excavation length), cm | Sample age | Type of boring machine used |
|---|---|---|---|---|
| 527-1 | IV | 15 (68.5) | | KURD |
| 528-1 | IV | 53 (78) | about 70 | KURD |
| -2 | IV | 70 | 39 | KURD |
| 529-1 | IV | 100 | | KURD |
| 604-1-A | IV | 170 | | SD-3A |
| -1-B | IV | 115 | | SD-3A |
| 529-2 | III or IV | 50 (50) | 22 | KURD |
| -3 | III or IV | 0 (70) | | KURD |
| 530-1 | III or IV | 43 (80) | 61 | KURD |
| 601-1* | III | 180 (196) | 138 | KURD |
| -2* | III | 185 (200) | 111 | KURD |
| -3* | III | 155 (155) | 168 | KURD |
| -4* | III | 173 (195.5) | | KURD |
| 602-1 | III | 28 (40) | | KURD |
| -2 | III | 10 (50) | 17 | KURD |
| -3 | III | 25 (40) | | KURD |
| 603-1 | III | 42 (50) | 26 | KURD |
| 606-1 | III | 45 (55) | about 44 | KURD |
| -2 | III | 70 (100) | 46 | KURD |
| 609-1 | II | 70 (90) | 85 | SD-3A |
| 20 holes | | Total 19 m 73 cm | | |

* Same colony.

*Rate of growth.* The thickness of growth in a year (uniform light and dark density stripes) for the three holes in sample 601 (-1, -2 and -3) was observed by x-ray transparency photographs and the mean value of 5 vernier

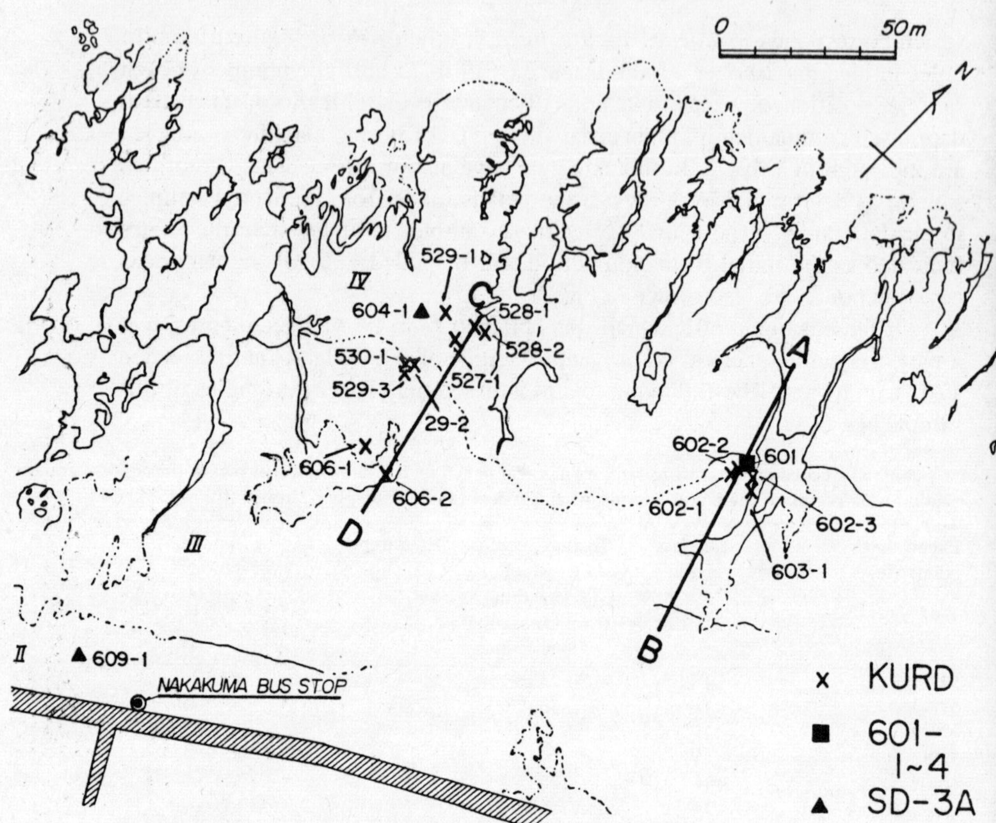


Fig. 3. Diagram showing the points of excavation in the Nakakuma region of the Kikai Islands used to obtain samples of a Holocene scleractinia colony from coral reefs for skeletal chronological studies, II, III and IV represent the same bench levels as in Fig. 1. × represents the use of a KURD-1 hydraulic hand-boring machine (Konishi and Niguchi, 1982), ▲ represents use of SD-3A sander hand-boring machine. Position of sample 601 is indicated by ■ which is discussed in greater detail in this paper. A–B, C–D are base lines of topographical cross sections but the cross-sectional view is slightly modified to suit the paper surface.

readings was determined. However, since the high density stripe pattern is disturbed in the region of proliferation they were purposely not included in the measurement. Subsequently the three cores were compared using the stress band, to be discussed later, and the mean value of growth (thickness) during the same age was determined. This value is assumed to be the yearly growth rate of the colony (mm/y). Lower part of Fig. 6 shows this value according to age. The dotted lines on both sides of this curve indicate those periods when a part of a colony was lost due to erosion and the core length was incomplete. Thus of three cores, one was lost and the mean value was obtained from one or two cores only.

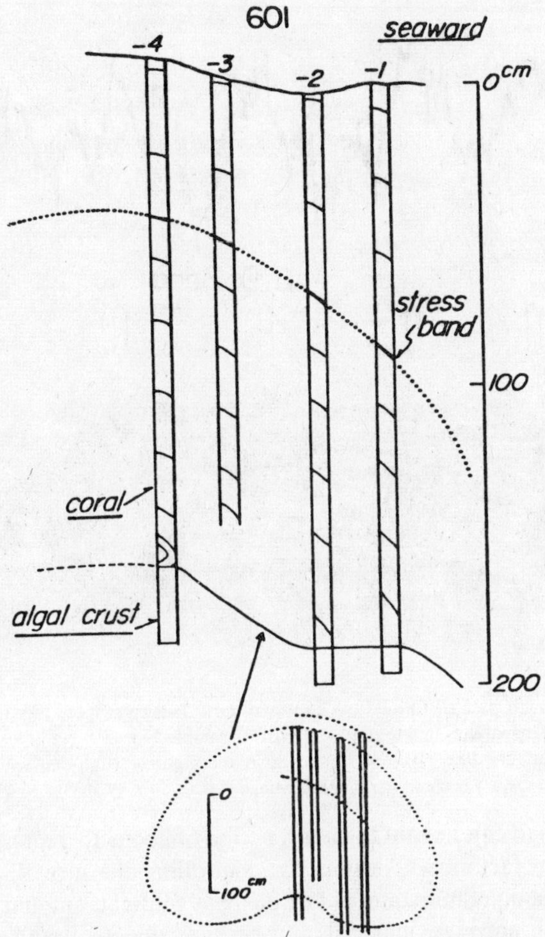


Fig. 4. This figure shows the shape and size of beach coral fossils recovered from 4 core samples at point 601. This is formed on top of an algal crust. A stress band during 90–91 years is also shown in the figure as a standard for comparison.

As can be clearly seen from this figure, the growth rate varies between 4.2 mm/y and 8.1 mm/y. In this variation the existence of a stress band can be determined at those positions where the growth rate is smaller than the previous or subsequent years. In this sample, assuming the top part of the colony as 0 years, ten stress bands were detected at 11th, 25th, 34th, 48th, 54th, 65th, 75th, 90th, 108th and 122nd year. Of these, the growth rate dring the 90–91st year is extremely small indicating the presence of a most significant stress band that plays an important role in the distinction between the four cores (Fig. 4 and Fig. 7). The period between these stress bands is 14–9–14–6–11–

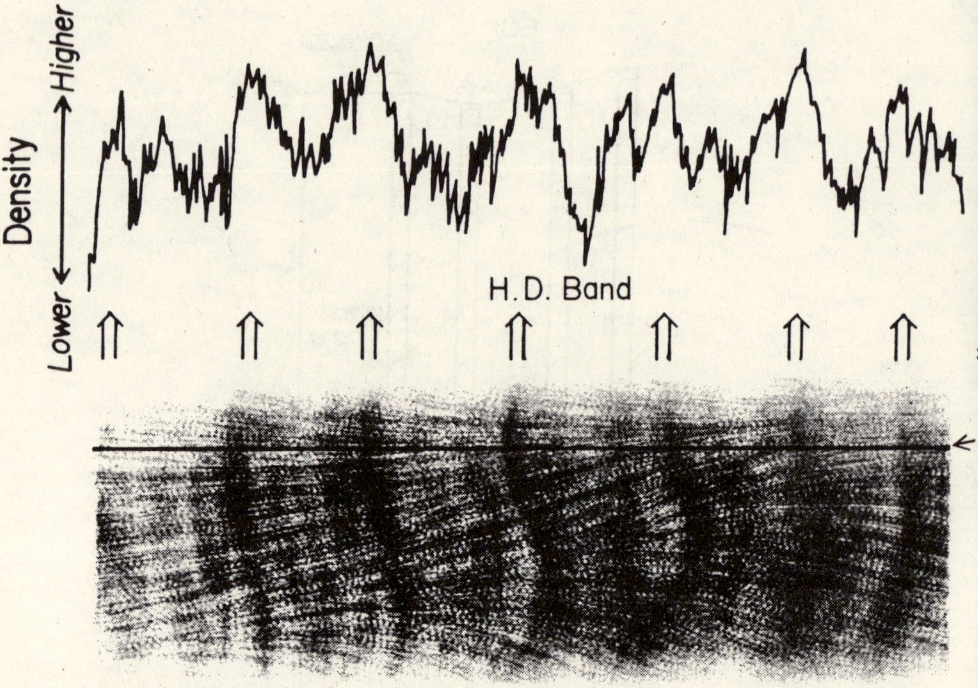

Fig. 5. Yearly growth pattern formed by light and dark density stripes. Figure shows a comparison of x-ray transparency photographs of part of sample No. 601-1 and a scanning from a densitometer. (Scanning line position is shown in the figure.) High-density points in x-ray photographs can be clearly seen on the scanned image.

10–15–18–14 and can be said to have a period of about 13 years.

Among the factors responsible for controlling the growth rate, we can think of water temperature, amount of incident sunlight, amount of precipitation, supply of nutritive materials, suspension due to mud, effect of hot streams, etc. The water temperature and amount of sunlight have a strong effect on a wide area of environmental variations and are also found to have a positive correlation with growth rate. Studies were conducted regarding the relation between the growth rate and weather factors for a colony of *Porites lutea* on the outskirts of Naha port between 1913 and 1979 AD. The results showed, although there is no correlation between the amount of precipitation and incident sunlight, a positive correlation was obtained with temperature. Looking at the stress band in sample No. 601 of the Kikai Islands, we can also attribute it to the periodic variation of such environmental factors as drops in water temperature which affect skeleton formation. In order to establish this fact independently, we subsequently carried out a geo-chemical analysis of skeletal materials.

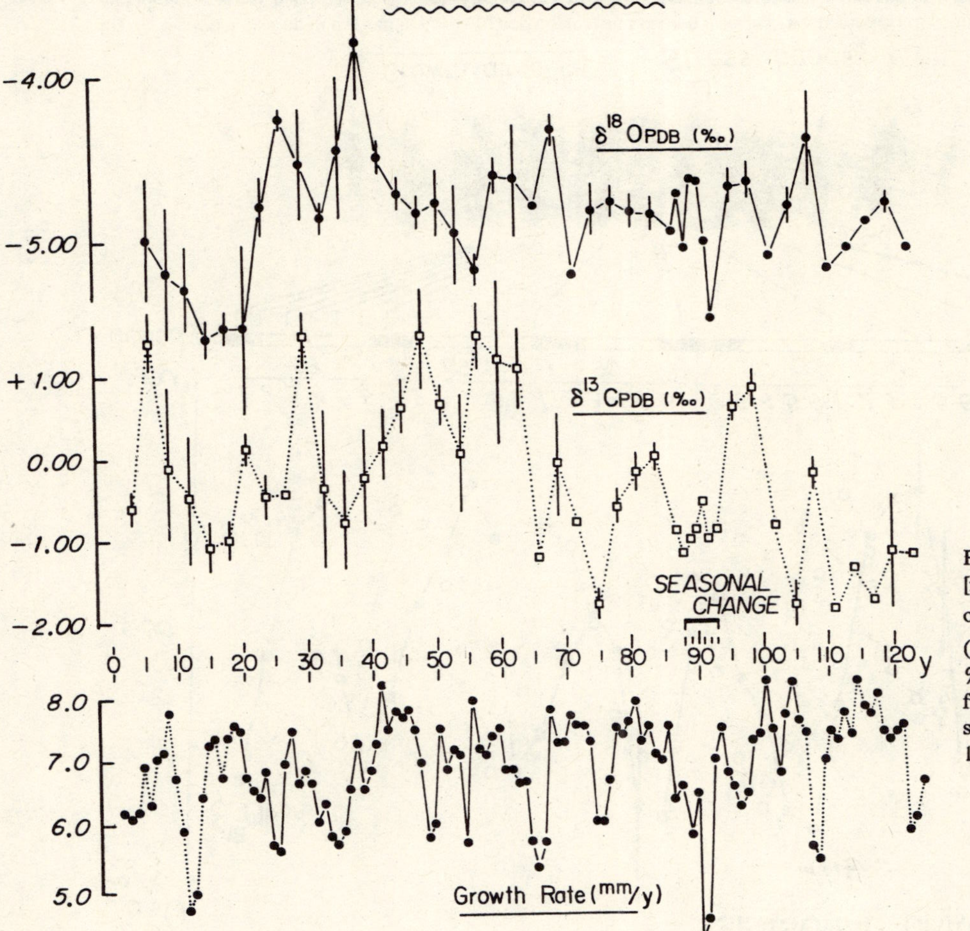


Fig. 6. Annual change in (1) rate of growth [mm/y (lower part)], (2) ratio of stable carbon isotopes [$\delta^{13}C_{PDB}$; % (middle part)] and (3) ratio of stable oxygen isotopes [$\delta^{18}O_{PDB}$; % (upper part)] over 120 years obtained from the analysis of core sample 601-1, from samples at 601 with a colony age of about 170 years.

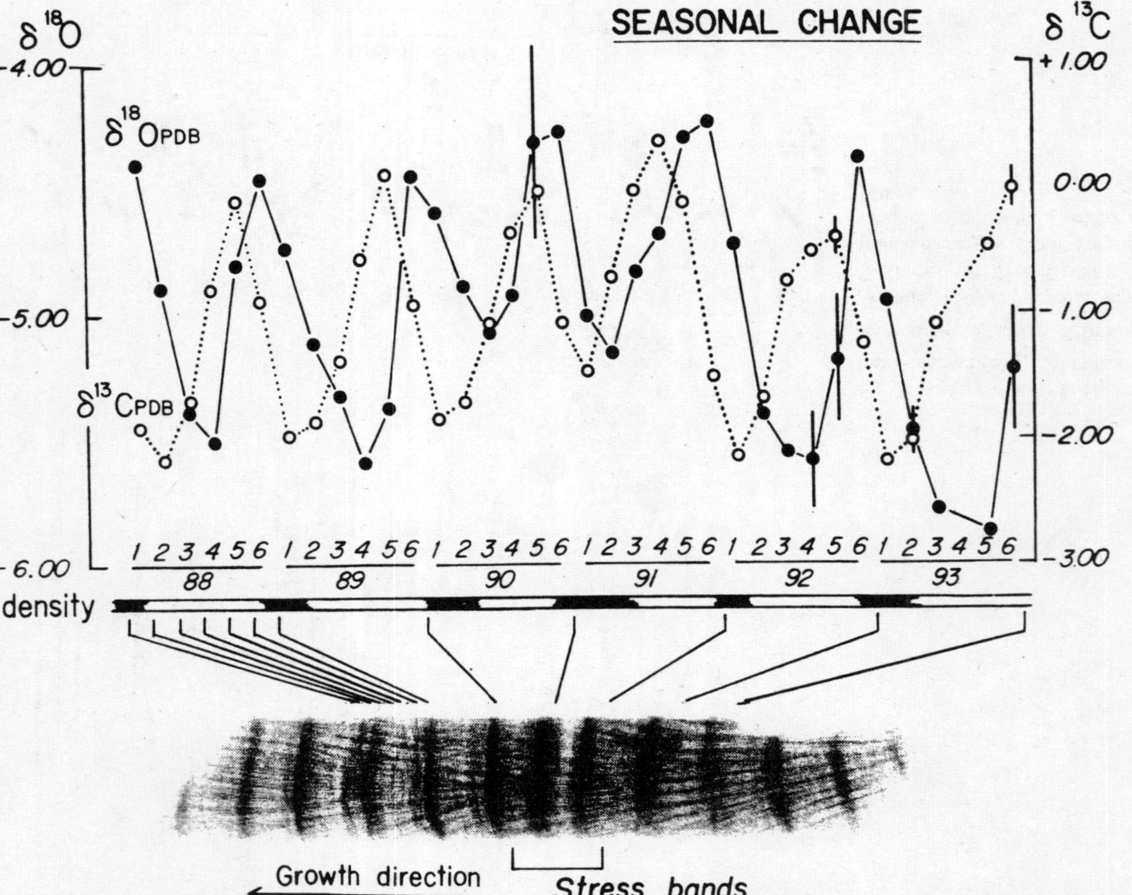


Fig. 7. Seasonal change in oxygen isotopes (black circles) and carbon isotopes (white circles) of sample 601 over a period of six years including the stress band (90-91st year). The lower part is an x-ray transparency photograph of a small sample wafer cut for isotope

*Stable isotopes of oxygen and carbon.* After a detailed examination of three cores, it was found that sample 601-1 is the most optimum sample since it satisfies the following three conditions, viz., 1) it has retained the longest growth history, 2) x-ray transparency photographs show clear density stripes and 3) fossils of scleractinia colonies do not conceal the continuous action of such factors as post emergence, leaching (eluviation), cementation, alteration, etc. Now, the 120th year part of the sample was selected.

As explained in Fig. 7, the pellet used for x-ray photographs was cut at right angles to the direction of growth to a width of about 1.3 cm and again x-rays were taken. These photographs were then transferred on tracing paper. The tracing paper was fixed on the sample and, based on that, the sample was drilled for every annual growth and then measurements were taken. However, during the six years between 88 and 93, the growth during each year was divided in six equal parts to study the seasonal changes. The powdered sample (about 30 mg) was first roasted in a helium atmosphere (heated for 30 minutes at 470°C) and then was allowed to react with 100% phosphoric acid in a bath maintained at 25°C. We added 10 ml (25°C : 1 atm) of $CO_2$ to this solution to measure the gaseous isotopes. For these measurements, we mainly used Hitachi's Nier-McKinney mass spectrometer model RMU-6 (Geology Dept., Faculty of Science, Kanazawa University); whereas sometimes we used a mass spectrometer for micro-measurements (Hot Spring Research Institute, Okayama University). Measured values of $\delta^{18}O$ and $\delta^{13}C$ are converted to PDB standard values and shown in Fig. 6 (upper and middle part of Fig. 6).

Let us first discuss seasonal change. A value of $\delta^{18}O$ shows considerable seasonal change over six years. Every year the minimum value appears around 3 or 4 (Fig. 7) which corresponds to a high water temperature period, viz., summer. Similarly the maximum value appears around 5-6 which is a low water temperature period, viz., winter. Looking at $\delta^{18}O$ during 90-91st year, during which a stress band was formed, we notice that compared to other years, the value during these two years is considerably large. Even the summer value is 0.5% higher than those during other years. It can therefore be concluded that years of stress band formation generally correspond to cool summers with a low water temperature. However, changes in $\delta^{18}O$ cannot be wholly attributed to water temperature variations, even when we compare the rate of growth and yearly experience (Fig. 6). It is thus necessary to take into account some other factors such as amount of precipitation, etc., in addition to water temperature, as factors controlling $\delta^{18}O$ value.

The $\delta^{13}C$ value, on the other hand, is known to vary with the photosynthesis of paragenetic algae [Zooxanthelles (Goreau, 1977)] and hence can be considered as a record of changes in the amount of sunlight those days.

Although a weak, but positive, correlation exists between the growth rate and $\delta^{13}C$ (Fig. 6). This should indicate that the growth rate corresponds, in addition to water temperature, to the amount of sunlight (cloudiness).

Seasonal change in $\delta^{13}C$, also clearly show an annual periodicity, same as $\delta^{18}O$, and shows a minimum value around one or two every year. This period can be considered winter or early spring from the values of $\delta^{18}O$ and hence can be considered a period with increased cloudiness (amount of precipitation).

*Concentration of radioactive carbon.* A tracing paper showing the x-ray measuring the growth rate was placed on the cross sectional surface of the core sample after dividing it into two equal parts. This is a guide for cutting more slices into samples for our analysis. In this analysis method, a minimum of 25 g of calcium carbonate in every sample is necessary for radioactive carbon analysis. As such the growth during one year is insufficient and we had to club together the growth over three years. Also to avoid mixing samples, a gap of one year was left (e.g., 4th year) between two samples by cutting that part and removing it from the sample. A fret saw was used for cutting the sample and cuts were taken along the lines shown by the tracing paper. An excess portion in each sample was removed with a file. In this way, we prepared 25 samples from the top of the scleractinia fossils. The separated samples were immersed in a solution containing sodium hypochlorate for 48 hours to remove any organic materials contained in the samples. It was then rinsed a few times with ultrasonic waves in distilled water and dried in a fur-

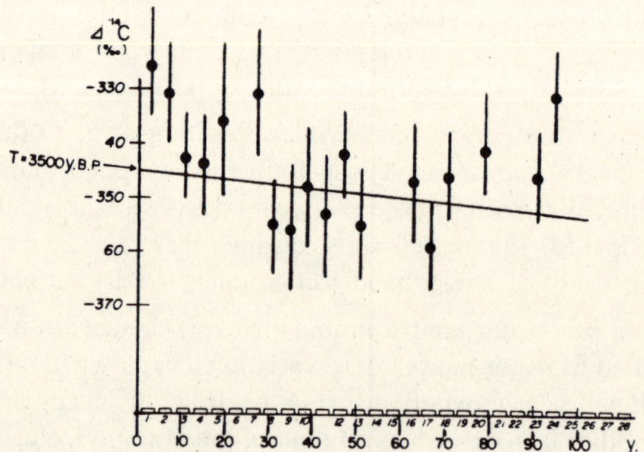


Fig. 8. Change in the concentration of radioactive carbon ($\Delta^{14}C$) in sample 601-1 over a period of 120 years. Measured value is for each three-year period. Vertical lines show the standard deviation. A line passing through the central part of this figure shows the expected age for each growth slice (of three years) calculated with the assumptions that the surface age of the colony is 3500 Y.B.P. and that there is no timewise change in the $\Delta^{14}C$ value of seawater.

nace maintained at 60°C. After this the sample was turned into a fine powder. Using this powered sample containing calcium carbonate, about 3 g of benzene was formed after the formation of carbide acetylene. It was then measured in a low background liquid scintillation counter (Aloca LSC-LB 1 type) placed in the low level radioactive experimental setup in the Faculty of Science of Kanazawa University.

Results of measurement are shown in Table 3. A graph showing the relationship between the value of $\Delta^{14}C$ and the age of the colony (actually representing the age for every three-year combination) is shown in Fig. 8. Of 25 samples collected, the measurement of 18 samples has been completed so far. The radioactive carbon age of sample 601-1 as calculated from the value of $\Delta^{14}C$ and happens to be 3,500 ± 80 Y.B.P. (half life period 5,568 years).

Now if we calculate, assuming that the age of the top of the coral colony is 3,500 Y.B.P. and that the $\Delta^{14}C$ value over the previous 100 years has decreased only due to the radioactive disintegration of radioactive carbon, then we get the line shown on the graph.

**Table 3. Measured values of carbon isotope in sample 601-1. The mean values for three-year growth are represented in each column**

$$\delta^{14}C(‰) = \left(\frac{As}{0.95\times Ao} - 1\right)\times 1000$$

$$\Delta^{14}C(‰) = \delta^{14}C(‰) - 2\left(\delta^{13}C(‰) + 25.0\right)\left(1 + \frac{\delta^{14}C(‰)}{1000}\right)$$

| Sample No. | $\delta^{14}C$ | $\delta^{13}C$ | $\Delta^{14}C$ |
|---|---|---|---|
| 1 | -294±9 | -0.63 | -328±9 |
| 2 | -296±9 | +1.40 | -333±9 |
| 3 | -309±7 | -0.57 | -343±9 |
| 4 | -311±9 | -1.18 | -344±9 |
| 5 | -303±13 | -1.04 | -336±13 |
| 7 | -296±11 | -0.48 | -331±11 |
| 8 | -319±8 | +1.46 | -355±8 |
| 9 | -323±11 | -0.85 | -356±11 |
| 10 | -314±11 | -0.28 | -348±11 |
| 11 | -319±11 | +0.14 | -353±11 |
| 12 | -305±8 | +1.44 | -342±8 |
| 13 | -322±10 | -0.59 | -355±10 |
| 16 | -311±11 | +1.00 | -347±11 |
| 17 | -327±8 | -1.33 | -359±8 |
| 18 | -310±8 | +0.90 | -346±8 |
| 20 | -307±8 | -0.73 | -341±8 |
| 23 | -325±8 | +0.50 | -346±8 |
| 24 | -297±8 | -0.96 | -331±8 |

It is quite clear from Fig. 8 that the measured values for sample 601-1 do not show a simple drop as seen in the line calculated with the assumption that the concentration of radioactive carbon in seawater is constant but there is a fluctuation of about 30%.

This kind of fluctuation occurs because the concentration of radioactive carbon in a liquid that is used for skeletal formation of scleractinia is not constant. Thus the above fluctuation can be considered to occur because the concentration of radioactive carbon in seawater at that time was not constant. Yearly changes in concentration of radioactive carbon in the atmosphere have been studied using the tree trunk (ring) based chronology up to the last 8,000 years [to 6,000 B.C. (Suess, 1978 and others)]. Similarly, using a 1,820-year old Yaku cedar *(Cryptomeria japonica)* on Yaku Island (30°40′N, 130°30′E) near the Kikai Islands, a record of fluctuations for the last 1,800 years has been reported (Kigoshi & Hasegawa, 1966). The factors responsible for the annual variation in the concentration of radioactive carbon in the atmosphere are known to be (1) variation in the earth's magnetic field which has a periodicity of as long as 12,000 years and an amplitude which is 8-10% higher than that of $\Delta^{14}C$ and (2) a variation which has a periodicity of 200 years (120-300 years) and which has a smaller amplitude (max. 20‰). This latter is supposed to have a relationship with the solar activity similar to blackspot activity with a shorter periodicity of about 11 years (Stuiver and Quay, 1980).

The analyzed data this time shows the variation in the concentration of radioactive carbon in the ocean over 120 years, about 1600 B.C. Now if we assume that the ocean variations also have a factor similar to atmospheric variations with a periodicity of 12,000 years, we can expect it to appear in the slope of rise on the right side of Fig. 8. However, actual calculations show that in just about 100 years, the inclination becomes almost 0 and can be ignored. On the other hand, the variation with a periodicity longer than 100 years, in Fig. 8 may indicate a possibility that these correspond to atmospheric variations with a periodicity of 200 years. Now, further detailed study shows a drop in the $\Delta^{14}C$ concentration, although within the error, in case of samples 9, 13, 17 and 23. When we look at the years of these drops, and compare them with the variation in growth rate, we notice that the points of drop in growth rate and points of drop in $\Delta^{14}C$ concentration match well. Also this drop shows a periodicity of 11 ± a few years.

There is no such short periodicity variation reported in the atmospheric concentration of $^{14}C$. However, a possibility has been indicated that the concentration of radioactive carbon may be related to solar black spot activity with a periodicity of 11 years (Stuiver and Quay).

Now, if the periodicity in radioactive carbon concentration within the scleractinia skeleton observed during this study is assumed to be due to variation in solar activity, we can also consider that the variation in growth rate is due to those factors which are directly affected by solar activity.

Studies using geochemical information on the ocean (such as metallic ion concentration or isotopes of oxygen and carbon, etc.) among the paleo-oceanographic analysis based on skeletal chronological studies of scleractinia were so far restricted to environmental sciences related to the social activities of human beings in the Quaternary-Holocene period and for the past few centuries. In Japan we also have examples of studies conducted to clarify the annual variation in carbon isotope composition in seawater during 1913–1979 A.D. using *Porites lutea* in the New Naha port of Okinawa and the Seitei Islands in that peninsula (Konishi, Tanaka and Sakanoue, 1981). When we compare the Suess effect as a result of an increase in fossil fuel consumption indicated by *Porites lutea* of Okinawa (when measured in terms of the drop in $\Delta^{14}C$ from 1900 to 1954 A.D., it is -13%) or the degree of pollution (when represented in terms of increase in $\Delta^{14}C$ concentration from the base line value of 1954–1956 A.D. to the peak value observed during 1972–1973 A.D. of + 217%) with the measurements in other oceans over the same period, we can illustrate the properties of ocean regions in which these scleractinia can grow, such as information on seawater cycles. The surface layer of seawater is actually formed by a mixure of two types coming from reservoirs, viz: (a) a surface water which is exposed to the atmosphere for a long period and with higher values of $\Delta^{14}C$ indicating a sufficient exchange and (b) the deep layer water which is away from the atmosphere for a long period with smaller values of $\Delta^{14}C$. Naturally the actual value of $\Delta^{14}C$ of the (combined) surface water will vary with the cycling of seawater. The composition of surface layer also varies according to its position and is recorded in terms of variation in the carbon isotopes in scleractinia skeleton. The surface water in Bermuda, which lies along the Mexican gulf stream (Nozaki, et al. 1978), Florida, and Belize (Druffel, 1980) or the surface water in Okinawa which lies along the Kuroshio all correspond to (a). On the other hand, the Galapagos Sea corresponds to that region where the deep water rises to the surface and has but a short exposure to the atmosphere (b). The Suess effect mentioned before is also -6%, about half compared with the other regions and thus strongly supports our above considerations (Druffel, 1981). In this region recently, the flow of El Ñino which is said to be a remote factor controlling abnormal weather conditions in Japan, has stopped but it is known to occur one or two times every ten years, causing abnormal weather conditions such as very high or very low surface layer temperatures. The growth rings formed in scleractinia in the year of El Ñino have a much higher concentration of radioactive carbon than the rings formed during normal weather years (Druffel, Ibid). If we can extend such findings to samples of long past, it could be very useful in the analysis of paleometeorology or paleooceanography.

Now, in order to further improve our understanding of the relationship between the concentration of radioactive carbon in the surface layer sea water (atmosphere) and solar activities in 1983, we have started collecting samples of *Porites lutea* colony fossils that grew when solar activity was weak—such as Maunder Minimum [from the middle of the 17th century to the early 18th century 1645-1715 A.D. (Eddy, 1976)] also known as the small glacier period or no black spot period. Although the same holocene *Porites lutea* of the Kikai Islands, they are formed much more recently than the samples excavated during 1982, which were formed around 1600 B.C. The problem water pump was changed to a new type (Yanmar dynamic sprayer cleaner for sea water Model GF) and the excavation went quite smoothly up to now. We have excavated 23 colonies from 27 holes with a total excavation depth of about 30 m. Pretty soon we should obtain results on the radioactive age determination of the excavated samples.

## Acknowledgment

Part of the measurements of oxygen and carbon stable isotopes was carried out with the help of a mass spectrometer to measure gaseous isotopes belonging to the Hot Spring Research Institute of Okayama University. It is, therefore, our duty to express our sincere thanks to Prof. Sakai who allowed the use of this equipment and to Prof. Kishima and other members of his staff who helped in numerous ways during the experiment. Prof. Takashima, Shimada, and Yamazaki of the Radiography Department of the Faculty of Medicine, Kanazawa University, helped us make x-ray transparency photographs. The actual excavation operation was carried out with the help of Mr. Niguchi (currently with Diamond Consultants Co. Ltd.), Mr. Ohara (2nd year graduate student) and Mr. Takahashi (4th year under-graduate student) of the Geology Department of Kanazawa University.

## References

Druffel, E.M. 1980. Radiocarbon in annual coral rings of Belize and Florida. *Radiocarbon, 22*(2), 363-371.

Druffel, E.M. 1980. MS Radiocarbon in annual coral rings of the Pacific and Atlantic Oceans. Ph.D. Dissert., U.C.S.D., 213 pp.

Druffel, E.M. 1981. Radiocarbon in annual coral rings from the eastern tropical Pacific Ocean. *Geophys. Res. Lett., 8,* 59-62.

Eddy, J.A. 1976. The Maunder Minimum. *Science, 192*(4245), 1189-1202.

Kigoshi, K. and H. Hasegawa. 1966. Secular variation of atmospheric radiocarbon concentration and its dependence on geomagnetism. *Jour. Geophys. Res., 71*(4), 1065-1071.

Konishi, K. 1983. Scleractinia skeletal chronology (sclerochronology) "Ocean Environments" (Ed. Hirano), 540-561.

Konishi and Niguchi. 1982. Analysis of paleooceanography during the Quaternary period using a coral reef boring. Part 1: Ocean Changes and Their Characteristics—*Newsletter* No. 7, 3-7.

Konishi, Tanaka and Sakanoue. 1981. Secular variation of radiocarbon concentration in sea water: sclerochronological approach. Proc. 4th International Coral Reef Symp. (Manila 1981) *1,* 181-185.

Konishi, Tsuji, Goto, Tanaka and Niguchi. 1983. Multiple shallow excavation of a Coral Reef—Example of Holocene system in the Kikai Islands. *Kaiyo Kagaku, 15*(3), 154-164.

Konishi, Ishihara, Iguchi and Sato. 1983. Course of carbon isotopes and seasonal changes in the sea during 1600 B.C. Outline of studies with a low-level, radioactive experimental setup in the Faculty of Science, Kanazawa University. Annual Report, 7, 12.

Nozaki, Rye, Turekian and Dodge. 1978. A 200-year record of Carbon-13 and Carbon-14 variations in Bermuda coral. *Geophys. Res. Lett. 5*(10), 825-828.

Suess, H.E. 1978 *in* Druffel, E.M. 1980 MS. Radiocarbon in annual coral rings of the Pacific Oceans. Ph.D. Dissertations, U.C.S.D. Fig. 2.

Stuiver, M. and P.D. Quay. 1980. Changes in atmospheric carbon-14 attributed to a variable sun. *Science, 207*(4426), 11-19.

# Report of the Second Meeting on the Future of Geosciences

The second "Meeting on Future of Geosciences" was conducted for two days on February 26 (Saturday) and 27 (Sunday) at the University Seminar House of Hachioji with the participation of 90 researchers. This time participants came from a wide region, from Kozen University in the north to Kyushu University in the south. Also all age groups and field interests were represented. The program was mainly divided in four sessions, viz., Geophysics, Paleontology, Geology, and Geochemistry. Discussions on the first day were about geophysics and paleontology, including two special lectures. Discussions on geophysics included such topics as the stress field of the subduction zone, earthquake, vertical movement, temperature structure of the lithosphere, etc. The special lecture was about the system to measure plate motion and a system to observe low layer currents. Discussions on paleontology considered such aspects as foraminifera, nanoplankton, algae, etc., and last, the problem of silica in sea water was discussed to establish a link with geochemistry. A social gathering was also arranged to establish the friendship between researchers of various fields and age groups. During the reception, young researchers who came a long distance explained their studies with the help of slides and it was followed by a lively discussion on the subject.

On the second day also a large number of people were present irrespective of the fact that there was a film show from 7:30 a.m. During the special lecture from 9:00 a.m. the top geoscientist of our time, Prof. Ueda, talked about the Present and Future of Geoscience. The lecture was extremely rich in contents for researchers engaged in this field as well as for probable researchers of future.

During the Geology session, the geology of the forefront of the island arc, the island arc itself, and the back arc basin were discussed from the point of view of marine geology as well as continental geology. Four comments in this regard were presented. During the session of geochemistry, discussions centered around strontium isotopes and helium isotopes along and across the island arc. Theories regarding the origin of three types of basaltic magma in the island arc were also presented.

In the brief discussion, representatives from each session briefly explained the proceedings of their group which was followed by free discussions. The response to a questionnaire indicated that a majority wanted this kind of meeting to be held again and hence it was decided to hold the next meeting mainly with young researchers from the Geological Survey Department in mind. The meeting ended with this resolution.

During this meeting, the contents were rich and although some restrictions had to be applied to the question hour and lunch break, the program generally went as per schedule. The discussions were serious and the meeting in general can be considered satisfactory. We would like to express our sincere thanks to the Administrative Department of the Special Research Project "Ocean Characteristics and Their Changes" for their assistance.

Reported by coordinator Kantaro Fujioka, Institute of Oceanography, Tokyo University.

# Report Regarding the Symposium on the Neotectonics of a Ridge Basin

This oceanographic symposium was held at the Institute of Oceanography of Tokyo University as per the program mentioned below. Large number of researchers, including those belonging to this Special Research Project, participated in the symposium with active presentations and opinion exchange. In addition to the presentation of some advanced ideas, the symposium also involved heated discussions about some provocative and unorthodox views presented—so much so that the scheduled time was not enough. Thus, for example, many speakers indicated subduction of the Sea of Japan while discussing the life cycle of a Back Arc Basin and, looking at this now, it is felt that it forecast the earthquake in the central region of the Sea of Japan (which actually took place) two months later. The discussions were thus completely orthodox.

*1983*

March 14 (Monday)

*Morning*

Welcome speech: Prof. Konishi (Faculty of Science, Kanazawa University)
Prof. Nasu (Institute of Oceanography, Tokyo University)

Chairman: Prof. Kimura (Faculty of Science, Ryukyu University)

Problems in the formation of a Back Arc Basin—Ueda (Seismological Research Institute, Tokyo University)

Crust tectonics and back arc openings—Yamaoka, et al. (Faculty of Science, Nagoya University).

*Afternoon*

Chairman: Prof. Isezaki

Mode of Back Arc expansion activity in the Izu Ogasawara Marianas Arc—Tamagi and Miyazaki (Geological Survey, Marine Survey)

Development of Senshima Arc—Kimura (Faculty of Liberal Arts, Kayawa University)

Formation of Okinawa Trough—Kimura

Seismotectonics of Back Arc Basin in the western Pacific Ocean—Eguchi (Disaster Relief Center).

March 15 (Tuesday)

*Morning*

Chairman: Kajihara (Department of Geoscience, Namba University)

Expansion of Back Arc Basin and formation of a hydrothermal sedimentary floor—Horikoshi (Faculty of Science, Toyama University)

The subduction cycle and the formation period of a Back Arc Basin—Kobayashi (Institute of Oceanography, Tokyo University)

*Afternoon*

Chairman: Prof. Konishi

Survey of heat flow in the Marianas trough (Mound area)—Fujii (Faculty of Science, Kobe University)

Chairman: Prof. Fujii

Back Arc Basin as seen from volcanic rock—Sato (Faculty of Science, Kanazawa University)

Volume of Sea of Japan—Wakida (Faculty of Science, Tokyo University)

Oil production in Back Arc Basin—in the seas surrounding Japan—Ishiwata (Hokyoku Oil Co.)

Has the Sea of Japan begun to close?—Nakamura (Seismological Research Institute, Tokyo University)

Subduction of the Sea of Japan—Kobayashi (Department of Natural Science, Namba University)

Discussions.

Reported by: Prof. Kenji Konishi.

# Report Regarding the Symposium on the Present Condition and Future Prospects of Chemical Oceanography

During the two days of May 20-21, 1983, the following symposium was held at the Institute of Oceanography of Tokyo University. Many researchers involved in Special Research Project "The Ocean Characteristics and Their Changes" took part in the symposium and, as a result, many papers were presented. The program of symposium was as follows:

May 20 (Friday)

Greetings and welcome: Nozaki Yoshiyuki (Institute of Oceanography)
Chairman: Okabe Shiro (Department of Oceanography, Tokai University)

1. Exchange of $CO_2$ between atmosphere and ocean—Kanamori and Ikegami (Marine Science Laboratory, Nagoya University)
2. Chemical properties of low layer waters in the western part of the north Pacific and their circulation—Gamo (Institute of Oceanography, Tokyo University)
3. Electrolytic mass spectrometric analysis of an isotope diluted surface and its application to Pacific Ocean waters—Murozumi, Nakamura, Yohoyama and Chagi (Muroran Institute of Technology)

Chairman: Kuwamoto Toru (Faculty of Science, Kyoto University)

4. Heavy metals in the ocean—Tsubota (Department of Science, Hiroshima University)
5. Behavior of heavy metals in estuaries and along the beaches—Maeda (Tokyo Fisheries University)
6. Development of an automatic monitoring system for trace elements in sea water using an anodic flocculometry—Okazaki (Faculty of Science, Kyoto University)

Chairman: Prof. Nagaya (Radiological Medical Research Institute)

7. $^{228}$Ra and $^{228}$Th at the seashore—Okubo (Kobe Mercantile Navy University)

8. Behavior of U/Th chains in the ocean—Nozaki (Institute of Oceanography, Tokyo University)

9. Submarine particles and chemical oceanography—Tsunogai (Department of Fisheries, Hokkaido University)

May 21 (Saturday)

10. Mass balance along the seashore—Matsumoto (Geological Survey Department)

11. Continuity of sedimentation as seen from intermediate waters—Kato (Department of Oceanography, Tokai University)

Chairman: Prof. Harada (Faculty of Science, Yamagata University)

12. Transport mechanism between deep-layer sediments—Osumi (Faculty of Science, Tokyo Institute of Technology)

13. Variation in the paleoenvironments of the Sea of Japan as seen from stable isotopes—Oba (Faculty of Liberal Arts, Kanazawa University)

Chairman: Ogura (Tokyo Agricultural University)

14. Transport of continental organic materials to a sea through atmosphere—Ota (Ocean Science Laboratory, Nagoya University)

15. Trends in the study of elemental stable isotopes found in the sea and the recent topics around $N_2O$—Wada and Yoshida (Mitsubishi Biological Research Laboratory).

General discussions

Chairman: Horibe (Tokyo University)

Debators: Kitano (Ocean Science Laboratory, Nagoya University), Yamamoto (Kyoto University of Education).

There were 62 participants on the first day and 53 on the second.

On the first day, discussions were about methods to measure carbonates, heavy metals and radioactive materials as the chemical tracers in sea and the actual results. First M/s Kanamori and Ikegami proposed a method to calculate the partial pressure of carbon dioxide when the sea water is in direct contact with the atmosphere, using the measured values of pH, alkalinity, nutrient salts, etc. This method is based on the chemical mass ratio according to Redfield-Ketchum-Richard's plankton composition and obtains the difference between the variation in total carbon due to the disintegration of organic materials and the variation due to the calcium carbonate solution with a different degree of alkalinity. As a result it was found that the partial pressure of carbon dioxide in sea water now at the surface is around 300-350 ppm, a value quite close to that of the atmosphere. Whereas the value at a 1,000 m or more in the Pacific was about 250 ppm. This may be considered to correspond to an increase in the $CO_2$ concentration in the atmosphere as a result of the use of chemical fuels, etc., and can therefore be assumed to be one mechanism by which atmospheric $CO_2$ is transferred to the ocean.

Mr. Gamo discussed the mixing and dispersion of low layer waters using the distribution of $^{222}$Rn. Thus, since $^{222}$Rn, with a half life period of 3.8 days is being dispersed from the sedimentary material of the ocean floor, the low layer water up to few hundred meters from the ocean floor contains more $^{222}$Rn than newly formed $^{226}$Ra. The distribution of this excess $^{222}$Rn generally reduces logarithmically with distance from the ocean floor. However, in some cases, the reduction may not be continuous. In this case, the rate of the reduction of excess $^{222}$Rn near the seabed is less and the rate of mixing is very high. Also such discontinuity in $^{222}$Rn matches in many cases with the temperature discontinuity observed during CTD measurements. It was made quite clear that $^{222}$Rn acts as an important tool in the study of mixing or dispersion near the seabed or for the study of boundary layers of the seabed.

These two presentations were followed by four papers by M/s Murozami, Tsubota, Maeda, and Okazaki regarding heavy metals. First, Mr. Murozami discussed in detail the actual case of micro-mass spectrum analysis using an isotope dilution method. Utmost care was taken to prevent contamination, e.g., using an extremely clean laboratory and purifying test samples, etc. With this, it became possible to measure the heavy metal contents in sea water correctly. It was also pointed out as a merit of isotope dilution analysis method that it is possible to measure the concentration correctly even though it is very low. Tsubota compared the results of measurements using the atomic absorption method with those obtained using the isotope dilution method for several heavy elements and ascertained that they both match. The results of the analysis of various heavy metals including Mn and Fe were introduced and the speaker pointed out its resemblance with the distribution of nutrient salts as well as discussed its meaning from an oceanographic perspective. It was impressed that this study of heavy metals in sea water should be taken to the top level at least as far as the collection and analysis of a sample is concerned. Considerable progress is expected in future.

Next, Maeda discussed the behavior of heavy metals in the estuary and along the seashore, particularly the importance of Fe elimination by condensation. Similarly, Okada introduced the state of the development of an automatic monitoring system for microheavy metals from the point of view of analytical chemistry.

Okubo and Nozaki discussed respectively the distribution of radioactive nuclei and their behavior along the seashore and in deep seas. Okubo made clear the nonequilibrium distribution of $^{228}$Ra and $^{228}$Th mainly in the Seito inland sea. $^{228}$Ra is supplied by the sediments at the sea floor and its concentration reduces due to radioactive disintegration as well as due to mixing with waters from the outer seas that have a low $^{228}$Ra concentration. In the central region of Seito backwaters, an extremely high concentration as much as 800 dpm/$10^3$ l was noticed. From the balance of this $^{228}$Ra, it is possible to estimate its mean residence time corresponding to the transfer between sea water from the Seito backwaters and the outer sea but then value of $^{228}$Ra flux from the sediments is quite important. Currently, the transfer time for sea waters is considered to be less than ten years. Now the concentration of $^{228}$Th, a kind of daughter nucleii, is much lower than that of $^{228}$Ra indicating that it is eliminated as a result of particle activity or seabed surface activation. The elements with good reactivity, such as $^{228}$Th, enter the sediments within a few months and most of these contaminants are thus accumulated in the sediments.

Mr. Nozaki discussed the difference in behavior of $^{230}$Th and $^{231}$Pa in the outer seas. The $^{230}$Th concentration increases almost linearly from the surface to the seabed, whereas $^{231}$Pa indicated a decrease beyond 3,000 m. This clearly indicates a vertical removal by the precipitation of particles and also the presence of other elimination operations. Thus in the outer seas, where the vertical particle flux is small, $^{231}$Pa is eliminated with more difficulty when compared to $^{230}$Th and is transported horizontally to regions of higher particle flux where they are eliminated. Now such a kind of horizontal removal should occur within a time span of a few tens of years. It is therefore necessary to study the operations that take place at ridge areas or continental shelf slopes and clarify the importance of horizontal transport.

Last, Tsunogai outlined the state of studies on granular materials using sediment trap experiments.

During the pre-noon session on the second day, there were four presentations regarding chemical studies of sediments. First, Matsumoto explained that measurements of heavy metal distributions in sediments in Tokyo Bay as well as the measurements of the sedimentation rate using the $^{210}$Pb method revealed that the extent of heavy metal contamination varies with time. He also calculated the balance of various elements in Tokyo Bay. Most of the heavy metals settle in sediments before they are taken outside as a result of water exchange.

Kato mainly discussed the initial continuity operation with reference to measurement on the distribution of various elements in intermediate water. He also explained the vertical distribution of Mn, $PO_4$, $IO_3$, etc., using the equation of diffusion, including chemical reaction. Osumi determined the coefficient of diffusion using the laboratory experiments on sediments. He also discussed the importance of transport processes due to rising transfer currents through intermediate water as explained by Sayles and Jenkins.

Oba explained the changes in the paleoenvironments of the Sea of Japan based on the changes in foraminifera in the Sea of Japan core and the changes in its oxygen isotopes. Accordingly, it was revealed that there were periods when fresh water from Yokogawa or Oya lakes entered the Sea of Japan and the ocean structure varied in many ways. The present structure of the Sea of Japan is considered to have formed about 8,000 years ago. This lecture was extremely interesting, particularly to persons specializing in different fields of chemistry as the speaker discussed ways of reading the past from the cylindrical sediments.

In the afternoon there were two lectures about organic materials. Ota discussed the terrigenous organic materials that are transported through the air using results obtained during the cruise of the *Hakucho-maru* and observations on islands. Studies in this field are very few but some development is expected in future. This was followed by a presentation of Mr. Wada who indicated that the concentration of $^{15}N$ is extremely high as a result of the food chain operation. Subsequently, Yoshida discussed the trends in studies on $N_2O$.

The symposium concluded with a general discussion under the chairmanship of Prof. Horibe during which the participants expressed their opinions on the future of oceanographic studies. This discussion went on for 30 minutes beyond the scheduled time and the symposium in general was quite meaningful.

(Reported by Yoshiyuki Nozaki).

# ADMINISTRATIVE REPORT

## Minutes of the 18th Meeting of Executive Committee

June 14, 1983 (Tuesday)
Institute of Oceanography, Tokyo University.

**Items reported**

The application for the research grant for fiscal 1983 was submitted to Mombusho (Ministry of Education).

Posters and programs for the General Symposium were mailed to concerned organizations and persons.

**Items discussed**

It was decided to hold the 7th Meeting of the Managing Committee at the end of the first day of the General Symposium (July 11) to discuss:

1) plans for publishing the final research report

2) propositions for future plans as decided by a committee for the Future Course of Studies, etc.

## Guidelines for Authors

This Special Research Project has now entered its last year. Since this may be the time to summarize the research conducted, Newsletter welcomes enterprising articles from all regarding the present status of studies and results obtained so far, etc., which will go a long way in finalizing this Special Research Project in general.

The contents are generally classified as:

| | |
|---|---|
| Highlights: | Quick reports of the results of this Special Research Project |
| Bulletin: | Domestic or international news reports closely related to this Special Research Project |
| Reports: | Reports on the activities of each section or working group |
| Announcements: | Announcements regarding the proposed meetings or conferences of each group |
| Bibliography: | Listing of papers published by the members on this Special Research Project |

Manuscripts should be written on standard 400 character size paper (handwritten material also accepted). The figures should be drawn in such a way that they can be reproduced as they are.

It is proposed to publish this Newsletter every two months.

**Note:**

When reproducing or citing the material published in this Newsletter, due acknowledgment should be given. Authors' names, wherever mentioned, should also be quoted. A copy of the publication containing such a reproduction may be sent to the Editorial Section of the Administrative Office.


| | |
|---|---|
| Published by: | Administrative Office, Special Research Project<br>"The Ocean Characteristics and Their Changes" |
| Published at: | Administrative Office,<br>Physical Oceanography Department,<br>Institute of Oceanography,<br>Tokyo University,<br>1-15-1 Minami-dai, Nakano-ku,<br>Tokyo 164<br>Tel: 03-376-1251, Ext. 251. |
| Editorial staff: | Taeko Kitano and Noriyo Kimura. |

Special Research Project

# The Ocean Characteristics and Their Changes

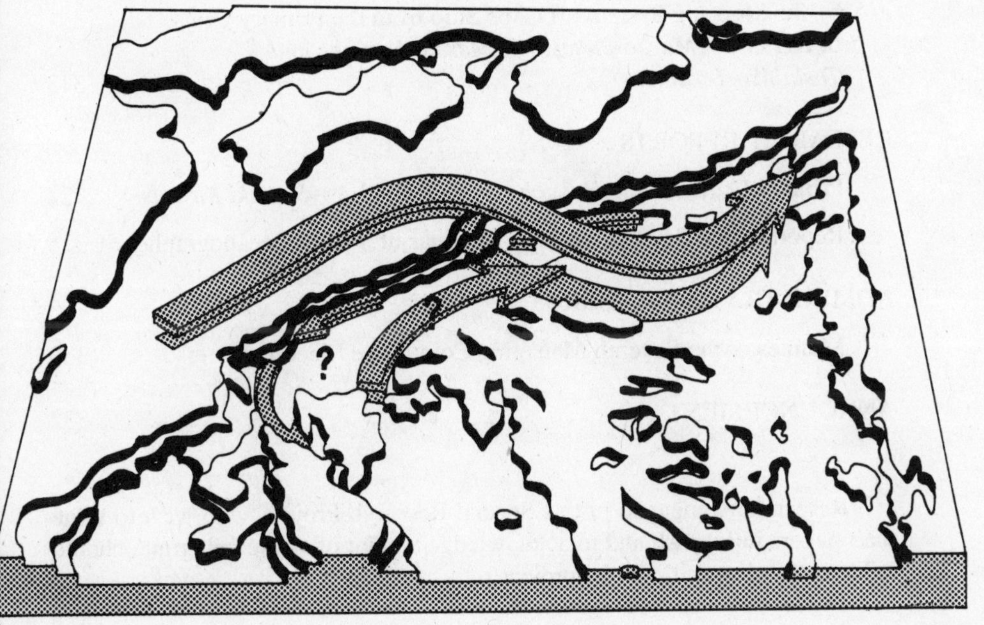


NEWSLETTER NO. 14, SEPTEMBER, 1983

# Contents



Researchers engaged in this Special Research Project, who wish to write their papers in English and to acknowledge receipt of a research grant, should indicate the above title of the project.

## Cover Page

We still cannot answer exactly how the deep layer currents in Shikoku Basin should be. However, we are nearing that answer with the help of data obtained by flow meters, CTD measurements, etc. The figure here represents an image before arriving at the exact picture and is a result of modeling the currents in the northern part of Shikoku Basin (Ref. Highlights).

The model preparation and this figure are provided by Kitagawa of the Institute of Oceanography, Tokyo University.

HIGHLIGHTS

# Deep-layer Circulation in Shikoku Basin: Flow Measurements by the Mooring System off Cape Shio from the Primary Cruise of KH83-1

*Masao Fukasawa, Masaki Kawabe and Toshihiko Teramoto*

(Institute of Oceanography, Tokyo University)

(Received: August 15, 1983)

## 1. Introduction

By deep-layer circulation we generally mean the entire geospheric circulation wherein sea water sinking occurs in the northern Atlantic region and its advances southward along the American continent. The water is supplied from both Polar currents and then flows to the Indian and Pacific oceans. Now the Philippine Sea, which lies in the western Pacific region, is separated from the Pacific in deep layers as a result of the Izu-Marianas ridge and was therefore excluded from conventional studies of deep-layer circulations. (This has been further explained in detail by Warren and Wunsch in their book *Evolution of Physical Oceanography.*) Certainly, the Philippine Sea is much smaller than the Pacific Ocean and it is difficult to assume that it contains a source of deep-sea water in itself. Therefore, naturally it was not considered an important region when studying the image of the deep current circulations on a geospheric scale. However, although smaller compared to the Pacific Ocean, it spans over 2,000 km in the north-south direction and 1,000 km east-west. So it is also difficult to assume that no deep-layer currents exist here. Actually the amount of dissolved oxygen at 3,000 m is much higher in the Philippine Sea than in the western Pacific. Thus, it may be possible that the deep-layer water of the Pacific flows from the marine canyons of the northern Caroline Islands into the Philippine Sea. The recent studies by Holland and Rhines (1980) have also proposed the concept of eddy-driven flow, indicating that a deep layer circulation exists even in isolated seas which do not have a deep water source of their own. This aspect is not much concerned with the circulation on a geospheric scale but establishes a link between the Philippine Sea, particularly its southern region, and the Pacific Ocean. It is quite interesting to study whether a deep-layer circulation exists in the Shikoku Basin, in

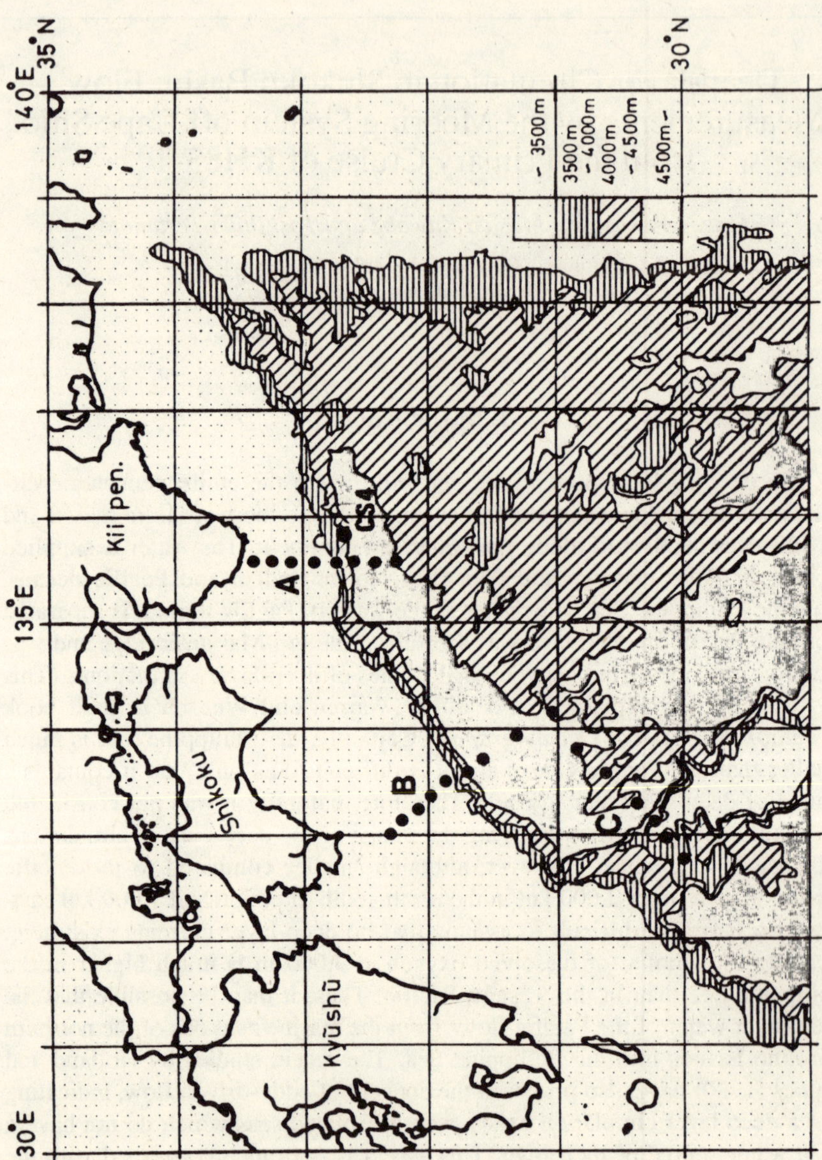


Fig. 1. Water depth distribution in the northern part of Shikoku Basin and the measuring points.

the northern part of the Philippine Sea, whether it contains a large energy source such as the Kuroshio and, if it exists, what could be its nature, etc.

The Teramoto Committee now carrying out studies under this special project is measuring the deep-layer currents off Cape Shio and since the results obtained, along with the results of CTD measurements by the KH83-1 cruise, could lead us to formulate an image regarding deep-layer circulation in the Shikoku Basin, we take this opportunity to present our observations here.

## 2. Data

Figure 1 shows the sea water depth distribution in the northern part of Shikoku Basin, locations of current measuring points, and locations of CTD measuring points. As far as current and CTD measuring points are concerned only those discussed here are shown in the figure. The current measuring point CS4 lies on the south slope of Cape Shio and gives data between April 29, 1981 and January 26, 1982. Two layers were measured, viz., 400 m and 2,200 m above the seabed. A moving average of flow velocity was obtained over 25 hours and used as the data base. The total record is 270 days long. The dotted lines A, B, C in this figure show CTD measuring points. At each point CTD-DO measurements were obtained to 15-16 m above the seabed. The period for this measurement is between April 21 and April 30, 1983.

## 3. Results of Flow Measurement

The flow velocity obtained at CS4 was subsampled every midnight and is shown in Fig. 2 as a stick diagram. The mean values over the entire period were for the upper layer, i.e., CS41: $\bar{u}$ (east-west component. Toward east is +ve) = -0.8 cm/sec and $\bar{v}$ (north-south component. Toward north is +ve) = 0.0 cm/sec. For the lower layer, i.e., CS42: $\bar{u}$ = -7.2 cm/sec and $\bar{v}$ = 0.8 cm/sec. This indicates a higher westward velocity at CS42. Also the variance of $\bar{u}$ in the case of CS42 shows a higher value of 124 $m^2/sec^2$ which means the standard deviation of $\bar{u}$, i.e., $U_{std}$ is 11 cm/sec. Thus statistically speaking it raises some doubts about the meaning of $\bar{u}$. However the ratio of $\bar{u}/U_{std}$ is 0.7 which is much larger than the $\bar{u}/U_{std}$ value obtained in regions where the existence of deep-layer circulations has been confirmed, such as the deep-layer currents in the gulf region. Now if we find the cross correlation between the fluctuations at CS41 and CS42, a higher correlation exists at lag 0 and the coherence of fluctuations of both measuring points exceeds 0.8 over the entire period. The phase (of such fluctuations) is also stable at 0°. All these indicate that fluctuations of CS41 and CS42 are of the same nature or at least the fluctuations having the same time scale as our observation period do not contribute to the difference in $\bar{u}$ of CS41 and CS42, i.e., to the large value of $\bar{u}$ at CS42.

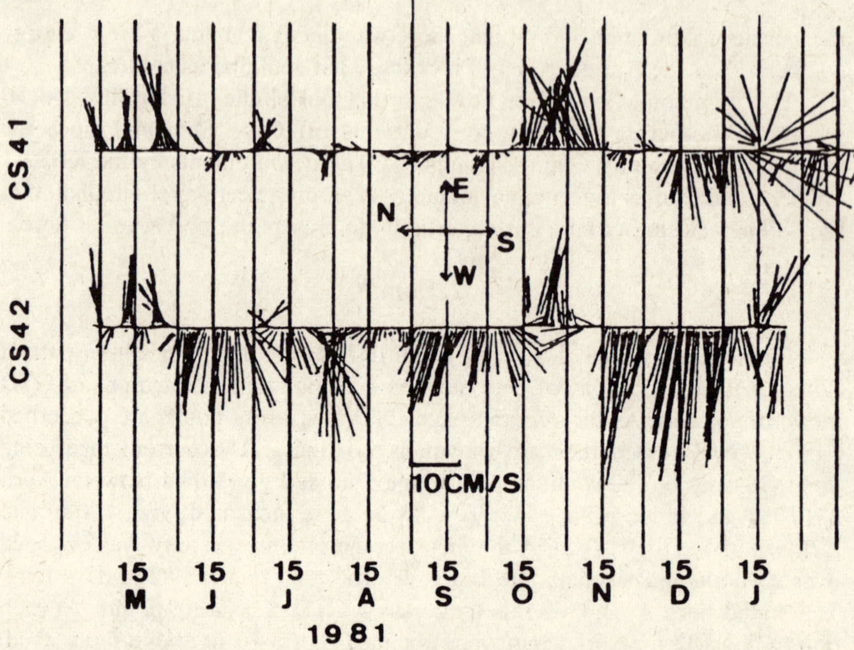


Fig. 2. Stick diagram of flow velocity at CS4. Upper part is CS41. Lower part CS42.

From this point of view, the value of $\bar{u}$ at CS42, i.e., -7.2 cm/sec, indicates that currents with a larger time scale than our observation period exist around CS42. Also comparing $\bar{u}$ for CS41 and CS42 we notice that the above currents are stronger in and around the lower layer.

## 4. Results of CTD Measurements

As shown in Fig. 1, CTD measurements were made all across the slopes off Cape Shio, off Cape Ashizuri and in the northern Kyushu-Palau ridge. The temperature profile at each measuring line is shown in Fig. 3(a)–(c).

Figure 3(a) shows the cross section off Cape Shio. The depth around 32°30'N is a Nankai trough. Attention is particularly invited to the isothermal lines below 1.4°. Thus at depths beyond 2,800 m, near the upper shelf toe, lying north of the South Sea trough, the iso-temperature lines are moving southward. On the other hand, still further to the Nankai trough, the isothermal lines turn northward. It should also be noted that at depths between 2,000 m and 2,500 m, either in the shelf region or off the Nankai trough, we do not see any particular direction for the iso-temperature line. Thus the inclination of the isothermal lines below 1.4° mentioned above may not be the result of the compensation of fast eastward currents (Kuroshio current) existing in the

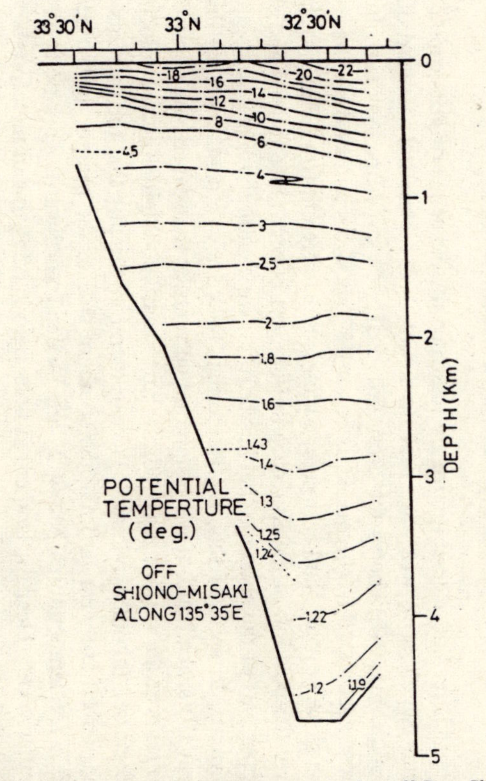


Fig. 3(a). Potential temperature cross section off Cape Shio.

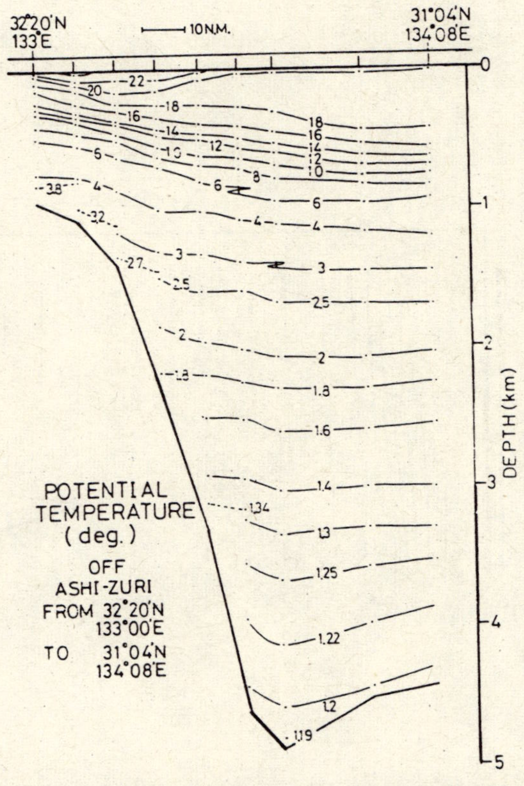


Fig. 3(b). Potential temperature cross section off Cape Ashizuri.

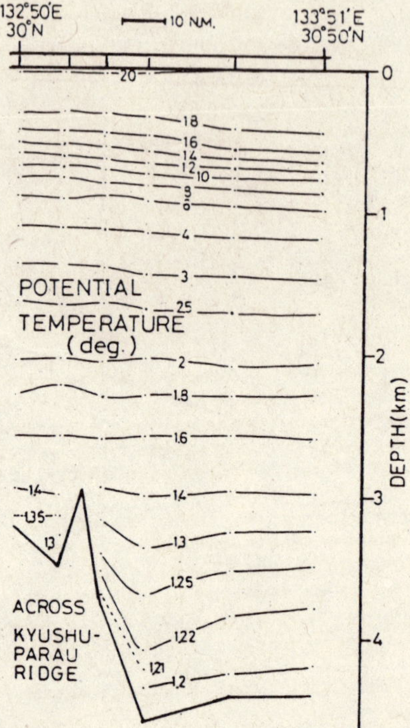


Fig. 3(c). Potential temperature cross section in the northern Kyushu-Palau ridge.

upper layer but instead it strongly indicates the existence of geostrophic currents which move westward near the lower shelf region and eastward off the trough. Fig. 3(b) shows the cross sectional view off Cape Ashizuri. In this cross section, all isothermal lines below 18°, show a southward inclination on the slope beyond Nankai trough, which is thus different from Fig. 3(a). However the position of the Kuroshio in Fig. 3(b) is very close to land and at the upper part of the shelf toe that lies beyond, we cannot think of strong surface currents. In spite of this, the isothermal lines show a southward inclination near the seabed and, if we assume no currents at the seabed, the geostrophic currents at the surface may indicate an excessively high eastward velocity of 90 cm/sec (about two knots). We can therefore assume that the currents near the seabed are westward. On the southern side of the trough, the isothermal lines for temperatures below 2° are inclined northward. This trend is particularly very clear for isothermal lines below 1.25°. It can therefore be concluded that an eastward geostrophic current exists at the lower layers of Fig. 3(b). Fig. 3(c) shows the cross section of the Kyushu-Palau ridge. This cross section differs from the earlier two in that it does not consider the Kuroshio.

In this cross section, the isothermal lines below 1.4° are all inclined away from the slope (eastward); but farther east, the same (the isothermal lines below 1.4°) lines change their direction to west. This condition of isothermal lines indicates that a southward (southeast) geostrophic current exists near the slope's lower layers, whereas a northward current (northwest) exists farther off.

So far we have discussed three cross sections and in all these cases, at a depth beyond 3,000 m (or below 1.4°), a narrow westward or southward current exists on the shelf toe or on the ridge slope, whereas a wider current in opposite direction is noticed on the farther side. As shown in Fig. 1, all these cross sections are obtained at the same depth (a continuous slope) and it may be safe to assume that the westbound current indicated on the upper shelf toe of the cross section off Cape Shio and the cross section off Cape Ashizuri is connected to the southbound current observed along the cross sectional slope of the Kyushu-Palau ridge. Similarly, the eastbound currents noticed off the Nankai trough side of the cross sections at Cape Shio or Cape Ashizuri can also safely be assumed to be a continuation of the northbound current observed off the trough side at the Kyushu-Palau ridge. The calculation of the velocities of geostrophic currents, assuming no existence of current at the 2,000 m layer, reveals a value of 5-10 cm/sec near the shelf toe or near the farther side of the ridge slope which changes to 2-5 cm/sec as we go further away. Although these values of current velocity are small, the calculation of water mass transport to a level of 2,000 m above the shelf toe, according to Fig. 3(a), indicates a value of 3 Sv (between two points on shelf toe), i.e., about 1/10 of the mass transport in the Kuroshio.

## 5. Discussion

In section 3, direct measurements of current showed a westward current whereas in section 4 we obtained a density structure (geostrophic current structure) which is not in contradiction with the results of section 3. However the difference in the period of current and the CTD measurements is a major problem when correlating these two results as the next step of our studies. Accordingly, at this point we cannot conclude that current measurements and CTD measurements point to the same phenomena. However, the westbound current with a velocity of 7 cm/sec obtained by direct current measurement can be assumed to have a sufficiently long time-scale. Also the measuring points are on the shelf toe itself and on the same locations where a westbound geostrophic current was indicated in Fig. 3(a). Therefore, it may be safe to conclude that the results of direct current measurement and those of CTD off the Kuroshio indicate the same phenomena. Also, as mentioned in section 4, since the density structure indicating deep-layer currents in each cross section are mutually connected, a narrow deep-layer current can be considered to exist at least in the northern part of the Shikoku Basin, beyond a depth of 3,000 m which has a northward component on the shelf toe north of the Nankai trough and a southward component flowing along the eastern slope of the Kyushu-Palau ridge. Beyond this region a deep-layer current flows in opposite directions. Also as shown in Fig. 1, if we assume that in deep layers, the Shikoku Basin is isolated from other seas, then these currents can be considered to form a circulation system within the Shikoku Basin. Now the temperature profile, as obtained in Fig. 3(a)-(c) cannot be viewed clearly from the density field measurement carried out by Taft. This may look like a contradiction to our conclusion but, as mentioned in Fig. 3(a)-(c), at lower layers, the logic applied in section 4 will be meaningless unless the measurements are carried out with a steady resolution of above 1/100°. When we consider the fact that Taft's observations were hydrographic-cast, it may not be difficult to discuss the density structure (temperature profile) of deep-layers using Taft's results.

## 6. Conclusion

From the results of current and CTD measurements, we carried an image that an anticlockwise circulation system exists in the deep-layers of the northern part of the Shikoku Basin. This circulation system can be assumed to consist of a narrow current on the shelf toe of the northern part of the Shikoku Basin (10-20 N.M.) and a wider current beyond. It is now necessary to carry out CTD and current measurements, particularly in the eastern and southern parts of the Shikoku Basin to confirm or modify this image.

## 7. Acknowledgment

During the moored current measurements and CTD measurements, we received great help from Capt. Henmi and other crew members of the *Keiten-maru,* a research vessel of the Department of Fisheries, Kagoshima University, Prof. Takahashi, Ichikawa, Maeda, Sakurai, and Sanjo of the same University and Capt. Tadema and other crew members of the *Hakuhomasu-maru,* the research vessel of Tokyo University as well as Prof. Taiva and Prof. Kitagawa of the Institute of Oceanography, Tokyo University. We would like to express our gratitude to all of them. Thanks are also due to Mr. T. Kanego of the Institute of Oceanography, Tokyo University who calibrated the CTD apparatus.

## References

Holland, W.R. and P.B. Rhines. 1980. An example of eddy-induced ocean circulation. *J.P.O.,* 10, 1010–1031.

Taft, B.A. 1978. Structure of the Kuroshio south of Japan. *J.M.R.,* 36, 77–117.

Warren, B.A. 1981. Deep circulation of the World Ocean. In Evolution of Physical Oceanography, B.A. Warren and C. Wunsch, Eds. The M.I.T. Press, 6–40.

# Primary Cruise of the *Hakucho-maru* KH-83-1

*Masaki Kawabe*

(Institute of Oceanography, Tokyo University)

This cruise of the *Hakucho-maru* was conducted for 29 days between April 21 and May 19, 1983, under the leadership of Prof. Teramoto of the Institute of Oceanography, Tokyo University. The cruise was divided into leg 1 and leg 2. In the first leg, we made CTD·XBT measurements, recovered and installed current meters, etc., along the south coast of Japan between Kii Peninsula and Cape Toi for 14 days until we arrived at Yokohama port on May 4. In the second part, similar operations were carried out at point B for 12 days from May 8. The CTD apparatus used this time has an oxygen sensor attached to it and, at almost all points of CTD measurements, we collected water samples to calibrate oxygen and salinity. In general there were no major troubles with equipment. The CTD, recovery, and installation of current meters was smooth and all operations were carried out just as planned. Particularly during leg 1, the weather was so nice that we could increase the number of CTD measuring points. During leg 2, however, the weather was rough for many days and we had to cancel two points for CTD measurements. Those were not, of course, the important measuring points and the objects of this time can be said to be achieved satisfactorily.

Figure 1 shows the locations of measuring points in leg 1. Under this Special Research Project, flow measurements are being conducted between the south of Cape Shio (Shiono-misaki) and southeast of Cape Toi (Toi-misaki) from April, 1981. Whereas flow measurements in the intermediate Shikoku Basin have been conducted since August, 1982. The object of leg 1 was to recover and install flow meters off Shikoku Basin (SB points) and off Cape Toi (CT points) as well as observe the flow cross section and the surrounding density structures very closely. CTD measuring points were selected so that they cross the Kuroshio three times and form a cross shape along the warm current vortex off Shikoku. The measuring range was approximately ten miles and at each point the data was collected to the ocean floor. In order to discuss the structure of the Kuroshio and compare the flow data with the density structure, an accurate observation of the density field to a considerable depth is essential. The operations were divided into CTD group, mooring

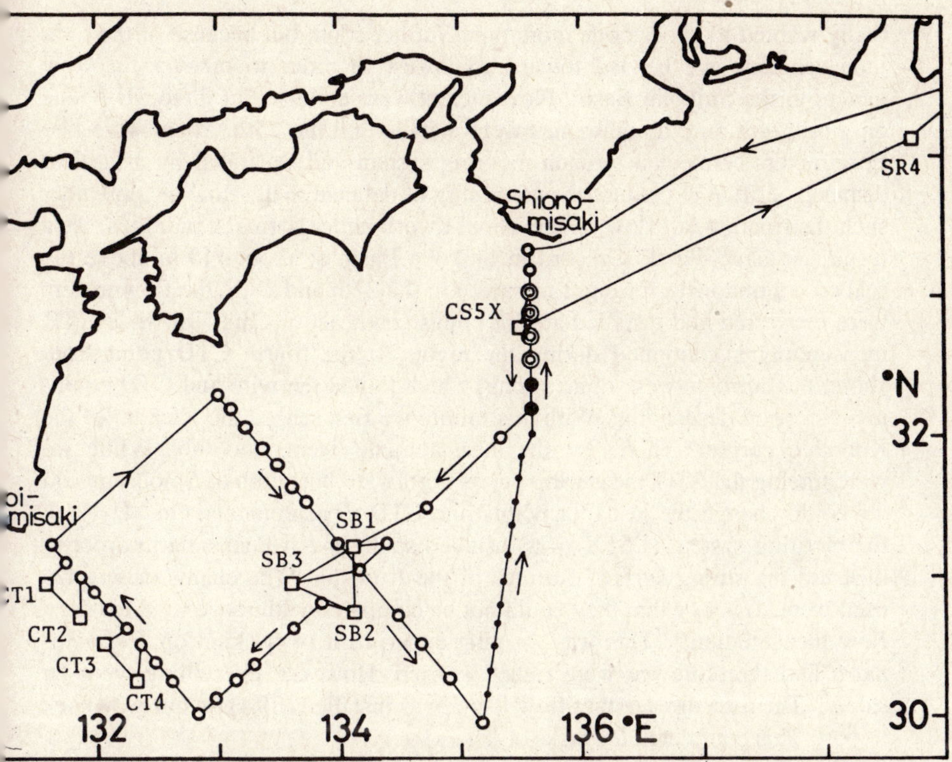


Fig. 1. Location of measurement points during leg 1 of the KH-83-1 cruise of the *Hakucho-maru*. ○ CTD; □ moorings for flow meter; • XBT measuring point; ● points where CTD were measured when going and XBT were measured when returning; ◎ points where CTD were measured twice, going as well as returning.

group, and water sampling group. Since the recovery and installation of flow meters can be carried out only in the daytime, this plays an important role in planning the schedule. CTD measurements can be made any time whether, day or night, and following their operations, the water sampling group can measure the salinity, oxygen, or concentration of various particles in sea water.

We reached the first CTD point at 12:40 on April 22. The real work starts from here. At this point we practiced CTD operations. Only here we could not collect water samples without the closing lid of the Rosette bottle. At other points operations were very smooth. During this time, the Kuroshio current was quite far from Shiono-misaki and it was therefore not possible to cross the Kuroshio current sufficiently at the southern side of Shiono-misaki. We really wanted to choose one more point further south but because of the tight time schedule we headed toward southwest in order to recover the flow meters in the Shikoku Basin. Flow meters were collected at three SB points on April 24th and the new ones were installed on the 25th. There were two flow meters connected to each mooring system and separated by a vertical distance of 50 m to counter the possibility of damage to the flow meter. Under such distribution, all flow meters should work under normal conditions. As a result, we have the flow record in Shikoku Basin at about 440 m above the seabed continuously for eight months. On the 27th and 28th, the flow meters were recovered and installed at four points southeast of Cape Toi while CTD measurements continued during the night. At the fourth CTD point from shore, the currents were quite strong, which pulled the wire and CTD equipment stopped descending. With this failure we first sensed the strength of the Kuroshio current. Thereafter the measurements went smoothly. While we were tracing the XBT measuring points northward but south of Shiono-misaki we could change the last four points into CTD measurements. On May 2nd, the mooring system (CS5X) was installed south of Kii Peninsula in order to measure the strong surface currents of the Kuroshio. The chains or weights used were so heavy that they could not be compared with those used for deep flow measurements. This was the first experiment of its kind on *Hakucho-maru* and therefore we were rather worried. However, everything went on safely. The next day another flow meter was installed (SR4) on our way back to port. This completed the tasks of leg 1.

Leg 2 started after a rest of three days. Almost 2/3 of the research members were replaced. At region B (30°N, 145–147°E) six flow meters were recovered and eight were installed; whereas, on the return journey one flow meter was recovered and installed west of Hachijoshima and two were recovered on the southern side of the port. In addition, CTD measurements, the acoustic testing of the Doppler flow meter, sediment trap installation, etc., were carried out at ten points in region B. While going as well as on the return journey, we conducted proton magnetometer tests. We did not have any significant trouble and returned to Tokyo port on May 19 just as planned. This completed our cruise of about one month.

However, Prof. Teramoto and five other members of the Physical Oceanography Department of the Institute of Oceanography were being awaited to join the Kh-83-8 cruise of the *Tansei-maru,* which can be called leg 3. One of the objects of this cruise was to recover and install flow meters at points off Cape Shio which were left aside during leg 1. During leg 1, this operation had to be abandoned in order to obtain the flow data corresponding to the observation period of the density field. However, in the last phase of observations which were smooth otherwise, our hearts skipped a few beats. Mooring system 1, which was being monitored closely, could not respond to frantic calls from the ship. Considering the 100% success obtained during the recent past regarding recovery of the mooring system, this failure may not be that serious. However, we felt it very much as this point was being monitored most closely. In general, the observations were satisfactory for the most part. It is quite interesting to find what kind of data we receive as a result of this cruise and what new knowledge it imparts regarding the ocean.

# Report on the Seventh Summer Seminar of Young Oceanographers

## 1. Introduction

The Association of Young Oceanographers has been formed to provide a forum for the exchange of information between young scientists and students who have a special interest in Physical Oceanography. Presently there are about 70 members of the association. The association conducts a residential summer seminar every year. This is the seventh year and this time the seminar was conducted at Shinjudo resort for three nights and four days between August 6 (evening) to August 9, 1983 (morning). There were 39 participants ranging from first year graduate students to professors and active discussions went on late each night. The composition of participants and programs are detailed below.

## 2. Parent Organizations of Participants and Their Classification

Faculty of Science, Hokkaido University—3
Faculty of Science, Tohoku University—7
Defense Academy—1
Faculty of Science, Tokyo University—7
Institute of Oceanography, Tokyo University—9
Department of Oceanography, Tokai University—6
Faculty of Science, Kyoto University—3
Department of Education, Kyoto University—1
Faculty of Engineering, Eihime University—2
Graduate course 1st year—12, 2nd year—11
Doctorate 1st year—2, 2nd year—2, 3rd year and above—7
Lecturers, Readers, Professors—5.

## 3. Program

August 6, 20:00-

Explanation of schedule, self introduction.

August 7, 09:15-12:00

Lecture 1: Numerical experiments about the Kuroshio. Prof. M. Ito (Faculty of Science, Tokyo University); the studies made so far regarding the changes in the path of the Kuroshio were introduced and, based on the results of numerical tests using a reduced gravity model, the speaker discussed the fluid mass dependence of the current path, etc.

Lecture 2: Numerical experiments regarding the Kuroshio, based on a general circulation model and an inflow-outflow model. N. Hosaka, I. Yasuda (Faculty of Science, Tokyo University); the results of two types of numerical tests, carried out to examine the dependence of the Kuroshio current on the shore profile or current velocity, were discussed as well as the characteristics of each model.

13:30-17:00

Lecture 3: Rossby waves in a stratified ocean and the seabed profile—Aoki (Institute of Oceanography, Tokyo University); the speaker discussed how an incident Rossby wave would reflect or pass through (transmit) when stratification or water-depth changes are taken into consideration.

Lecture 4: Long-period waves of the Sanin shore—Jikida (Faculty of Engineering, Eihime University); the fluctuations in water level or flow velocity were in response to changes in winds or atmospheric pressure. Characteristics of such fluctuations were discussed and the reasoning based on this phase velocity was presented.

Lecture 5: Structure of temperature fluctuation phenomena in Yadoge Bay—Akiyama (Faculty of Engineering, Eihime University); during the season, the water temperature of the Yadoge Bay region was found to fluctuate with a period of ten days. The factors responsible and the mechanism of these changes were studied and, as a result, the two-dimensional structure—horizontal and vertical—was presented.

Lecture 6: Problems involved in Weiner-Elmit development when applied to a two-dimensional turbulent flow—Zewamura (Faculty of Science, Kyoto University); the basic considerations in applying the Weiner-Elmit development to a two-dimensional turbulent flow and the problems involved therein were presented. The speaker also introduced part of his studies to find a solution for these problems.

20:00

General meeting of the Association of Young Oceanographers. It was decided to appoint Tokai University member as the Secretary for next year (October 1983 to September, 1984).

August 8, 09:00-12:00

Lecture 7: Circulation in the tropical belt—Kitamura (Faculty of Science,

Tokyo University); the speaker introduced the results of measurements as well as of numerical experiments on the eastward undercurrent at lower latitudes and discussed the possibility of explaining those results with a linear theory.

Lecture 8: Movement of particles according to the growth of a vortex formed as a result of tides passing through a strait—Kurota (Faculty of Science, Kyoto University); the speaker presented the results of model experiments carried out to study the shift of particles with a vortex and the growth of a vortex in a strait region during tides.

Lecture 9: Fundamentals of a nonlinear long wave theory and its applications to oceanographic phenomena—Kubogawa (Faculty of Science, Tohaku University); the speaker discussed a nonlinear long wave theory particularly the behavior of a KdV solitary wave and a shock wave and presented the results of applying this theory to the problem of fluctuations in a shore density current.

13:00-15:45

Lecture 10: Indications on the deep-current circulations in the northwest Pacific from material distribution—Kanego (Institute of Oceanography, Tokyo University); the speaker discussed basic considerations in postulating the sources of water from the $DO_2$, Si distribution and presented part of his studies regarding the cross currents between the Philippines Sea and the northwest Pacific Basin.

Lecture 11: Behavior of discharged low-density water in a rotation field—Michida (Faculty of Science, Tokyo University); the speaker outlined the phenomenon observed in a rotating water tank experiment carried out to study the behavior of warm discharge waters released from the shore and discussed the dependence of this phenomenon on parameters.

Unscheduled lecture: Qualitative analysis of inflow-outflow models of the central latitudes—Sakai (Department of Education, Kyoto University); the speaker presented a qualitative image obtained by a low-order model regarding the normal currents in the circular ocean region in the central latitudes resulting from inflow-outflow conditions.

16:20-17:30
Recreation (softball)

20:00-
Party

August 9, 09:00
Dispersal.

### 4. Invitation to Association of Young Oceanographers

The details of proceedings of the Seventh Summer Seminar of the Association of Young Oceanographers are proposed to be published in Vol. 5 of the journal of this association. It will be published sometime around October–December, 1983. Persons desirous of obtaining a copy of Vol. 5 may send their requests to Mr. Michida at the following address.

We welcome the participation of many people, not necessarily restricted to the academicians. Inquiries regarding the Association of Young Oceanographers may also be directed to the following address:

Physical Oceanography, Chair of Geophysics,
Faculty of Science, Tokyo University 2-11-16,
Bunkyo-ku Tokyo-113.
Tel.: 03-812-2111, Ext. 4283.

# Minutes of the Seventh Managing Committee Meeting, July 11, 1983 (Monday)

*Aoi Kaikan*

(Toranomon, Minato-ku, Tokyo)

**Items announced**

The progress of the project after the sixth steering committee meeting (February 28, 1983) was presented.

(For details, refer to the minutes of the 16th, 17th, and 18th Executive Commitee meetings.)

**Items discussed**

Prof. Teramoto presented discussions of the first and second meeting of the committee on future plans.

Among the problems to be handled as an extension of this special project, major importance is attached to deep-layers, ridge and trench area. The following six topics are currently under consideration:

1) Structure of deep-layer circulation

2) Warm water activity

3) Are Basin

4) Dynamic behavior of substances under deep layers or at the ocean floor

5) Flow conditions of sea water in a trench and the behavior of materials there

6) Compound opening technique, development of tomography.

Thus there is a close connection between the different fields and topics therein. Our first proposal for studying these aspects will be considered in the next meeting and the details will be prepared by autumn.

In continuation with this announcement, our relationship with WESTPAC while studying the deep layers or the social importance, etc., were also discussed.

**Preparation of final research report**

This will contain all the studies carried out under this special project.

More pages can be allotted to each research group if more copies are made available for sale and the report would then gain social importance.

By autumn the executive committee will decide the authors and their extent and will also prepare a budget for its publication.

After the details have been worked out, requests will be made for manuscripts. As far as possible, everything will be ready by the end of this fiscal year and actual printing is planned for next summer-autumn.

**Future schedule of the administrative group**

The eighth meeting of the Managing Committee will be held in autumn; during which the various aspects related to the preparation of the final research report will be discussed. The last meeting of the Managing Committee is planned in February-March, after which the last research conference will be held.

## ANNOUNCEMENT

# Seventh Symposium of Young Chemical Oceanographers

This symposium will be held in the following manner, in continuation with the previous two years. This symposium is mainly for young scientists enagaged in chemical studies of ocean or earth. It provides a forum for free discussions on subjects ranging over a wide field without restrictions and should provide an opportunity for opening a new viewpoint or considerations.

This time there is no particular main theme but it is planned to have lectures on subjects mentioned below followed by discussions. For registration (your suggestions for topic selection are welcome) and other information regarding the symposium, please contact the coordinator.

**Program**

1. Time and date: From 17:00 of October 19 (Wednesday) to 12:00 of October 20 (Thursday). One night and two days. It will be held after the conclusion of the annual meeting of the Japan Geochemical Association.

2. Place: University Seminar House (1987-1 Hachioji-city, Tokyo).

3. Registration fee: About 5,000 yen (including lodging and boarding).

4. Lectures (proposed):

Oba (Faculty of Education, Kanazawa University): Paleoenvironments in seas surrounding Japan.

Ebihara (Institute of Oceanography, Gumma University: How the elements exist in the solar system.

Fujioka (Institute of Oceanography, Tokyo University): Sedimentary structure of sediments at the ocean floor.

5. Coordinators:

Gamo (Department of Inorganic Chemistry, Institute of Oceanography, Tokyo University), 1-15-1 Minami-dai, Nakano-ku, Tokyo 164. Tel: 03-376-1251, Ext. 274.

Suzuki (Meteorological Research Laboratory).

Ozumi (Faculty of Science, Tokyo Institute of Tehnology).

Kato (Department of Oceanography, Tokai University).

## Guidelines for Authors

This Special Research Project has now entered its last year. Since this may be the time to summarize the studies conducted so far, Newsletter welcomes enterprising articles from all regarding the present status of studies, results, etc., which will go a long way to finalize this special research in general.

The contents are classified as:

| | |
|---|---|
| Highlights: | Quick reports of the results of this Special Research Project |
| Bulletin: | Domestic or international news reports closely related to this Special Research Project |
| Reports: | Reports on the activities of each section or working group |
| Announcements: | Announcements regarding the proposed meetings or conferences of each group |
| Bibliography: | Listing of papers published by the members on this Special Research Project or other reports. |

Manuscripts should be written on standard 400 character size paper (handwritten material also accepted). The figures should be drawn in such a way that they can be reproduced as they are.

It is proposed to publish this Newsletter every two months.

**Note:**

When reproducing or citing the material published in this Newsletter, due acknowledgment should be given. Authors' names, wherever mentioned, should also be quoted. A copy of the publication containing such a reproduction may be sent to the Editorial Section of the Administrative Office.


| | |
|---|---|
| Published by: | Administrative Office, Special Research Project<br>"The Ocean Characteristics and Their Changes" |
| Published at: | Administrative Office,<br>Physical Oceanography Department,<br>Institute of Oceanography,<br>Tokyo University,<br>1-15-1 Minami-dai, Nakano-ku,<br>Tokyo 164<br>Tel.: 03-376-1251, Ext. 251. |
| Editorial staff: | Taeko Kitano and Noriyo Kimura |

Special Research Project

# The Ocean Characteristics and Their Changes

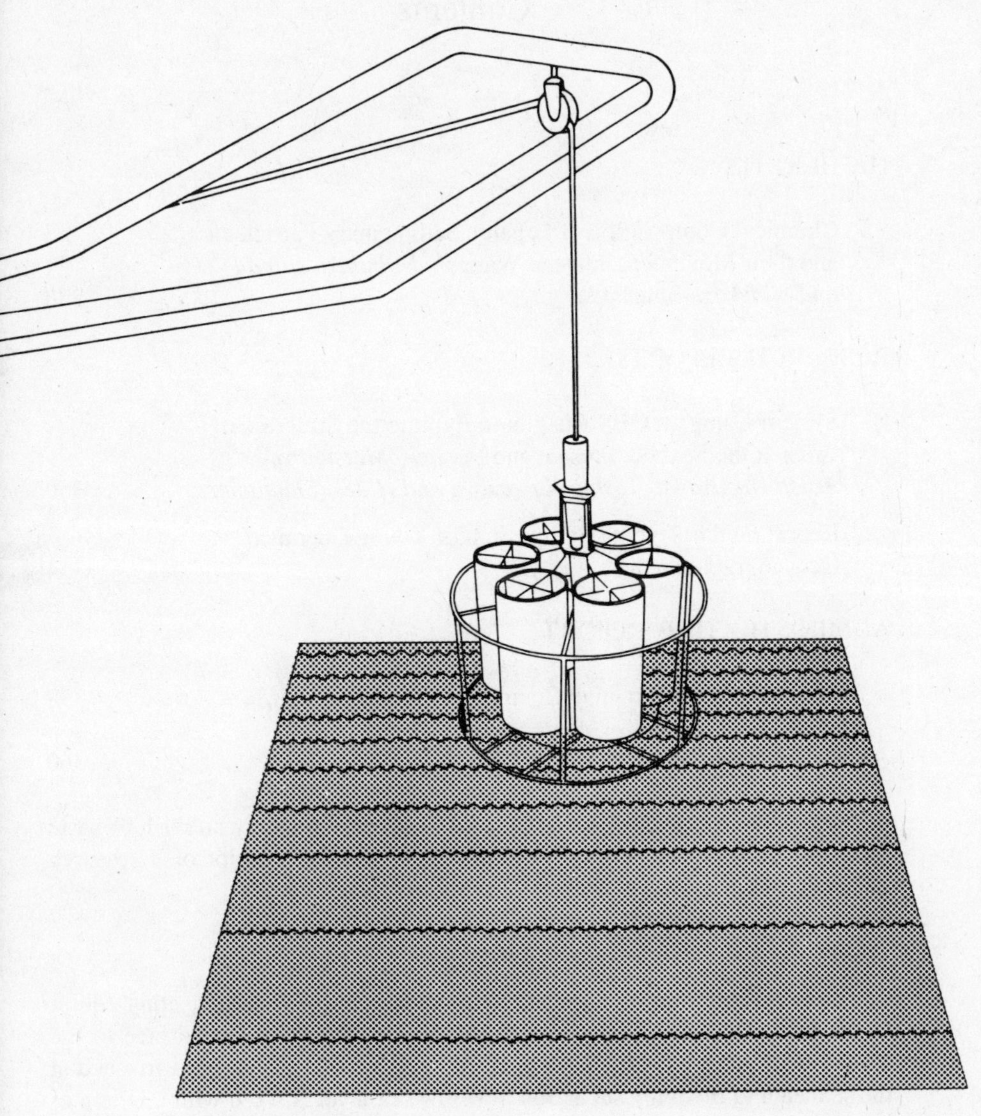

**NEWSLETTER NO. 15, NOVEMBER, 1983**

# Contents



Researchers engaged in this Special Research Project, who wish to write their research papers in English and to acknowledge receipt of a research grant, should indicate the above title of the project.

## Cover Page

This cover shows a sediment trap (sedimentary particle collecting equipment) to measure the flow velocity of particles from the sea surface to the seabed while they are precipitating. Generally, a 500 kg weight is attached at the seabed and the buoy set at 300 m below sea level is fixed with the help of nylon rope. There are five sediment traps, like the one shown here, connected in this interval in such a mooring system. At the time of recovery, each cylinder is covered internally to avoid scattering the sample through turbulence. This sediment trap is the improved version of the system developed by Mr. Mochigai, Department of Fisheries, Hokkaido University.

The figure has been prepared by Mr. Kimura (Institute of Oceanography, Tokyo University) based on photographs by Mr. Handa (Oceanographic Laboratory, Nagoya University).

HIGHLIGHTS

# Chemical Composition of Organic Sedimentary Particles and their Movement in Deep Waters

*Nobuhiko Handa and Hidekazu Matsueda*

(Oceanographic Laboratory, Nagoya University)

(Received: October 31, 1983)

## 1. Introduction

Organic compounds are formed by the photosynthetic reaction of plant plankton in the surface-layer water of the ocean and that forms a major component of the suspended particles. Such particles increase in size as a result of mutual aggregation or by the positive accumulative action of animal plankton and are transformed into particles which precipitate with high speed (sedimentary particles). Such sedimentary particles are supposed to transport, at a high velocity, the organic compounds or radioactive nuclei of the ocean from its surface layer to the deep-water layer and eventually to the ocean floor. These organic materials play an important role in material recycling in the ocean since they eliminate the dissolved oxygen in the bio-activities and at the same time generate nutrient salts and carbon dioxide. They are also important since they impart fertility to the deep-layer waters.

The object of this study is to clarify the vertical transport mechanism of organic materials following the sedimentation of particles in the ocean and evaluate the role of the organic material constituting the sedimentary particles in the material recycling system of the ocean, particularly in the intermediate and deep layers.

With these objects in mind, the sediment trap experiment and the on-site filtering experiment were conducted at Sagami Bay as well as in the northern and eastern parts of the north Pacific Ocean to collect sediment and suspended particles. We analyzed the composition of the organic materials, mainly the fats, and tried to estimate the sources of organic materials which form sediment particles as well as their ability to consume the oxygen dissolved in water. We also calculated the vertical flow rate of organic materials in the intermediate or deep ocean layers, following the sedimentation of particles, and evaluated the role of organic materials and sediment particles in the material recycling of deep water.

## 2. Chemical Properties of Organic Sedimentary Particles in the Ocean

Sedimentary organic materials generally consist of hydrocarbons, amino acids, proteins, fats, etc. Among these constituents, there are many reports available regarding the fats as a source of sedimentary organic materials. We, accordingly, carried out a comparative study of the fat composition of precipitated layers at the ocean floor, suspended particles and sedimentary particles.

Samples of sedimentary particles, suspended particles and precipitations were treated with chloroform-methanol (2 = 1 V/V) to extract the fats. The extract was washed with distilled water and then a chloroform phase was separated and condensed. It was then dissolved into a small quantity of hexane and allowed to be absorbed in a silica gel column (Merck, Silica gel 60). A solution containing a mixture of hexane, benzene, ether, ethyl ether and methanol, etc. was flowing through the silica gel column and by changing the polarity of the medium, different components of fats were separated.

The organic materials of each group were then subjected to gas chromatographs equipped with a capillary column (length 25 m, inner diameter 0.25 mm) and each constituent was identified. In this case fatty acids were assumed to be in the form of a methyl ester after they were made to react with $BF_3$-methanol; whereas, fatty alcohols were assumed to be in the form of trifluoroacetate esters after they were made to react with anhydrous trifluoroacetic acid.

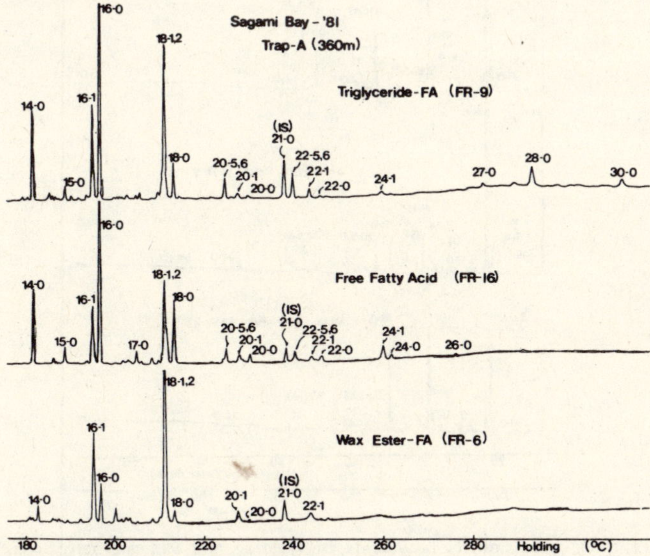


Fig. 1. Gas chromatograms of free fatty acids, triglyceride and wax ester fatty acids in sedimentary particles, suspended particles, and precipitates in Sagami Bay.

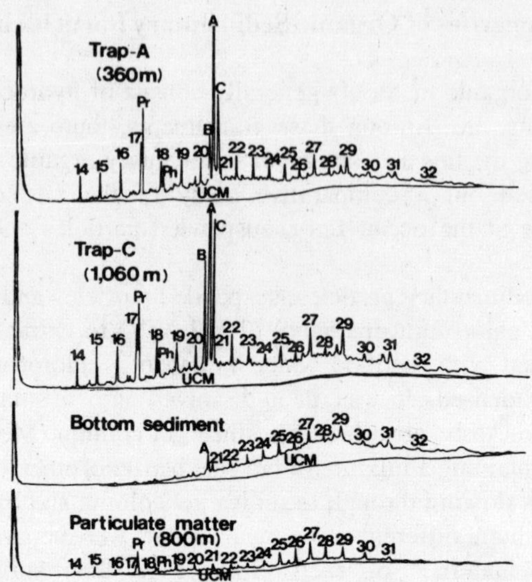


Fig. 2. Gas chromatograms of hydrocarbons in sedimentary particles, suspended particles, and precipitates in Sagami Bay.

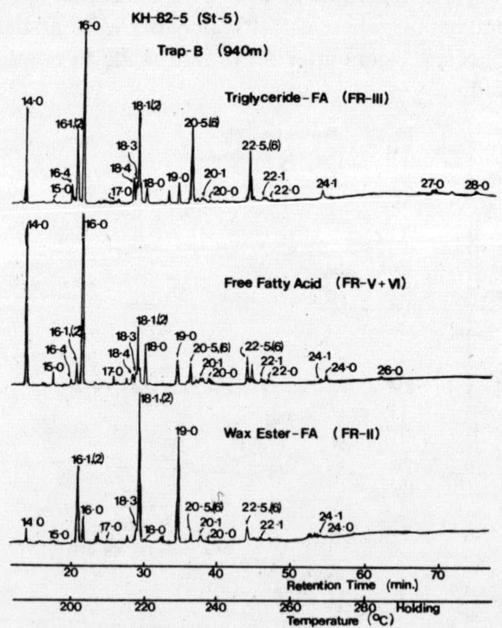


Fig. 3. Gas chromatograms of free fatty acids, triglyceride, and wax ester fatty acids in sedimentary particles of the eastern region of the north Pacific Ocean.

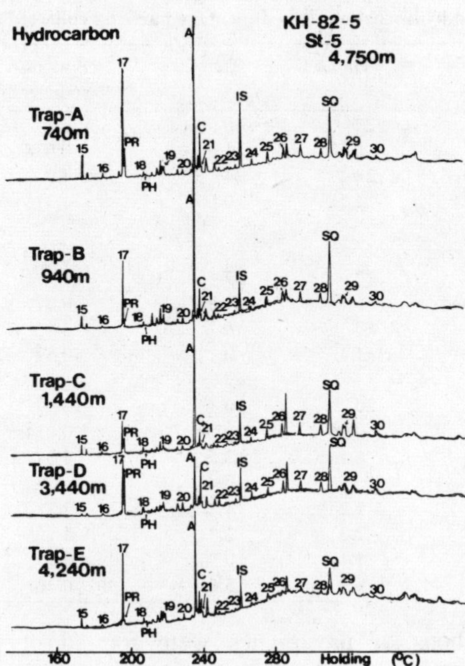


Fig. 4. Gas chromatograms of hydrocarbons in sedimentary particles of the eastern part of the north Pacific Ocean.

The gas chromatograms of fatty acids and hydrocarbons are shown in Figs. 1-4 as representative fat components. Table 1 shows an outline of the composition of each component constituting the fatty groups, using the samples from Sagami Bay as an example.

Compared to sediments and suspended particles, the CPI in precipitates is considerably less, meaning that the level of disintegration of organic materials (initial continuous degeneration operation) is considerably advanced. This fact is also supplemented by the ratio of unsaturated/saturated fatty acids and fatty alcohols. Generally, the ocean bacterias contain large amounts of free fatty acids occupying more than 50% of the total (Leo and Parker, 1966). A similar fact has also been ascertained by Tanoue and Handa (1980). It can therefore be concluded that the degeneration of fatty acids and fatty alcohols in precipitates depends on the activity of micro-organisms.

The hydrocarbon composition of the sedimentary particles, suspended particles, and precipitates varies considerably from each other. According to the method adopted this time, the number of carbon was found to vary from 14($C_{14}$) to 32($C_{32}$) in each sample.

**Table 1. Properties of each component constituting free fatty acids, triglyceride fatty acids, fatty alcohols, and hydrocarbons in sedimentary particles collected at Sagami Bay**

| | Trap A | Trap C | Bottom sediment | PM—800 m |
|---|---|---|---|---|
| Free fatty acids | 22 | 40 | 6.4 | 42 |
| CPI | 0.73 | 1.3 | 0.078 | 1.8 |
| Unsaturated/saturated | 28 | 24 | 4.0 | 12 |
| $C_{12-23}/C_{24-30}$ | 1.1 | 0.66 | 4.31 | 0.36 |
| Branched (%) | | | | |
| Fatty acid of triglycerides | | | | |
| CPI | 27 | 37 | 11 | 19 |
| Unsaturated/saturated | 1.1 | 0.83 | 0.21 | 1.0 |
| $C_{12-23}/C_{24-30}$ | 13 | 6 | 3 | 39 |
| Branched (%) | 1.1 | 0.12 | 0.59 | 0.83 |
| Fatty alcohols | | | | |
| CPI | 20 | 70 | 13 | 51 |
| Unsaturated/saturated | 0.26 | 0.79 | 0.014 | 0.47 |
| $C_{12-23}/C_{24-30}$ | 6.1 | 3.5 | 2.3 | 13 |
| Hydracarbons | | | | |
| CPI ($C_{14-23}$) | 1.2 | 1.2 | 2.1 | 1.3 |
| CPI ($C_{24-30}$) | 1.3 | 2.0 | 2.1 | 1.1 |
| $C_{14-23}/C_{24-30}$ | 1.1 | 0.94 | 0.15 | 0.33 |
| Pristane/Phytane | 4.9 | 3.6 | 0.46 | 5.6 |

The hydrocarbons of precipitates were formed of long chains of hydrocarbons $C_{21}$–$C_{32}$ only and the distribution pattern for each component was almost identical to that for soil particles. As against this, hydrocarbons for sedimentary particles were made from $C_{14}$–$C_{32}$. In this case particularly, the distribution pattern for short chains of hydrocarbons ($C_{14}$–$C_{22}$) mainly involving $C_{17}$ and the presence of Pristane (Pr) as well as Phytane (Ph), which causes the degeneration of chlorophyl, indicates that hydrocarbons in sedimentary particles are a mixture of short chained hydrocarbons coming from plant plankton and long chained hydrocarbons coming from soil particles. Thus during the sedimentation process of particles in the ocean, the short-chain and long-chain hydrocarbons are not much affected by microbiological degeneration but after the particles precipitate at the ocean floor, the hydrocarbons from plant plankton are rapidly disintegrated, leaving behind only the long-chain hydrocarbons which have their origin in soil particles.

The hydrocarbon composition of suspended particles varies considerably from that of sedimentary particles. Here the short-chain hydrocarbons are in extremely small quantities. Looking from this aspect, we can conclude that differing from sedimentary particles, the suspended particles follow the same disintegration process as precipitates.

Another characteristic of sedimentary particle hydrocarbons is the

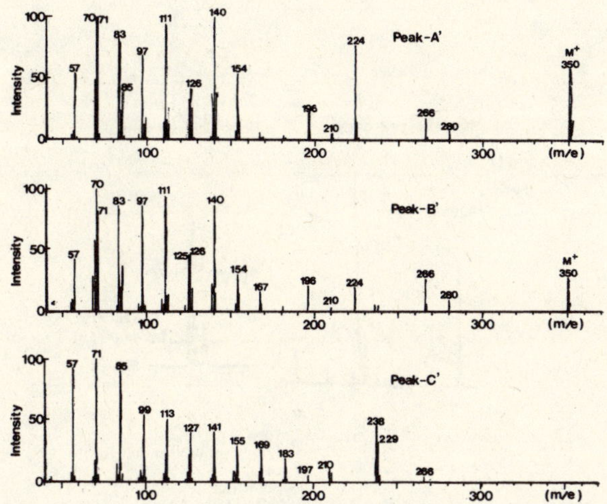


Fig. 5. Mass spectrum of hydrocarbon peaks A, B, and C, constituting the sedimentary particles in Sagami Bay.

presence of compounds corresponding to peaks (traps) A, B, C. The structure of these compounds was analyzed by gas chromatography and mass spectrometer.

The EI mass spectrum of compounds at peaks A, B, and C is shown in Fig. 5. The molecular ions of each compound are respectively 348, 346 and 346 and compared with saturated hydrocarbon with 25 ($C_{25}H_{52}$, $C_{25}$), the mass number is short by 4 (2 × 2H, degree of unsaturation 2) and 6 (3 × 2H, degree of unsaturation 3). In addition, the molecular fracture pattern indicates that there are large number of methyl radicals present within the molecule and therefore the structure was concluded to be a multibranched, unsaturated hydrocarbon.

In order to obtain detailed information about the unsaturated bonding in molecules, these compounds were reduced by $PtO_2$/EtoAC/$H_2$ system and the structure of the resultant compounds was analyzed. The retention time for unsaturated compounds after reduction varied as shown in Fig. 6 during gas chromatography. The EI mass spectrum of these compounds (Fig. 7) indicated that $M^+$ for peaks A′, B′, and C′ stand at 350, 348 and 352 respectively. This means that even after reduction, the compounds of peaks A′, B′, and C′ have a mass number less by 2 and 4 than the saturated $C_{25}$. It can therefore be concluded that compounds of peaks A, B, and C are respectively multibranched—with one degree of unsaturation—one ring type hydrocarbon ($C_{25}$ = 1 + 1), multibranched—with two degrees of unsaturation—one ring type hydrocarbon ($C_{25}$ = 2 = 1), and multibranched—with three degrees of unsaturation hydrocarbon ($C_{25}$ = 3 = 0).

These compounds are of great interest since their structure is still dif-

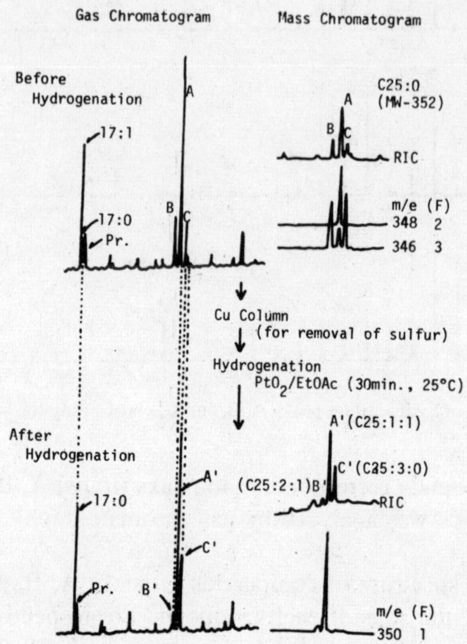


Fig. 6. The variation in retention time of peaks A, B, and C after reduction and mass spectrogram.

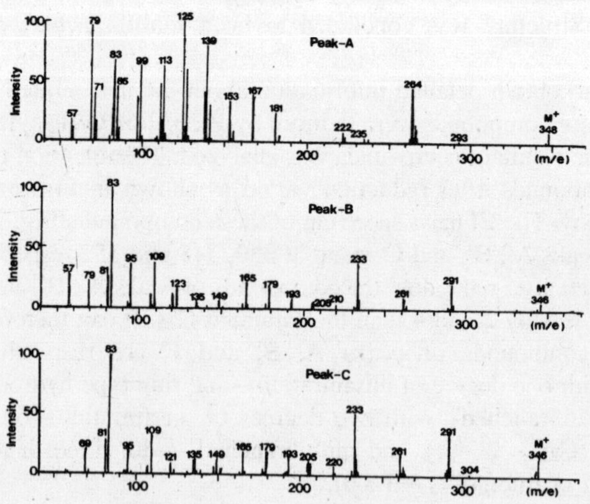


Fig. 7. Mass spectrum of reduced products of hydrocarbon peaks A, B, and C constituting sedimentary particles of Sagami Bay.

ferent from isoprenoid which is widely present in the atmosphere. Materials similar to these compounds have been reported to be present in the surface layer of seashore precipitates (Barrick, et al., 1980). Similarly, Prahl, et al. (1980) has ascertained the presence of hydrocarbons with carbon numbers of 20($C_{20}$), 25($C_{25}$), and 30($C_{30}$) and with similar structure in the sedimentary particle samples collected along the seashore. He has estimated that as a typical material in the excretion of animal plankton.

During a cruise on the eastern part of the north Pacific region, conducted during November, 1982-February, 1983, the sediment trap experiment was carried out a few times. Here, during a sediment trap experiment carried out 200 miles off San Francisco, the sediment particles were mostly the excretions of sulpha (0.5-1 × 1-2 cm). This is an extremely precious sample to discover information regarding the role of animal plankton in sedimentary particles.

These excretions mainly consisted of pebbled seaweed in addition to diatomaceae, radiolaria and foraminifera. A gas chromatogram, using a capillary hydrocarbon column, is shown in Fig. 4. Thus the composition of hydrocarbons is almost the same over all layers. Also similar to the samples from Sagami Bay, these sediments were also a mixture of short hydrocarbon chains around $C_{17}$ and long hydrocarbon chains. In this sample we also detected the presence of multibranched, unsaturated, ringed hydrocarbons. Gas chromatograms and mass spectrums of the compounds after reduction revealed that these compounds were the same as compounds from peaks A and C observed in sedimentary samples of Sagami Bay.

Thus, multibranched, unsaturated, ringed hydrocarbons are characteristic compounds of animal plankton excretions and can be considered indicative materials of excretions.

The fact that this material flows through the seashore and outer seas into the sedimentary particles also indicates that animal plankton excretion undergoes considerable changes while forming large particles such as sedimentary particles.

Now, in either Sagami Bay or the eastern part of the north Pacific, there was no change in the component pattern constituting fatty acids, fatty alcohols, and hydrocarbons, according to the depth of the sample. This means that the vertical motion of sedimentary particles occurs very smoothly.

### 3. Vertical Flux of Organic Carbon and Carbon Recycling

The vertical flux of organic carbon, following particle sedimentation in the outer seas has been measured since around 1970. Table 2 shows some of those efforts. It can be seen from the table that although the value of flux is rather high along the seashores—such as Sagami Bay sample—in the outer seas, flux does not change much according to the seas. Also organic carbon being transported to depths beyond 1 km, is less than 10 mg C/m$^2$ · day corresponding to only a minor percentage of the basic quantity produced at the surface.

Let us now discuss qualitatively the direction of flow of organic carbon into deeper waters based on sediment trap experiment and rate of precipitation.

**Table 2. Vertical flux of organic carbon measured in the sediment trap experiments carried out in different seas**

| Region | Depth (km) | Flux of organic carbon (mg C/m$^2$/day) | % with respect to basic production | Author |
|---|---|---|---|---|
| Tongue of Ocean Bahama | 2.15 | 6.4 | — | Wiebe, et al. (1976) |
| | | 4.7 | | |
| Gulf of Alaska | 0.9 | 7.6 | 3.8* | Iseki (1977) |
| (49°57′N:144°29′W) | 1.5 | 4.7 | 2.4* | |
| | 1.875 | 3.3 | 1.7* | |
| Sargasso Sea | 5.367 | 1.23 | 1.6** | Spencer, et al. (1978) |
| (31°34′N: 55°03′W) | | | | |
| North Pacific Gyre | 1.05 | 12 | 9 | Knauer, et al. (1979) |
| (32°47′N: 144°26′W) | | | | |
| Sargasso Sea | 0.976 | 2.43 | 4.1 | Honjo (1980) |
| (31°32.5′N: 55°55.4′W) | 3.694 | 0.87 | 1.5 | |
| Tropical Atlantic | 0.389 | 6.76 | 3.4 | Honjo (1980) |
| (13°32.2′N: 54°00.1′W) | 0.988 | 3.95 | 2.0 | |
| | 3.755 | 1.73 | 0.9 | |
| | 5.068 | 1.70 | 0.9 | |
| North Central Pacific | 0.378 | 3.56 | 5.1** | Honjo (1980) |
| (15°21.1′N: 151°28.5′W) | 0.978 | 0.55 | 0.8** | |
| | 2.778 | 1.09 | 1.6** | |
| | 4.280 | 0.88 | 1.3** | |
| | 5.582 | 0.66 | 0.9** | |
| Northern North Pacific | 1.1 | 7.27 | 2.2 | Tanoue and Handa |
| (47°57.7′N: 176°28.8′E) | 2.2 | 7.33 | 2.2 | (1980) |
| | 4.4 | 2.70 | 0.8 | |
| | 5.25 | 2.38 | 0.7 | |
| Sagami Bay | 0.36 | 88 | | Matsueda and Handa |
| | 0.56 | 73 | | (1983) |
| | 1.06 | 64 | | |
| Eastern North Pacific | 0.74 | 23 | | Matsueda and Handa |
| | 0.94 | 18 | | (1983) |
| | 1.44 | 6.7 | | |
| | 3.44 | 6.6 | | |
| | 4.24 | 8.7 | | |

*Calculated from the data of 200 mg C/m$^2$/day of primary production (Koblenz-Mishke, et al., 1970).

**Calculated from the data of 70 mg C/m$^2$/day of primary production (Koblenz-Mishke, et al., 1970).

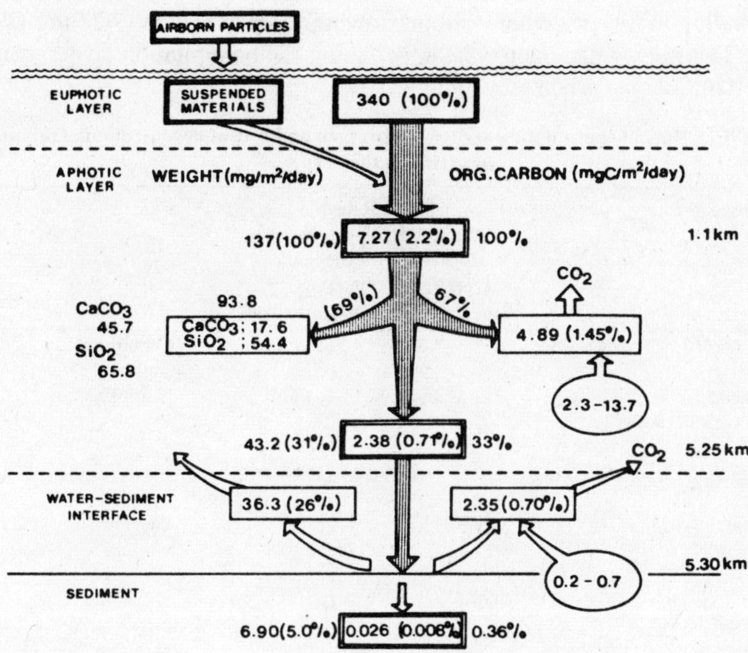


Fig . 8. Organic carbon flux and carbon recycling following the sedimentation of particles in the deep seas of the northern North Pacific region. Figures in ☐ indicate the basic quantity produced at the surface layer; at a depth of 1.1 km and 5.25 km they indicate the vertical flux of organic carbon at those levels. In the case of sediments, the figure indicates rate of decrease of the organic carbon.

Let us use the values of organic carbon flux measured in the northern North Pacific region (Tanoue and Handa, 1980). The vertical flux of organic carbon at a depth of 1 km and 5.25 km (50 m above the seabed) was 7.27 and 2.38 mg C/m$^2$ · day respectively (Fig. 8). This means, the loss of organic carbon in deep waters (1–5.25 km layer) is 4.89 mg C/m$^2$ · day.

The rate of consumption of dissolved oxygen in the deep waters of various seas has been calculated with a diffusion-advection model and is estimated to be around 0.6–4 µl/l · yr (Table 3). This corresponds to 2.3–14 mg C/m$^2$ · day of organic carbon. Thus at depths beyond 1 km, the organic carbon destroyed by the routine daily activities of organisms is returned to those waters in the form of organic particle sedimentation.

The organic carbon which flows toward precipitates and remains there is calculated from the rate of precipitation and the contents of organic carbon in the precipitates to be 0.026 mg C/m$^2$ · day. This means the amount of organic carbon that disintegrates near the ocean floor is 2.35 mg C/m$^2$ · day corresponding to 1/3 of organic carbon flowing to deep waters (7.27 mg C/m$^2$ · day). This shows that deep waters, including the precipitation layers, play an important role in carbon recycling.

**Table 3. Rate of consumption of dissolved oxygen and rate of degeneration of organic materials in the deep seas**

| Author | Rate of consumption of dissolved oxygen*, μl/yr | Rate of degeneration of organic materials, mg C/m$^2$ · day | Seas |
|---|---|---|---|
| Munk (1966) | 4 (2.7–5.3) | 13.5 (1–4 km) | Central Pacific |
| Arons and Stommel (1967) | 2–2.5 | 7.6 (1–4 km) | North Pacific |
| Tsunogai (1972) | 1–3 | 13.7 (1–4.5 km) | North Pacific |
| Kroopnick (1974) | 0.6–3 | 2.3–11.4 (1–4.5 km) | East Pacific |
| Range | 0.6–4 | 2.3–14 | |
| Sediment trap | | 4.89 (1.1–5.25 km) | Northern North Pacific |

*Determined by the diffusion-advection method.

## 4. Conclusion

During the special research project "Dynamic Characteristics of the Ocean," we mainly collected samples of sedimentary particles in a sediment trap experiment and measured the vertical flow rate of the organic carbon through these sedimentary particles. It was possible to specify the sources of sedimentary organic particles from their composition and chemical properties. Also we could stress the importance of animal plankton excretion in the transport of organic materials. A qualitative analysis was carried out regarding the vertical flux of organic carbon in deep waters and the material recycling system. It can be seen from this analysis that the rate of disintegration of organic materials in deep waters was of the same order as that calculated from the diffusion-advection model. It was also revealed during this discussion that the disintegration of organic materials in a sedimentary layer or in deeper waters plays an important role in oceanic material recycling.

## Bibliography

Barrick, R.C., J.I. Hedges and M.L. Peteron. 1980. *Geochim. Cosmochim. Acta,* **44,** 1349-1362.

Leo, R.F. and P.L. Parker. 1966. *Science,* **152,** 649-650.

Prahl, F.G., J.T. Bennett and R. Carpenter. 1980. *Geochim. Cosmochim. Acta,* **44,** 1967-1976.

Tanoue, E. and N. Handa, 1980. *J. Oceanogr. Soc. Jap.,* **36,** 231-245.

RESEARCH REPORTS

# System Using an OBS Air-gun to Explore the Structure of Crust at the Seabed—Present and Future

*Matsuhiro and Azusa Nishizawa*

(Faculty of Science, Tohoku University)

*Toshio Kanazawa*

(Faculty of Science, Tokyo University)

*Hideki Shimamura*

(Faculty of Science, Hokkaido University)
(Received: October 27, 1983)

## History

Surveys on the structure of the earth's crust have been carried out for ages but the survey methods have changed.

These methods were the two-ship method using an explosive-hydrophone-array, the sonobuoy method which analyzes seismic vibrations, the hydrophone-steamer method using reflection waves, methods using air-guns, etc.

At present there is no best method available. However, the OBS, OBH method which was successfully developed commercially during the '70s has an improved S/N ratio compared with other seismological equipment. It can be installed at any specified point and records even S waves with OBS. Also its recovery is now simple. Although its use is restricted to off-line conditions, the OBS air-gun method became common. The course of its development is shown in Table 1. The later part lacks details. It also excludes the work of Soviet scientists, Neprotinov, Sedoff, and Ricknoff.

Using this method, it is possible to explore to the top mantle region of the ocean floor.

## Survey Using Tohoku University's Present System

In 1980, experiments were started using the self-propelled OBS

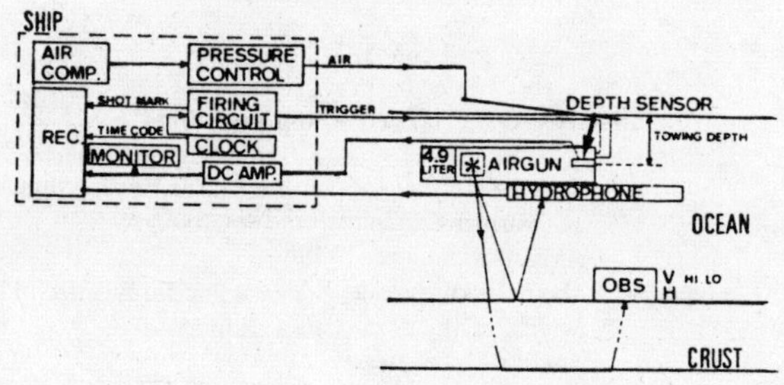


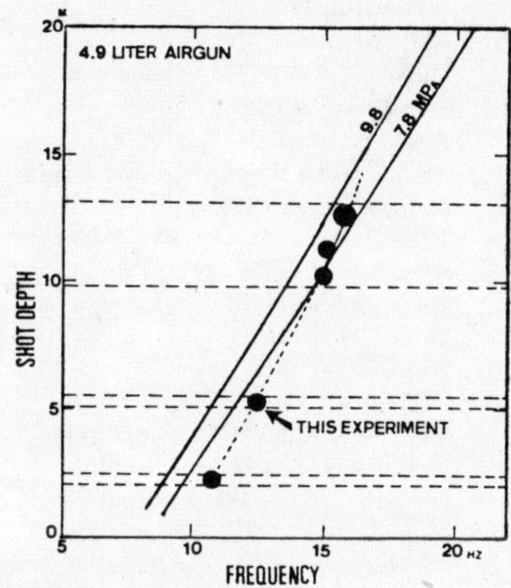


Air-gun depth preparation and waveform sounding.

developed by Tokyo University and Tohoku University as well as using a BOH 1500 C air-gun (9, 4.9, 2.4 l) under exclusive use of the Faculty of Science, Tohoku University. During the 1980–1981 cruise of the *Tansei-maru*, it was experimentally proved that depth preparation is important for the air-gun's waveform sounding (see Figure).

During 1980, 1981 and 1982 few cruises were conducted off the Tohoku region and a large volume of data was collected including long distance data using explosives. Similarly, a survey was also conducted at Amami Plateau and for the first time the structure below the plateau was observed.

## Present and Future Problems

During the experiment: The present efficiency of the compressor is not sufficient. We would like to develop a compressor with double the present efficiency which can discharge 18 l in 50 s so that it can be effectively used in ship time without any restrictions.

**Table 1.**

**Air-gun—OBS (H) refraction profiling**

| | | | |
|---|---|---|---|
| 1971 | C.O.B., | France | Bolt 1900 B 0.7 l, Sodera 4.1 l, off Brest compared with drifting and anchored sonobuoys. Avedik & Renard, 1973 |
| 1972 | KH-72-2 | Japan | about 20 km profile, Western Pacific Basin. Asada, et al., 1975 (Cruise Report) |
| 1973 | C.O.B. | France | 5.9 l (× 2), off Brest Avedik, 1975 |
| | I.O.S. | G.B. | 16 l, FAMOUS, OBH Whitmarsh, 1975 |
| 1974 | U.T. | Japan | 28 l, Okhotsk Shibuya & Miyashita, unpublished. |
| 1975 | C.O.B. | France | 9 l × 2, Bay of Biscay Avedik, et al., 1978 |
| 1976 | KH-76-2 | Japan | 4.9 l, Shikoku Basin, IPOD site survey Nagumo, et al., 1980 |
| | C.O.B. | France | 16 l × 2, Bay of Biscay; 45 l, Norwegian Sea Avedik, et al., 1978 |
| | GDP-22 | Japan | 4, 5l, northern Philippine Sea Takeda, et al., 1977 |
| | U. Texas | USA | 24.6 l, Texas Shelf zone Ibrahim & Latham, 1978 |
| 1977 | KH-77-1 | Japan | 4.9 l, Daito Ridge, IPOD site survey Nagumo, et al., 1977 |
| | KH-77-1 | Japan | 4.9 l, off Sanriku, IPOD site survey Nagumo, et al., 1980, Tokuyama, et al., 1981 |
| | GDP-23 | Japan | 4 l, Northwestern Pacific Basin, PmP observed Takeda, 1978 |
| 1978 | U. Hamburg C.O.B | | Eastern Mediterranean Weigel, Makris, Avedik, WHOI, et al. |
| 1979 | I.O.S. C.O.B. | | English Channel Roberts, Whitmarsh, Avedik, et al. |
| | Chiba U. | Japan | 4 l, Suruga Bay, OBH |
| 1980 | KH-80-3 | Japan | 2, 4.9 l, Northwestern Pacific Basin Urabe & Imahori, 1981 (Cruise Report); Kinoshita, et al., 1981 (Cruise Report) |
| | Chiba U. | Japan | 10 l, Suruga Bay, OBH Murauchi, et al., 1980 |
| 1981 | Chiba U. | Japan | 20 l, Suruga Bay, OBH JMAOBS |

**Air-gun—OBS Refraction profiling (Tohoku U. related)**

| | | | |
|---|---|---|---|
| 1980 | May | 23-24 | KT-80-7 off Miyako |
| | Jul | 24-25 | Ocean Discoverer off Sanriku |
| | | 28-30 | |
| 1981 | May | 30 | KT-81-7 off Joban |
| | Jun | 11-16 | Ocean Discoverer off Sanriku |
| | | 18, 21-23 | |
| 1982 | Jun | 18-19 | Sakura off Sanriku |
| | | 22-26 | |
| | Jul | 15-17 | Tokyo off Fukushima |
| | Aug | 1-2 | KH-82-4 Amami Plateau |

Data processing: Digital processing became possible with the introduction of air-gun data processing and analysis equipment this year but software support is also necessary. From the processing aspect, we are still lagging behind Europe or America. Although analysis is mainly based on refractive waves, the structural constraints due to reflected waves are also important.

# Report on the Seventh Symposium of Young Chemical Oceanographers

## 1. Aim and Course

The main object of this symposium was to study the viewpoints and considerations of a wide range of new concept that have been developed in modern geoscience. The symposium was held as mentioned below. The participants were young scientists engaged in chemical studies of the earth, including the ocean. The symposium saw a lively question-answer session and free discussion on the seven specified themes.

## 2. Place and Dates

University Seminar House (II Seminar Room)
October 19, 1983, 5:30 p.m.-October 20, 3:30 p.m.

## 3. Participants

Total 33. The names and sponsoring organizations of these participants are appended at the end of this report.

## 4. Outline of Program

October 19.

5:30-6:00 p.m. Opening: Self introduction, etc.

7:30-9:00 p.m. Address 1: "How the elements exist in a solar system". Speaker: Ebihane, Lecturer, Department of Liberal Arts, Tokyo University. He explained the basic features of what a meteorite is and what the chemical studies of meteorites mean. He also discussed the form of the original solar system, based on a condensation model and the subsequent behavior of elements using the results from the chemical analysis of meteorites. He introduced the research activities going on in Chicago University as well as the positions corresponding to recent solar abundance.

(Chairman: Ohsumi Takashi).

9:00–10:30 p.m. Address 2: "The present meaning of entropy and its involvement in geochemistry". Speaker: Ebihane, Lecturer, Department of Liberal Arts, Tokyo University. The speaker explained in detail the transition of the term entropy in different fields such as nonequilibrium thermodynamics, economics, or environmental science. As an example of the application of the entropy concept, he offered problems related to the self-purification process of the ocean and entropy. The lecture was followed by a lively discussion.

(Chairman: Yoshimi Suzuki).

10:30 a.m.–12:00 noon. Free discussion in small groups.

October 20

9:00–10:30 a.m. Address 3: "Construction of the original (first) cell under ocean environments". Address 4: "Chemical evolution reaction under high pressure". Speaker: Yanagikawa, Chief Researcher, Bioscience Research Laboratories of Mitsubishi Chemical Co. He introduced the results of studies carried out to clarify the course taking place during protein synthesis in the ocean or atmosphere. In this context he discussed the mechanism of amino acids formation determined experimentally under various temperatures or pH values, or the form and properties of artificial proteins. He also indicated the examples and problems involved in the chemical evolution of the warm water system at the ocean floor.

(Chairman: Yoshimi Suzuki).

11:00 a.m.–12:20 p.m. Address 5: "Geochemical cycling of $N_2O$ as seen from the stable isotope ratio of Nitrogen". Speaker: Yoshida, Special Research Member, Bioscience Research Laboratories of Mitsubishi Chemical Co. He introduced the studies undertaken to clarify quantitatively the atmospheric $N_2O$ cycle by applying the $^{15}N/^{14}N$ ratio. He also discussed how the reduction activation of $N_2O$ in river water varies from forest land to estuary region and mentioned the experimental findings that $\delta\ ^{15}N$ in $N_2O$ existing in sea water changes rapidly by a low oxygen water.

(Chairman: Shinichi Watanabe).

1:00–2:30 p.m. Address 6: "Paleoenvironments in seas surrounding Japan". Speaker: Oba, Associated Professor, Department of Liberal Arts, Kanazawa University. He mentioned the results of recent studies conducted to find the variation in sea level from the last glacier period to the presentday and the subsequent environmental changes by analyzing the data regarding the $^{14}C$ period of core sample in the Sea of Japan and the Pacific Ocean surrounding Japan, the distribution of foraminiferal fossils, and the ratio of oxygen isotopes.

2:30–3:00 p.m. Address 7: "The sedimentation structure and bioturbation of columnar sediments at the seabed". Speaker: Yoshika, Applied Geomorphology, Department of Humanities and Science, Japan University (co-researchers: Tanaka, Haruna, and Fujioka, Institute of Oceanography, Tokyo University). He explained the structure of sediments at the seabed as seen from soft x-ray photographs. It was mentioned that quite frequently, the layers were mixed together as a result of turbidities or bioturbation.

(Chairman for 6, 7: Prof. Gamo).

3:00–3:30 p.m. Summary discussions: Views were exchanged on the manner of conducting the symposium or the selection of topics. It was decided to hold the next symposium around 1984. Annual Conference of the Geochemical Society (Nagoya University) and the symposium were declared over after it was decided to request Mr. Sakugawa of the Oceanographic Laboratory of Nagoya University to act as coordinator for the next symposium.

**Names of Participants**

| Name | Sponsoring Organization |
|---|---|
| Watanabe | Department of Fisheries, Hokkaido University |
| Yamada | Department of Fisheries, Hokkaido University |
| Suzuki | Department of Fisheries, Hokkaido University |
| Kairohara | Department of Liberal Arts, Tokyo University |
| Suzuki | Meteorological Laboratory |
| Hirose | Meteorological Laboratory |
| Yanagikawa | Bioscience Research Laboratories, Mitsubishi Chemical Co. |
| Yoshida | Bioscience Research Laboratories, Mitsubishi Chemical Co. |
| Fujiwara | Faculty of Science, Tokyo University |
| Akagi | Faculty of Science, Tokyo University |
| Hashimoto | Faculty of Science, Tokyo University |
| Tomita | Faculty of Science, Tokyo University |
| Yasumagoko | Department of Liberal Arts, Tokyo University |
| Gamo | Institute of Oceanography, Tokyo University |
| Yang | Institute of Oceanography, Tokyo University |
| Tanaka | Institute of Oceanography, Tokyo University |
| Fukushima | Faculty of Science, Tokyo Prefectural University |
| Okuma | Faculty of Science, Tokyo Institute of Technology |
| Oba | Department of Liberal Arts, Kanazawa University |
| Yoshika | Department of Humanities & Science, Japan University |
| Masuzawa | Oceanographic Laboratory, Nagoya University |
| Ikegami | Oceanographic Laboratory, Nagoya University |
| Sakugawa | Oceanographic Laboratory, Nagoya University |
| Matsueda | Oceanographic Laboratory, Nagoya University |
| Sato | Oceanographic Laboratory, Nagoya University |

| | |
|---|---|
| Ogasawara | Oceanographic Laboratory, Nagoya University |
| Ichiiro | Faculty of Science, Kyoto University |
| Tsuji | Faculty of Science, Kyoto University |
| Okumura | Faculty of Science, Shumane University |
| Okubo | Kobe Mercantile Marine College |
| Kadotani | Faculty of Agriculture, Kagawa University |
| Hayase | Faculty of General Science, Hiroshima University |
| Shimonoshima | Faculty of General Science, Hiroshima University |

(Convener and Reporter: Gamo).

## ADMINISTRATIVE REPORT

Outline of the Research Organization and Plans for Fiscal year 1983: The printing was over and copies were distributed to concerned persons.

Holding of General Symposium on "Dynamic Structure of Ocean": The general symposium under this special project was held for three days on July 11-July 13, 1983 at Aoi Kaikan (Torano-mou, Minato-ku, Tokyo). It was attended by a large number of people concerned with oceanographic studies. The proceedings of the symposium were printed and distributed to concerned persons.

### Minutes of the 19th Meeting of the Executive Committee

September 27, 1983 (Tuesday), Conference Room, Institute of Oceanography, Tokyo University.

**Items discussed**

It was decided to hold the eighth managing committee meeting on October 27 (Thursday) between 13:30-17:00 hrs at the Conference Room of the Institute of Oceanography, Tokyo University. The agenda for that meeting will include:

Preparation of final report
Future plans for study
Others.

**Preparation of final report**

It has decided to prepare a report that will consist of results of various studies under this project, background and a review of the studies, etc. The target reader is not just an expert; a more wider readership is assumed. The report will be 450 pages of B5 size.

Each planned study group will be allotted about 30 printed pages. The results of public groups not included in the planned groups will also be appended at proper places. The report will generally be divided into the following Chapters (tentative plan) and authors of each Chapter should offer their text at the time of the Managing Committee meeting.

### General Conclusion (Responsibility: Kajiura)

Chapter 1: Flow and density structure along the west coast of the

| | |
|---|---|
| | North Pacific Ocean. (Responsibility: Teramoto). |
| Chapter 2: | Sea water cycling and material distribution in the western North Pacific Ocean. (Responsibility: Teramoto and Horibe). |
| Chapter 3: | Water cycling and fluctuations in the continental shelf region. (Responsibility: Kajiura). |
| Chapter 4: | Material cycling in the ocean. (Responsibility: Kitano). |
| Chapter 5: | Structure of the ocean floor and its paleoenvironments. (Responsibility: Takayanagi). |
| Chapter 6: | Development of measuring methods to study recycling sea water and other materials. (Responsibility: Mochizuki). |

# ANNOUNCEMENT

## Announcement of Symposium on "Present and Past Ocean Environments"

This symposium will be held as stated below.

The object of the symposium is to explain the progress made in each topic to the geochemists and geologists engaged in the study of past and present ocean environnments, list the problems involved, identify the direction of future studies by first exploring the present state of ocean environmental studies and then the developments desired in future so that both geochemists and geologists can jointly attack the problems.

In addition to those who have agreed to speak at the symposium, there are many people who are closely connected with studies on the ocean environment. We invite the participation of all those people and also of other persons interested in this symposium so that there will be a lively discussion. For further information about the symposium and for suggestions regarding topics of discussion, please contact the coordinator.

We would like to thank Prof. Horibe and Prof. Teramoto, who encouraged us by stating that the symposium would be considered a part of the Special Research Project, "The Ocean Characteristics and Their Changes."

Time: December 9, 1983 (Friday), 10:00-17:00. December 10, 1983 (Saturday), 09:30-16:00

Place: Assembly Hall of the Institute of Oceanography, Tokyo University.

**Program**

December 9 (Friday), 10:00-11:45

1. Ocean environments, using chemical components as tracer—Gamo (Institute of Oceanography, Tokyo University).
2. Absorption by the ocean of the carbon dioxide developed as a result of human activity—Ikegami (Oceanographic Laboratory, Nagoya University).
3. Cycling of carbon in the ocean—Nozaki (Institute of Oceanography, Tokyo University).

13:00-14:45

4. Basic ocean productivity—Ishimaru (Institute of Oceanography, Tokyo University).

5. Nitrogen and carbon isotope separation by marine organisms and its meaning in terms of marine ecology—Wada (Mitsubishi Bioscience Research Laboratory).

6. Vertical transport of organic materials in the ocean—Ota and Handa (Oceanographic Laboratory, Nagoya University).

15:15-17:00

7. Habitat segregation of carboniferous nanoplankton in the present ocean and its effect on the recovery of the paleoenvironments of the ocean—Okada (Faculty of Science, Yamagata University).

8. Analysis of past ocean environments using coral reef skeletal chronology—Onishi (Faculty of Science, Kanazawa University).

9. Small ice age records left on the icebed—Kato (Oceanographic Laboratory, Nagoya University).

December 10 (Saturday), 09:30-11:50

10. CCD variations over the last 30,000 years in the southern Sea of Japan—Kitaumi (Faculty of Science, Shizuoka University).

11. Geochemistry of sediments and intermediate waters—Transition and diagenesis of elements—Kato (Institute of Oceanography, Tokyo University).

12. Paleoenvironments of the Sea of Japan and variations in the oxidation reduction cycle—Masuzawa (Oceanographic Laboratory, Nagoya University).

13. Changes in ocean conditions after the last glacier period in the Pacific Ocean east of Honshu—Chinze (Faculty of Science, Tokyo University).

13:00-14:45

14. Studies on the ocean environments during the last glacier period—Oba (Department of Liberal Arts, Kanazawa University).

15. Reversal of the earth's magnetic field and variation in biomass—Niizuna (Faculty of Science, Shizuoka University).

16. Analysis of paleoenvironments from the Late Miocene period to the Pliocene period based on deep seabed core samples—Saito (Faculty of Science, Yamagata University).

15:00-16:00

General discussion.

Coordinators:

1) Oba (Department of Liberal Arts, Kanazawa University) 1-1, Marunouchi, Kanazawa City, Ishikawa Prefecture (Tel.: 0762-62-4281. Ext. 655).

2) Nozaki (Inorganic Chemistry Department, Institute of Oceanography,

Tokyo University), 1-15-1, Minami-dai, Nakano-ku, Tokyo. (Tel.: 03-376-1251, Ext. 272).

### Publications Concerned with this Special Project

The following data book—"Distribution of Trace Elements in Marine Algae: Comparative Biogeochemical Data"—was printed in connection with this Special Research Project (Material elimination process in the ocean—group). Please send your request for copies to the following address:

Mr. Yamamoto (Department of Chemistry, Kyoto College of Education), Kyoto City, Fukumi-ku, Sokafuji, Morimachi.

Also some copies of the following publications are still available. Interested persons can send their requests so that they can be mailed along with the next issue of the Newsletter.

Newsletter (each issue except No. 1)
Outline of Research Organization and Plan for Fiscal 1981
Research Report for Fiscal 1981
Outline of Research Organization and Plan for Fiscal 1982
Research Report for Fiscal 1982 (1)
Outline of Research Organization and Plan for Fiscal 1983
Proceedings of Annual General Symposium 1983.

## Guidelines for Authors

There are only two more issues of the Newsletter to be published. Therefore, all are requested to report the activities undertaken under this Special Research Project or the results of studies as highlights or a research report. The proposals for submission should be sent to the administrative office at the earliest.

The contents are generally classified as:

| | |
|---|---|
| Highlights: | Quick report of the results of this Special Research Project |
| Reports: | Reports on the activities of each section or working group |
| Announcements: | Announcements regarding the proposed meetings or conferences of each group |
| Bibliography: | Listing of papers published by the members on this Special Research Project or other reports. |

Manuscripts should be written on standard 400 character size paper (handwritten material also accepted). Figures should be drawn in such a way that they can be printed as they are.

Newsletter issue No. 16 will be published in January 1984 and issue No. 17 will be published in March, 1984.

**Note:**

When reproducing or citing material published in this Newsletter, due acknowledgment should be given. Authors' names, wherever mentioned, should also be quoted. A copy of the publication containing such a reproduction may be sent to the Editorial Section of the Administrative Office.

| | |
|---|---|
| Published by: | Administrative Office, Special Research Project "The Ocean Characteristics and Their Changes" |
| Published at: | Administrative Office,<br>Physical Oceanography Department,<br>Institute of Oceanography,<br>Tokyo University,<br>1-15-1 Minami-dai, Nakano-ku,<br>Tokyo 164<br>Tel.: 03-376-1251, Ext. 251. |
| Editorial staff: | Taeko Kitano and Noriyo Kimura. |

Special Research Project

# The Ocean Characteristics and Their Changes

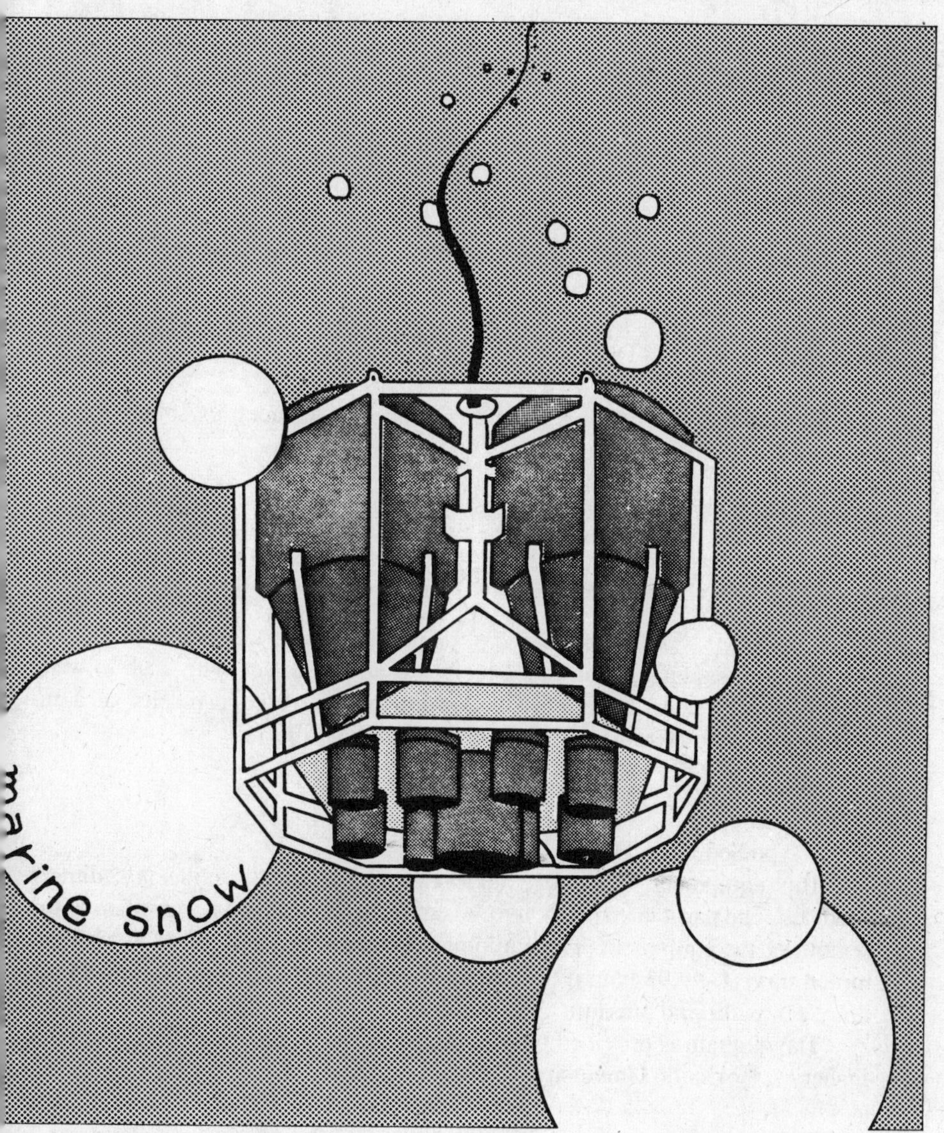


NEWSLETTER NO. 16, JANUARY, 1984

# Contents



Researchers engaged in this Special Research Project, who wish to write their research papers in English and want to acknowledge receipt of a research grant, should indicate the above title of the project.

Cover Page

D type Sediment Trap:

This equipment is installed under sea and it directly collects sediment particles. The particles enter in two 40 cm dia cylinders and accumulate at the bottom of the equipment. The equipment is turned through 36° at a predetermined time (1-99:99 hours) four times, thereby collecting eight samples. The drive can withstand pressure up to 6,000 m.

The diagram is provided by Mr. Taguchi and Handa of the Department of Fisheries, Hokkaido University.

HIGHLIGHTS

# Studies Regarding the Elimination Process of Materials in Seawater, Using a Sediment Trap

*Shizuo Tsunogai*

(Department of Fisheries, Hokkaido University)

(Received: November 14, 1983)

## 1. Origin of Sediment Trap Experiment

In October, 1975, I was received at Boston airport by Dr. D.W. Spencer and Dr. S. Honjo of Woods Hole Oceanographic Institution and taken to the residence of Dr. Honjo. There I was asked about sediment trap experiments in Japan (results of the Nishizawa group of Tohoku University and our group). There we discussed until late night plans for conducting actual sediment trap experiments in the outer seas. It was decided to carry out these studies under grants from the NSF jointly with Dr. Honjo, Dr. Spencer, and Dr. Brewer. In July, 1976, tests were conducted for the prototype trap which was then installed in the Sargasso Sea in the Atlantic Ocean in October. This was the beginning of an interesting experiment on the American side. However, when I returned to Japan in September, the data was not sent to me. At that point I decided to install the sediment trap equipment in the northern part of the Pacific Ocean with the help of the *Hakucho-maru* in August, 1978, although no sufficient funds were available. We were successful in collecting a few samples. This was how we got involved in sediment trap experiments.

Dr. Honjo has observed that the structure of the calcareous plate (coccolith) of a plankton containing carbonaceous material is quite similar to the layer of their residence or the one immediately below. If the coccolithophorid of a few micron size is not to be washed away by horizontal sea currents and if it is to settle at the seabed from the seawater not saturated with $CaCO_3$, it is necessary that they descend (vertically) at least 50–60 m per day. Here we assumed that the transport is provided by plankton excretions (fecal pellets) and the first object of our study was to prove that point.

In our laboratory, we have studied the mechanism that eliminates the chemical components from seawater by analyzing the variation in ratios of radioactive nuclides with a long life, that have been dissolved in seawater (U-238, U-234, Ra-226, etc.) and daughter nuclides arising from them, which are not easily soluble (Th-234, Th-230, Pb-210, etc.). The American chemical oceanographers (Dr. Craig, Broecker, and others) have proposed a scavenging model wherein these nonsoluble components are eliminated from seawater in a nonreversing manner while they form a stable solid phase.

As against this, our results indicated that the materials eliminated from the top layer have to be regenerated after they precipitate in the lower layer. However, since the pH value or oxidation-reduction conduction of normal seawater containing oxygen does not vary considerably at the top or bottom (upper or lower layer), we cannot assume that the solid phase formed, which is thermodynamically stable, would easily dissolve (in the lower layers). Therefore we considered that "regeneration" would mean that either the particles just become smaller and the sedimentation rate becomes low enough to be ignored or forms a big particle after numerous small particles with a considerable sedimentation rate cluster together. Based on these considerations, a settling model was proposed where these two processes alternate (Tsunogai and Minagawa, 1978). Taking into account the distribution of radioactive isotopes, we estimated that the nonsoluble components generally precipitate at a rate of 50–100 m per year. This value means that if only 1% of the materials existing in seawater are assumed to precipitate and the remaining 99% are assumed to remain in seawater, then the average sedimentation rate of precipitating particles would be 5,000–10,000 m per year. If we take into consideration that the mean velocity of tourists (chemical component) toward a target site, as they alight and get into the train, and the mean velocity of the train (the transport mechanism) are different then we can explain the above phenomena. We feel that sea ice (large suspended particles) plays a better side role as a transport medium than the excretory particles (of plankton). This is because, excretory particles are mostly formed at the surface level and therefore it is difficult to explain the removal of material from the intermediate- or deep-layer.

This is one of the main reasons we felt sediment trap experiments were necessary to arrive at the solution of the above problem; the samples collected in the trap can also be used for various other studies. In the future, the samples collection by the trap should become as ordinary as collecting water with a water sampler or collecting seabed samples with a mud collector. Hence, in our study group, the sediment trap was developed after taking into account, the future use aspect. Using this equipment, dynamic aspects of materials in the ocean were studied in terms of recycling materials and the mechanism of removing chemical components in the ocean.

## 2. Sediment Trap and Mooring System

These days there are different types of sediment traps being used for studies, viz., the H type which we improved to the NH type, the D type which was used in this special project, and the NU type which is used along the seashore. Since they are explained in detail elsewhere (e.g., Tsunogai, 1981), we shall only mention salient points here.

The NH type sediment trap has grouped together six PVC tubes 25 cm in dia and 60 cm long. The bottom of the tube is throttled where the sample is collected. At the time of recovery, the lid is closed and the equipment is floated. Since there are six tubes, it is easy to measure the variation in particle flux or study antiseptic (preservative) materials, distribution of sample, etc. If the sea has a high productivity, then one tube can collect a few hundred mg dry weight of chemical elements in one month.

The D type sediment trap (time-plan method) has one cylindrical tube 40 cm in diameter and 110 cm long as well as two funnel-shaped samplers. The receiver at the lower part of the equipment rotates at a predetermined interval and by dividing the total period into four sections, it can collect eight samples at a time. It can be installed for maximum period of one year at a depth of 6,000 m.

The NH type sediment trap (for seashore) has grouped together six tubes with a diameter of 6 cm and 60 long and the construction is such that it can be recovered by a simple lid. It is light in weight and can be carried even in a small ship and as such is useful for a seashore area where particle flux is large.

*Mooring systems in the outer seas.* This is almost similar to the mooring systems used in deep layer flow measurements. Tsunogai and Harada (1983) have mentioned one such example. This is suspended within the ocean and at the time of recovery is floated on the surface by activating the acoustic release equipment. Since it accumulates much heavier materials than in flow measuring experiments, the size of rope or number of floats etc. are slightly larger.

## 3. Discussions on the Results Obtained in Sediment Trap Experiments

Results obtained during this special study conducted in the eastern Pacific region, in addition to the results of studies conducted in the northern North Pacific region (Tsunogai, et al., 1982) or Atlantic Ocean (Brewer, et al., 1980) or along the seashore (Tsunogai, et al., 1980) made the following aspects about the material cycling in ocean water quite clear. We will discuss those aspects while arriving at some conclusions.

*(1) The particle flux varies by more than one digit even in outer seas, depending on the area*

The particle flux (on a dry weight basis) in deep sea waters is shown in

Table 1. It can be seen that it varies considerably depending on the area of observation. The northern North Pacific region and Panama Basin are regions with large particle flux. In either case, as a result of mixing the upper and lower layers, the nutrient salts of the lower layers are brought to the upper layers and the surface layer productivity is very high in this region. As against this, slight particle flux is observed in less nutritive seas of subtropical regions (including tropical regions but excluding the equatorial area) which are separated from the continent, such as the seas around Hawaii. Thus, it can be concluded that the amount of particles being sedimented in the deep sea layer reflects the bio-activity at the surface.

**Table 1. Mean value of particle flux in deep sea water (0.7–4.5 km deep)**

| Region (symbol of measuring point)* | Location | Total particle flux $g/m^2/yr$ |
|---|---|---|
| Northern North Pacific (KH-78) | 47.9°N 176.3°E | 67.1 |
| Panama Basin (PB) | 5.4°N 81.9°W | 51.8 |
| Eastern Pacific** (St. 5) | 37.0°N 127.6°W | 23.0 |
| Tropical Atlantic (E) | 13.5°N 54.0°W | 17.4 |
| Eastern Pacific** (St. 11) | 17.5°N 117.0°W | 8.4 |
| North Atlantic (S) | 31.5°N 55.9°W | 7.0 |
| Eastern Pacific ** (St. 7) | 31.7°N 124.6°W | 5.8 |
| Pacific (Hawaii) (P) | 15.4°N 151.5°W | 5.0 |
| Pacific (Hawaii)** (St. 27) | 26.8°N 146.8°W | 4.8 |

*The same symbols and sequence are followed in subsequent tables.

**Measured values during this special study.

*(2) In spite of the fact that the total particle flux varies considerably, the concentration of heavy metals or radionuclides born and eliminated in seawater does not vary so much*

As we can see from Table 2, the concentration of Copper or Th-234 in sedimented particles is not related to the total particle flux of Table 1. Th-234 is produced from U-238 which is dissolved uniformly throughout the seawater and, therefore, the supply of Th-234 is the same in all waters. Accordingly, the above figures indicate that there is no dilution effect where the concentration of metallic elements decreases as the particle flux increases but instead, with the increase in particle flux, the removal of chemical elements proportionally increases. Thus the ocean seems to be clearly divided in two regions: one, chemical elements are positively removed as a result of the bio-generation activity at the surface and another region which does not respond as well.

**Table 2. Mean concentration of Copper and Th-234 in sediment particles (mean value of all measurements at each measuring point)**

| Measuring point | Cu, µg/g | Th-234, $10^3$dpm/g |
|---|---|---|
| KH-78 | 163 | 3.8 |
| E | 79 | 7.3 |
| St. 11 | 293 | 5.1 |
| S | 238 | 3.6 |
| St. 7 | 97 | 10.2 |

*(3) Measured value of sedimentary particle flux for nonsoluble radionuclides such as Pb-210 or Th-230 is smaller in subtropical regions when compared with the quantity that precipitates from the upper layer, but it is higher in sub-arctic zones*

Particle flux of these nuclides can be directly calculated from the non-equilibrium quantity of radioactivity when both types of parent nuclides and thin concentration in seawater are measured. (In case of Pb-210, the quantity coming from the atmosphere is also added). The value of sedimentary particle flux of Pb-210 obtained in this manner happens to be (2–3) $\times 10^3$ dpm/m$^3$/yr in the North Pacific region. However, the measured values of Table 3 show that they vary 1/3–3 times that central value, and it is higher in places where the total particle flux is higher. Thus, combining (1) and (2) means that chemical elements, as a result of ocean currents, may be driven to the areas which are active in biogeneration where they are removed in the form of particles (i.e., they precipitate below). Such a flow of material (cycling) exists both at the surface and in the deep layers.

**Table 3. Sedimentary particle flux of Pb-210 (mean value in deep waters between 0.7–4.5 km area)**

| Measuring point | Particle flux, $10^3$dpm/m$^2$/yr |
|---|---|
| KH-78 | 7.4 |
| E | 5.0 |
| St. 11 | 1.1 |
| S | 2.4 |
| St. 7 | 1.0 |

*(4) There is not much difference in the concentration of Th-234, which has a half life of just 24 days, in suspended and sedimentary particles*

There are not many examples where concentrations of Th-234 (dpm/g) in suspended particles have been measured by filtering the seawater (the measured values reported for Th-234 concentrations in a suspended state are in terms of dpm/l of seawater). When measured at a depth of 100 m, we found it to be $2.2 \times 10^3$ dpm/g (St. 27) and $5.4 \times 10^3$ dpm/g (St. 11). Also, at the surface layer in the Atlantic, a value of $(1.4–39.5) \times 10^3$ dpm/g has been reported (Krishnaswami, et al., 1976), but it is per unit weight of ash. This value also coincides with the fluctuations in particle concentration of Table 2.

Born from U-238, Th-234 is in atomic or ionic condition in the beginning and adheres to particles in seawater (mostly of biological source) through some action. At first the concentration of Th-234, in smaller particles (but with larger specific surface area) can be expected to be higher. Generally, since the suspended particles are smaller than those which precipitate rapidly, the concentration of Th-234 in suspended particles should be higher. However, the concentration of Th-234 in sedimentary particles is also higher. We can therefore assume that not only sedimentary particles are formed from suspended particles but the transformation between these two is quite fast (considering the 24-day time scale).

*(5) Aluminum particle flux increases with depth*

Aluminum is supposed to represent the soil particles. As shown in Table 4, at all measuring points, the aluminum sedimentary particle flux increases with the depth of water. This point has been further discussed by Tsunogai, et al. (1982) and can be explained as follows. Small soil particles, when clustered together to form big particles like marine snow, precipitate easily but, since their density is not very different from seawater, they follow the (up and down) wave motion of the sea currents (it is not scattering) and when they go down they enter the trap. Such mixing is assumed to become more prominent as they approach the seabed. However, since the above effect acts in conjunction with the effect mentioned in (3), it may not be as large as thought by Tsunogai, et al. (1982). Also as we go deeper, the ocean area becomes smaller that is why the sedimentary particles may cluster together.

*(6) There is considerable difference in the composition of sedimentary particles and suspended particles as far as the main component is concerned*

Table 5 shows estimated values of particle composition of sedimentary particles classified according to their sources and based on their chemical analysis. It can be seen from this table, that sedimentary particles are rich in calcium carbonate (foraminifera) or silicates (diatom) which are formed by bioorganisms. As against this, organic material contents are extremely higher

**Table 4. Variation of aluminum particle flux with depth of water (indicated in brackets)**

| Measuring point | Aluminum particle flux, mg/m$^2$/yr (water depth, km) | | | |
|---|---|---|---|---|
| KH-78 | 52 ? (0.11) | 129 (1.04) | 164 (2.16) | 308 (4.38) |
| E | 143 (0.39) | 211 (0.99) | 565 (3.76) | 658 (5.09) |
| St. 11 | 48 (0.47) | 49 (0.69) | 80 (1.23) | 87 (3.35) |
| S | — | 22 (0.98) | — | 162 (3.69) |
| St. 7 | 33 (0.49) | 37 (0.73) | 69 (1.26) | 74 (3.39) |

in suspended particles. This is probably because carbonates or silicates formed by bioorganisms are large, high in density, and therefore precipitate rapidly (young); whereas organic particles excreted by them are light and cannot easily precipitate (old). Now the problem is to determine what percentage of organic materials in the sedimentary particles comes from young particles and what percentage comes from old particles via suspended particles.

**Table 5. Structural composition of sedimentary particles (mean values observed between the water depth of 0.7–4.5 km) and suspended particles**

| | Terrigenous | Biogenous | | |
|---|---|---|---|---|
| | Soil, % | Organic substances, % | Carbonates, % | Opal, % |
| KH-78 | 5.2 | 15.8 | 19.3 | 57.5 |
| E | 34.8 | 11.8 | 46.9 | 7.0 |
| St. 11 | 13.2 | 29.4 | 36.4 | 12.0 |
| S | 17.6 | 14.9 | 64.1 | 3.3 |
| St. 7 | 18.6 | 23.6 | 41.5 | 14.8 |
| Suspended particles: Surface waters in the North Pacific Ocean | 10.5 | 72.4 | 5.5 | 11.6 |
| Sediments: KH-78 surface layer | 75.4 | 5.0 | 3.0 | 14.8 |

Compared with others, the total particle flux at KH-78 was very large and this may be mainly because it is rich in silicates (opal) formed by bioorganisms.

*(7) Concentration of aluminum is higher in sedimentary particles along the seashore, whereas it is higher in suspended particles near the surface layers of the outer seas*

As can be seen from Table 6, the aluminum concentration or the soil content in sedimentary particles along the seashore is higher than that in suspended particles. This may be because the soil particles at the shore area have just entered from rivers, etc. and they precipitate at comparatively higher speeds. It would mean that these particles are similar to the biogenous silicate or carbonate particles of the outer seas mentioned in (6) above. As against this, the soil particles in the outer seas are mostly supplied by the atmosphere and are a few μm in size, which can therefore be considered to be similar to the "old" organic particles mentioned in (6). Thus if we assume that marine snow in the outer seas is formed as a result of condensation of old organic particles, this process can be considered to take place while removing small soil particles from the seawater without much contradiction. Also this condensed marine snow can be considered to accelerate the sedimentation of large "young" particles.

Now the process described here may appear to be in contradiction with the process described in (4), but since we do not really have a detailed explanation of the extent of fluctuations or the types of particles that can easily adhere to Th-234, it is difficult to arrive at any conclusion at this point.

**Table 6. Concentration of aluminum in particles near the surface layers of the outer seas and along the seashore**

| | Sedimentary particles, % | Suspended particles, % |
|---|---|---|
| Outer seas | | |
| KH-78 (around 100 m) | 0.26 | 1.2 |
| St. 11 (around 500 m) | 0.52 | 1.2 |
| St. 7 (around 500 m) | 0.48 | 3.0 |
| Funka Bay (92 m deep) | | |
| Summer (around 40 m) | 2.0 | 0.66 |
| Winter (around 40 m) | 3.8 | 1.2 |

*(8) Cadmium particle flux reduces considerably at a depth of 1,000 m*

Bruland (1980) and subsequent researchers have ascertained that the concentration of heavy metals in the Pacific Ocean increases with depth in the same way as nutrient salts. Particularly, cadmium increases rapidly with depth, same as the phosphates, and shows extremely high concentrations around 1,000 m. This is much different than the behavior of copper or nickel which have distributions similar to silicates. It is quite evident from Table 7 that this is a result of regeneration from the particles which have been transported from the upper layer in the same way as phosphates. This is because the concentration of cadmium in sedimentary particles goes down rapidly around 1,000 m. The fact that cadmium can regenerate easily has also been ascertained at the seashore (Noriki, et al., under preparation).

**Table 7. Concentration of cadmium and copper in sedimentary particles (figures in bracket indicate depth)**

| | KH-78 | St. 11 | St. 7 |
|---|---|---|---|
| Cadmium concentration, μg/g (depth of water, km) | 2.95 (0.11)<br>0.61 (1.04)<br>0.28 (2.16) | 4.55 (0.47)<br>0.66 (0.69)<br>0.52 (1.23) | 1.45 (0.49)<br>0.36 (0.73)<br>0.41 (1.26) |
| Copper concentration, μg/g (depth of water, km) | 144 (0.11)<br>64 (1.04)<br>176 (2.16) | 170 (0.47)<br>460 (0.69)<br>184 (1.23) | 45 (0.49)<br>66 (0.73)<br>125 (1.26) |

*(9) Manganese particle flux and its concentrations in sedimentary particles increase considerably with depth*

Concentrations of inorganic elements such as aluminum in sedimentary particles generally increases with depth since organic parts dišintegrate in this region. However, the increase in manganese concentration is much more than that (Table 8). This certifies the facts that manganese from the seabed circulates between deep layer waters and the seabed (Tsunogai and Kusakabe, 1982) and also the sedimentary particles precipitate jointly with the particles in deep layer waters (Tsunogai and Minagawa, 1978).

**Table 8. Manganese concentration in sedimentary particles (figures in bracket indicate depth of water)**

| Measuring point | KH-78 | E | S |
|---|---|---|---|
| Concentration, μg/g (depth of water, km) | 33 (0.11) | 42 (0.39) | |
| | 67(1.04) | 72 (0.99) | 54(0.98) |
| | 158 (2.16) | — | |
| | 397 (4.38) | 322 (3.76) | 768 (3.69) |
| | 1850 (5.25) | 462 (5.09) | |

## 4. Conclusion

During the IUGG-sponsored symposium on "Chemical Fluxes in the Water Coloumn" held in August, 1983 at Hamburg, the importance of marine snow or fluff was recognized and many of the above conclusions should be accepted. However, there is also an element of uncertainty in the above conclusions and particularly there are many shortfalls in the quantitative understanding of material cycling. In order to arrive at satisfactory conclusions in this respect, it is necessary to install time based sediment traps in four regions, the subarctic, subtropic, equatorial, and trench regions of the Pacific Ocean at a minimum of five layers and then measure particle fluxes every month over a period of two years.

## 5. Appendix

The unpublished measured values mentioned here have been obtained by Mr. Noriki, Harada, Ishimori, and others of the Analytical Chemistry Wing of the Department of Fisheries, Hokkaido University.

## Bibliography


Brewer, P.G., Y. Nozaki, D.W. Spencer and A.P. Fleer. 1980. Sediment trap experiments in the deep north Atlantic: isotopic and elemental fluxes. *J. Mar. Res.*, 38, 703–728.

Bruland, K.W. 1980. Oceanographic distributions of cadmium, zinc, nickel, and copper in the North Pacific. *Earth Planet. Sci. Lett.*, 47, 176–198.

Krishnaswami, S., D. Lal and B.L.K. Somayajulu. 1976. Investigations of gram quantities of Atlantic and Pacific surface particulates. *Earth Planet Sci. Lett.*, 32, 403–419.

Tsunogai, S. 1981. Sediment traps and mechanisms of particle sedimentation. Sedimentation-precipitation process in a bay-coastal area. (Japan Marine Resources Preservation Society), pp. 7–26.

Tsunogai, S. and K. Harada. 1983. Local and vertical variations in sedimentary particles in the Pacific Ocean and the process of material removal. Proceedings of General Symposium under Mombusho Special Project "Ocean Characteristics and Their Changes", pp. 43–46.

Tsunogai, S. and M. Kusakabe. 1982. Migration of manganese in deep sea sediments. The dynamic environments of the ocean floor (D.C. Heath). pp. 257–273.

Tsunogai, S. and M. Minagawa. 1978. Settling model for the removal of insoluble chemical elements in seawater. *Geochem. J.*, 12, 47–56.

Tsunogai, S., M. Uematsu, N. Tanaka, K. Harada, E. Tanoue and N. Handa. 1980. A sediment trap experiment in Funka Bay, Japan: "Upward flux" of particulate matter in seawater. *Mar. Chem.*, 9, 321–334.

Tsunogai, S., M. Uematsu, S. Noriki, N. Tanaka and M. Yamada. 1982. Sediment trap experiment in the northern North Pacific: Undulation of settling particles. *Geochem. J.*, 16, 129–147.

# Proceedings of the Symposium on Ocean Environment: Present and Past

The above symposium was held under the joint auspices of the Geochemists Society and the Geological Society on the two days of December 9, 10, 1983 at the Assembly Hall of the Institute of Oceanography, Tokyo University. A large number of researchers, including those involved in this special project, participated in the symposium and an active discussion took place in continuation with 16 lecturers. The total number of participants was 50 and the program was as follows.

**Program**

December 9, 1983, 10:00-17:00

Welcome address—Oba (Department of Liberal Arts, Ozawa University).

1) Cycling in the ocean using chemical elements as tracers—Gamo (Institute of Oceanography, Tokyo University)

2) Absorption of anathropogenic carbon dioxide by ocean—Ikegemi (Oceanographic Laboratory, Nagoya University)

3) Carbon cycling in the ocean—Nozaki (Institute of Oceanography, Tokyo University)

4) Basic productivity of ocean—Ishimaru (Institute of Oceanography, Tokyo University)

5) Vertical transport of organic substances in the ocean—Ota and Handa (Oceanographic Laboratory, Nagoya University)

6) Habitat segregation of carbonaceous nanoplankton in the present ocean and its application to estimate old ocean environments—Okada (Faculty of Science, Yamagata University)

7) Rate of reduction of sulfuric acid in the warm water system of the ocean floor and the isotope exchange reaction between sulfuric acid—$H_2S$—Sakai (Institute of Oceanography, Tokyo University)

This lecture was supposed to be delivered by Prof. Onishi of the Faculty of Science, Kanazawa University but since he was recalled, Prof. Sakai obliged with short notice.

8) Records of the small ice age remaining on the ice floor—Kato (Oceanographic Laboratory, Nagoya University)

9) Segregation of nitrogen-carbon isotopes by bio-organisms and its significance in marine life—Wada (Bioscience Research Laboratory, Mitsubishi Chemical Co.)

December 10, 1983 (Saturday), 09:30-16:20

10) CCD variations in the southern part of the Sea of Japan in the last 30,000 years—Kitaumi (Faculty of Science, Shizuoka University)

11) Geochemistry of the intermediate waters in the sediments—transition of elements and the succession process—Kato (Institute of Oceanography, Tokyo University)

12) Old environments in the Sea of Japan and changes in oxidation reduction environments—Masuzawa (Oceanographic Laboratory, Nagoya University)

13) Changes in the ocean after the last ice age—in the Pacific Ocean on the eastern side of Honshu—Shizunishi (Faculty of Science, Tokyo University)

14) Discussions on ocean environments during the last ice age—Oba (Department of Liberal Arts, Kanazawa University)

15) Reversal of the earth's magnetic field and variations in biomass—Niizuna (Faculty of Science, Shizuoka University)

16) Analysis of old environments between the Late Miocene period and the Pliocene period, using the deep seabed core—Saito (Faculty of Science, Yamagata University).

**General discussions**

Chairman: Saito (Faculty of Science, Yamagata University)

There were nine lectures on the first day mainly regarding the present ocean environments. Each lecturer introduced the general features of the studies conducted with reference to his research topic and added his original data. Those lectures were quite interesting and easy to understand even for participants specializing in geology. An outline of each lecture is mentioned below.

1. Gamo: He discussed the movement of water just above the seabed and the speed of vertical mixing in the western North Pacific Ocean region using $^{222}$Rn coming from sedimentary materials at the sea bottom. The vertical vortex diffusion coefficient calculated from the vertical distribution of $^{222}$Rn at 40 points, indicated a wide range of variation between 1-300 cm$^2$/S and it became clear that a benthic boundary layer of 10-100 m thickness existed at a majority of the measuring points. It was also indicated with actual measured values that water movement in a horizontal direction following the localized topography of the seabed plays an important role in the formation of a benthic boundary layer.

2. Ikegami: If, from the total carbon dioxide existing in the deep layer of seawater $\sum CO_2$ (partial pressure of carbon dioxide, $PCO_2$), we subtract the changes due to the disintegration of organic matter and the dissolution of calcium carbonate, then it is possible to determine $\sum CO_2^\circ$ ($PCO_2^\circ$) when the same water existed at the sea surface. Thus by knowing the vertical distribution of $\sum CO_2^\circ$ ($PCO_2^\circ$) we can estimate what part of $CO_2$ released in the atmosphere by human activities would be absorbed by seawater. When the distribution of $PCO_2^\circ$ in all oceans was studied, using GEOSECS data, it was found that the anthropogenic carbon dioxide penetrates to a depth of 1,000 m in all oceans. This corresponds to the 40–70% of $CO_2$ quantity coming from fossil fuel sources. Now, of the total carbonic acid data used from GEOSECS, the data regarding alkalinity is supposed to have some statistical errors. The absolute value of $PCO_2^\circ$ before the industrial revolution was found to be around 250 ppm during our studies. In a model based on the analysis of air bubbles in the water bed core or a variation of $^{13}C$ in annular rings, the same value is said to be 240–270 ppm. Considering these facts, it seems necessary to modify the pre-industrial revolution value of $PCO_2 = 290$ ppm being assumed presently.

3. Nozaki: These days the project "Transient tracers in the ocean" is in progress in nations led by the United States. The final object of the project is to determine more accurately the $CO_2$ absorption by the sea. The project is trying to determine the three dimensional distribution of $^3H$, $^3He$, $^{14}C$, Freons, $^{85}Kr$, $^{39}Ar$, $^{222}Rn$, $PCO_2$, Alk, $\sum CO_2$, etc. in all oceans and so far the measurements in equatorial and northern Pacific are complete. The speaker compared part of these results with those obtained in the GEOSECS experiment carried out ten years ago and at the same time explained the proposed plans for studying the south Pacific Ocean during 1984–1989.

4. Ishimaru: Photosynthesis of phytoplankton in the ocean is regulated in addition to the actual photo quantum by the concentration of nitrogen and phosphorus. Phytoplankton is preyed upon by animal plankton and part of the N or P assimilated by plants is lost from the photosensitive layer in the form of excretion. For this reason it is assumed that the productivity of the ocean can be calculated from the quantities of N and P being supplied to the photosensitive layer as a result of rise or mixing, the vertical diffusion, or the transition currents in a horizontal direction. Among the basic produces, the presence of micron sized planktons such as nanoplankton (2–20 μ) or ultraplankton (below 2 μ) has attracted attention recently.

5. Ota: The organic substances produced by phytoplankton at the ocean surface form large particles in conjunction with animal plankton and are removed by precipitation. When we studied the relation between the variation in organic material flux as accumulated in a sediment trap and the consumption of dissolved oxygen, in addition to studies of composition of organic

materials and microscopic observations, it was revealed that these rapidly precipitating particles provide an important means of transport for organic materials in the sea.

6. Okada: The speaker introduced various examples of determining the origin of old environments using a limited distribution of living organisms in carbonaceous nanoplankton or substances with different forms of nanori's (those which are termed nanoplankton fossils, such as cocoris or discoaster, etc.) depending on the factors responsible for those environments and discussed the ways in which it can be used. In the case of the nanoflora of *F. profunda*, the increase in relative ratio indicates an increase in water depth whereas the same for *G. oceanica* indicates a transition to environments such as a reduction in water depth or closed environments. Also the increase in *C. pelagicus* is supposed to indicate the prevalence of cold water.

7. Sakai: In the warm water system at the seabed, the sulfuric acid in seawater is reduced as a result of the reaction of seawater with basalt or hyper basic rocks and only sulfates remain in the system. This speed of reduction and its relation with the speed of isotope exchange between sulfuric acid—$H_2S$ was measured experimentally. The reduction speed of sulfuric acid strongly depends on pH in the same way as the speed of isotope exchange but the results from our experiments revealed that, over a wider range, the latter is much higher than the former. Accordingly, it is expected that if $H_2S$ is formed from $SO_4$ in the warm water system at the seabed, then equilibrium could be achieved for mutual isotope exchange.

8. Kato: With reference to the boring core obtained from the South Pole or the Greenland ice sheets, we tried to read the records of the small ice age, using oxygen isotope composition as an index of temperature and cosmic generic nuclides as an index of solar activity. After ascertaining that proper mutual correlation exists between the temperature fluctuation of Showa base and the fluctuation in oxygen isotope composition in a core obtained from Showa base, we could find records of temperature drops corresponding to the small ice age in a Greenland core which is supposed to be 8,000 years old. Now, the small ice age is supposed to be a low activity period for solar activity known as maunder-minimum, but in the Greenland ice sheet core, we obtained contradictory records.

9. Wada: 1) Summarized the factors affecting $\delta\,^{15}N$, $\delta\,^{13}C$ of phytoplankton in the sea

$\delta\,^{15}N$: Chemical form of inorganic nitrogen forming the substrate ($NO_3^-$, $NH_4^+$, $N_2$)

$\delta\,^{13}C$: Bio-velocity.

2) By measuring $\delta\,^{15}N$ and $\delta\,^{13}C$ in bioorganisms and organic material, it is possible to analyze the following phenomena:

i) contribution of continental organic material in sediments along the shore,

ii) analysis of food cycle and feeding habits.

3) He also reported on living organisms, indicating maximum and minimum values of $\delta^{15}N$ and $\delta^{13}C$.

On the second day there were seven lectures about the recovery of ocean paleoenvironments from the last 50-60 thousand years to the last few hundred years, using a seabed core as a sample. A particularly active question-answer session took place when the same sample was analyzed by geologists and geochemists. The summary of those lectures is given in the following part.

10. Kitaumi: Core samples were collected from the Sea of Japan at different depths and the level-wise distribution of carbonaceous fossils (foraminifera) was studied to determine the CCD variation in the Sea of Japan in the past. As a result, we noticed, there were four major changes in CCD over the last 30,000 years.

| | |
|---|---|
| 30,000-20,000 years | deeper than 2,500 m |
| 20,000-10,000 years | 1,500 m |
| 10,000-6,000 years | around 700 m |
| 6,000-present | 2,000 m |

This CCD variation is in good agreement with the historical transformation in the Sea of Japan as concluded by Oba (1983).

11. Kato: The organic materials generated at the surface of the ocean subsequently precipitate in sediments. Such organic materials degenerate microbiologically. At this stage, the chemical elements are either flowing into the intermediate waters or they are transformed into a solid phase. The behavior of that part which dissolved in the intermediate water then becomes a good index of such continuous operation. At this point we used samples of sediments from the Philippine Sea, Okinawa Basin, and the Sea of Japan and studied the behavior of the distribution of all carbonates, sulfate ions, phosphates, ammonia, manganese, calcium, etc. in the intermediate waters. We also estimated the mass transport of carbonates from sediments to seawater.

12. Masuzawa: Deep-layer core samples were collected from the Sea of Japan and the intermediate water as well as sediments were chemically analyzed to see whether the lower-layer water contained any hydrogen sulphide during the last ice age. The results revealed that the core from the south-west Sea of Japan with solidified sulfur compounds or sulfur materials is formed as a continuous process. In the core from the Japan Basin, where no reduction of sulphates occurs, a large quantity of phyrite sulphur is observed in layers corresponding to the last ice age. This shows the existence of low-layer water containing hydrogen sulfide.

13. Chinzei: In a joint research by a team of ten scientists, surface layer sedimentary cores were obtained from three seas, Kagoshima, Abashiri, and

Enshu and details of planktonic fossils (such as planktonic foraminifera, radiolaria, seaweeds, carbonaceous nanoplankton, etc.) as well as the composition of benthic foraminifera fossils were studied. The ratio of warm currents and cold currents was also determined. For the same sample, we also measured the ratio of oxygen isotopes in planktonic and benthic foraminifera. The results indicated that it could be around 14,000–11,000 y. B.P. when the continental ice had melted substantially. Since then the rise in water temperature is noteworthy but around 10,500 y. B.P., a severe cold spell occurred in northeastern Japan. Thereafter the water temperature again rose rapidly and by 7,000-5,000 y. B.P. we had reached the high temperature era. All these facts became clear during our studies.

14. Oba: The value of $\delta\ ^{18}O$ was studied in all the foraminifera found in seabed cores measured so far to explain the following four facts about the environments during the last ice age (18–20 ky), viz., 1) atmospheric $CO_2$ concentration was 200 ppm, 2) supply of aerial dust was large, 3) rate of sedimentation at the seabed was higher, and 4) lysocline became deeper around 18–11 ky. At the same time, the drop in sea level over 18 ky was determined from the $\delta\ ^{18}O$ curve for the cores from the Sea of Japan and Pacific Ocean; whereas, the density distribution was determined from contemporary salinity estimates and water temperature. As a result, we could say that the vertical mixing in the ocean was more prominent 18 ky ago than at present and the bio-productivity was also higher. We should also take into consideration the low $CO_2$ concentration in the atmosphere.

15. Niizuna: In order to discover the physical phenomena that took place on earth when its magnetic field was reversed, studies were conducted on geology, extinct plants and animals, isotopes, etc. Studies on the old terrestrial magnetism in the sediments from the seabed and oxygen isotope ratio revealed the relation between the reversal of the earth's magnetic field and the glacial age and interglacial age. However, since the records of old terrestrial magnetism get solidified about 50-60 cm below the sedimentary surface (seabed), it is not possible to establish a direct correspondence between them. This time we studied the sediments from the Abashiri Peninsula, where the rate of sedimentation is said to be 170 cm/1,000 years. We found that the reversal of the earth's magnetic field at the interface of the Brunhes and Matsuyama periods corresponds to the interface between stage 20 and stage 19 as explained by Emiliani (1978). At this level, a large reduction in bio-organisms was found to have occurred from the measurement of $\delta\ ^{13}C$. Mesurements of $\delta\ ^{18}O$ indicate that the temperatures were warm only during this period which could be a warm room effect following a large-scale expulsion of $CO_2$ gas.

16. Saito: A method to determine the old ocean environments using deep seabed cores was studied with an example of the ocean between the Central American Panama Isthmus and that on the northeast side of Japan. Saito

(1976) has made it clear that the separation between the Atlantic and Pacific oceans due to the bulging Panama Isthmus dates back to more than 3.5 million years. If we analyze the environmental changes in the eastern Pacific following that process of bulging, using the DSDP core, then we can ascertain 1) a temporary drop in bioproductivity as an effect of the bulging process which starts about ten million years ago, 2) recovery of that high productivity in the eastern Pacific due to an increase in both salinity and temperature on the Caribbean side around six million years ago and 3) the subsequent recession (in productivity) after the formation of the Panama Isthmus. During the Pliocene epoch in northeast Japan, the sea extended from the northern part of Fukushima prefecture to Hokkaido and then to Sakhalin. If we analyze the regression of this sea with the help of the polarity changes of the earth's magnetic field and the changes in isotope ratio in the north Pacific core, we can find that it occurs during the Gauss-chron period and it matches with the period during which glaciers appeared in the central latitude region.

Last, general discussions were conducted under the chairmanship of Prof. Saito of Yamogata University and the active question-answer session had to be extended for 20 minutes after which the symposium was adjourned. The general discussion consolidated the view that in future a closer cooperation is necessary between geochemists and geologists. We would like to extend our thanks to the Administrative Section of the Special Research Project "The Ocean Characteristics and Their Changes" who rendered assistance during the symposium.

(Coordinator: Tadamichi Oba and Yoshiyuki Nozaki)

## ADMINISTRATIVE REPORT

Eighth Managing Committee meeting: October 27, 1983 (Thursday), Assembly Hall, Institute of Oceanography.

**Items reported**

The Committee on Future Plans is trying to expedite the study plans in multiple meetings since it was instructed by the seventh meeting of the Managing Committee. There were slight changes and additions to the group.

New members: Tsunogai (Department of Fisheries, Hokkaido University) replaces Prof. Horibe, Mr. Yamazaki (Faculty of Engineering, Tokyo University) replaces Mr. Ogami. Mr. Takematsu (Applied Research, Kyushu University) is a new member.

Coordinators for the preparation of the Plan Report (Teramoto, Kananari, Takematsu, Tsunogai, Kagami, and Mochizuki) were selected and it is proposed to prepare a draft plan as well as the proposed organization for this year.

**Items discussed**

Preparation of Final Research Report:

The contents will generally be as decided in the 19th Executive Committee Meeting. An editorial committee will be formed next year. However, until that time, this task will be entrusted to the Executive Committee. After checking with printers, etc. the schedule will be prepared and the authors will be requested to send their papers through their respective group leaders.

| | |
|---|---|
| First deadline for draft submission | June 1984 |
| Second deadline | August–September 1984 |
| Completion | March 1985. |

It is proposed to apply for a grant of eight million yens to cover the cost of preparing the final research report, conduct the symposium, etc.

20th Meeting of the Executive Committee: December 5, 1983 (Monday), Assembly Hall, Institute of Oceanography.

**Items reported**

There was a hearing on November 30 at the Scientific Research Grant Commission under the Ministry of Education, regarding this Special Research Project. On this occasion we prepared a "Report on the course of studies undertaken under the Special Research Project financed by a Science Research

Grant" and explained various points.

The Committee on Future Plans is currently busy preparing a draft with the following four topics under consideration:

1) Seawater and material cycle in the deep sea
2) Formation and behavior of the boundary layer at the seabed
3) Movement of the seabed
4) Information communication from deep seas and remote measurements.

**Items discussed**

When preparing the final report, the following changes will be made. Index will be written as:

Chapter 1. Seawater cycle and material distribution in the northwest Pacific.

Chapter 2. Currents in the Kuroshio and connected region and the effects of topography.

The editorial committee will be formed as follows with the help of persons responsible for each Chapter:

Chief Editor: Kajiura

The printer and publisher will be decided at the next meeting in January next year and then the printer will be hired.

**Note:**

When reproducing or citing material published in this Newsletter, due acknowledgment should be given. Authors' names, wherever mentioned, should also be quoted. A copy of the publication containing such a reproduction should be sent to the Editorial Section of the Administrative Office.

| | |
|---|---|
| Published by: | Administrative Office, Special Research Project "The Ocean Characteristics and Their Changes" |
| Published at: | Administrative Office,<br>Physical Oceanography Department,<br>Institute of Oceanography,<br>Tokyo University,<br>1-15-1 Minami-dai, Nakano-ku,<br>Tokyo 164<br>Tel: 03-376-1251, Ext. 251. |
| Editorial staff: | Taeko Kitano and Noriyo Kimura. |

Special Research Project

# The Ocean Characteristics and Their Changes

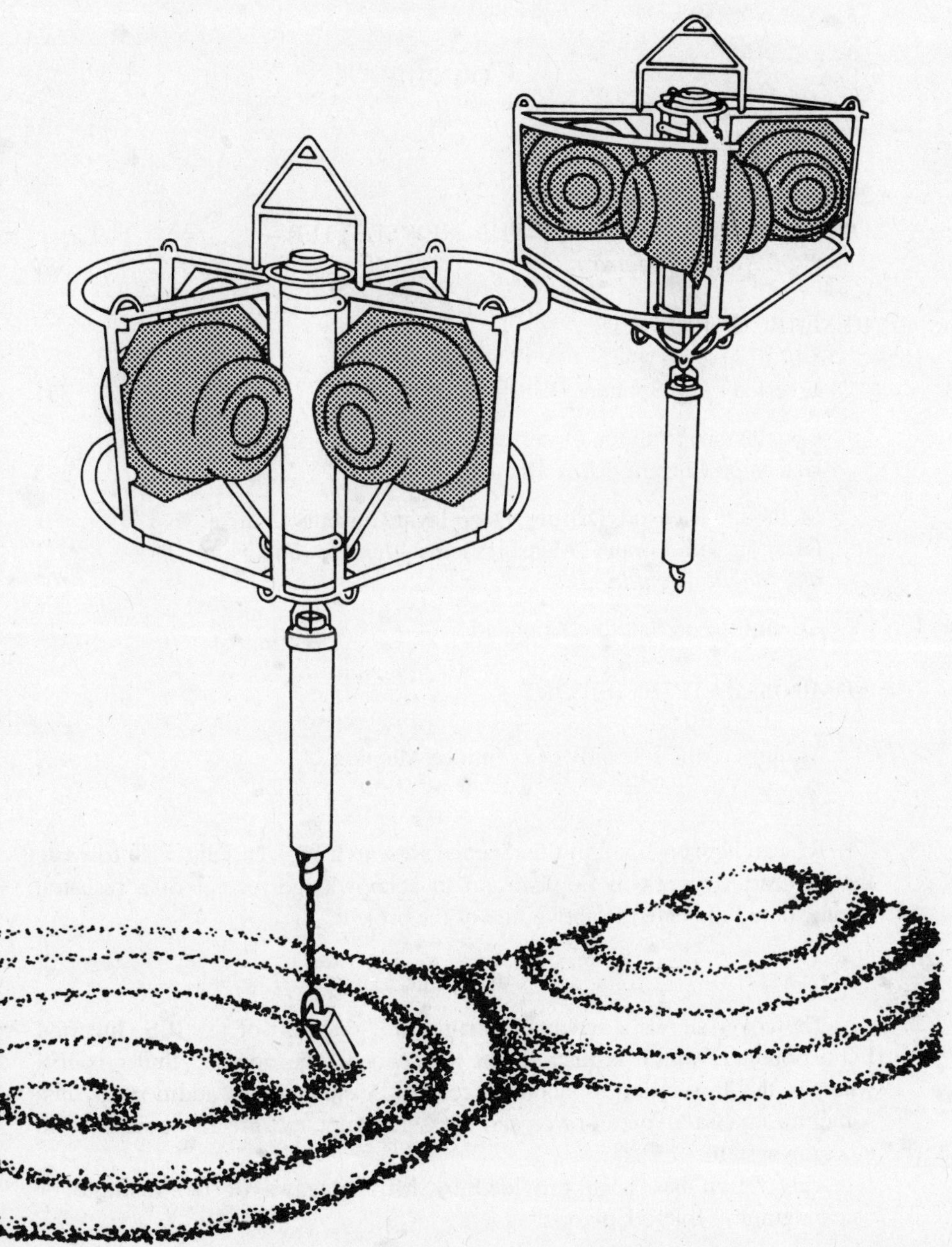

NEWSLETTER NO. 17, MARCH, 1984

# Contents



Researchers engaged in this Special Research Project, who wish to write their research papers in English and to acknowledge receipt of a research grant, should indicate the above title of the project.

## Cover Page

The cover shows a schematic diagram of that part of the IES (Inverted Echo Sounder) which is installed at the seabed. The upper cylinder is IES whereas the lower part is acoustic release equipment. In addition to this, sometimes a current meter or crystal pressure meter, etc. are also added to the mooring system.

This figure has been provided by Mr. Kitagawa of the Institute of Oceanography, Tokyo University.

# On the Final Issue of the Newsletter

*Prof. Kinjiro Kajiura*

(Coordinator, Special Research Project)
(Seismological Research Institute, Tokyo University)

At the end of this fiscal year, the Mombusho Special Project "The Ocean Characteristics and Their Changes" completes its three-year program and this issue of the Newsletter will also be the last.

As mentioned in the first issue also, recent developments in oceanography, attempting to ascertain new dynamic ocean phenomena, are quite eye-catching worldwide. The 1980s had as their background the results of the International Decade of Ocean Exploration (IDOE) of the 1970s and all countries are trying to execute various research plans, singly or jointly, to analyze different ocean phenomena. For example, we can quote various research projects dealing with problems of the fluid dynamic structure of seawater along the shore or on the continental shelf, the ocean structure of the western part of Mediterranean Sea, the mechanism of mixing seawaters on an oceanic scale using chemical tracers, deep and shallow boundary layer currents, etc. While the whole world is thus studying the ocean, in our country, there are special considerations with reference to this research project and we would like to congratulate the researchers for arriving at splendid results which will greatly contribute to the world's oceanographic studies.

During the course of this study, different kinds of basic data were obtained. This was extremely difficult, mainly because the northwest Pacific Ocean using the long term and continuous observations of changes in deep current flux, accurate measurement of heavy metal traces or elemental isotopes, etc. and by recovering the sedimentary particles in ocean water. This is quite a significant achievement. The project also included studies of methods to estimate the old environments by a multifaced analysis of sediments at sea floor, structural analysis of the appendages found in plateau subduction zones and new oceanographic measurement methods, etc. On the top of this data, many new ideas for the understanding of dynamic ocean structure have been evolved, such as deep layer circulation model of south Sea of Japan or vertical transport model, formation of appendages in plateau subduction zone, etc., which are introduced in Newsletter features like Highlights, etc.

Prof. Yoshida has written a small paper titled "Cold water mass of Kuroshio—the experiment of this century in American hands" expressing the sense of danger with respect to Japanese oceanographic studies. However, during this special study, quite interesting ideas came up about the dynamic structure of ocean including the problem of serpentine motion of Kuroshio and a convincing explanation of mechanism behind this serpentine motion were offered.

This special study was planned as a first step to increase the overall understanding of various ocean phenomena and from very beginning it was designed as a joint research scheme between different fields such as physical oceanography, chemical oceanography, marine geology, geophysics and engineering, etc. However, the studies were multidisciplinary rather than interdisciplinary. This was unavoidable considering the current state of development in each field, but even then, during the course of these three years of study, there was a good understanding about the studies being conducted in comparatively unrelated fields and some conversation became possible. We feel this should have a good effect on future developments in oceanography.

The studies undertaken to understand various natural phenomena including the ocean cannot be completed in three years. It need not be mentioned that it is necessary to observe (the phenomena) for a long period continuously, collect the basic data, and analyze it. For this purpose, a new research project has been planned to study further the new ideas that evolved from the analysis of data collected during these three years and to analyze the new problems that have newly surfaced. We request all concerned to participate with full energies in the new research plan.

More than five years have passed since the Mombusho Science Commission prepared its report on "Progress of Oceanographic Studies" around the end of 1978. During this period we received warm support from a number of concerned persons, even those not directly involved in the studies and we express our gratitude to all of them. We also take this opportunity to express our gratitude to Mr. Kitano and other staff members who have taken a lot of pains in editing this Newsletter.

## RESEARCH REPORTS

# Inverted Echo Sounder (IES)

*Keisuke Taira*

(Institute of Oceanography, Tokyo University)

(Received: January 21, 1984)

In Japanese this is also called an inverted depth measuring apparatus, acoustic sea level height measuring apparatus, etc. But IES seems to be a proper and popular abbreviation. The acoustic depth measuring equipment fitted to the bottom of an observation vessel sends sound pulses in the direction of the seabed and by measuring the time it takes the reflected sound waves to return, it calculates the water depth. Even when the depth is the same, the traverse time of the sound wave may vary with the vertical distribution of sound velocity at that place. By using direct measured values of sound velocity distribution or using the density distribution with CTD measurements, etc., the sound velocity is properly evaluated and then the time delay is converted to water depth. On the other hand, IES is installed at the sea bottom. It sends sound pulses in an upward direction and measures the time required for these pulses to travel to the the sea surface and return. With this, we can measure the depth at a given point. The distance between the sea surface and the seabed varies as the water level changes due to waves, tides and tsunamis or due to ocean currents, etc. These changes are ignored by conventional acoustic depth measurements. Changes in the vertical distribution of sound velocity also have a considerable effect. CTD measurements reveal that a difference of about eight milliseconds can occur in the traverse time of sound waves as a result of changes in the vertical distribution of sound velocity across the Kuroshio cross section. This, in terms of distance, would be 12 m if a constant sound velocity of 1,500 m/sec is assumed. In the outer seas, quite frequently the waves have greater amplitude than this level, but since their period is less than 20 seconds they can be eliminated from the averaging process. We hope to study the level of the Kuroshio using IES.

IES was proposed by T. Rossby (1969) to measure changes in the temperature profile with depth and was used in the 1973 MODE experiment. Bitterman (1976) has reported the construction of that IES. There are two types in IES. In one the acoustic release equipment is put together with glass balls developed by Bentos Co. In another it is mounted in the cylindrical container. At Rhode Island University, to which T. Rossby or R. Watts belong, they have made more than ten glass ball type IESs and they have measured three dimensional changes in the position of the Gulf stream since 1979 with this equipment. Cylindrical type IES is being used at the Woods Hole Oceanographic Institution. The plans to observe position of the Kuroshio using IES were first proposed by R. Watts in July, 1977 through Dr. H. Solomon to Prof. Teramoto, but it could not be actually carried out due to problems of research grant, etc. During 1980, the author worked at the Woods Hole Oceanographic Institution and since he was given the same room as Bitterman, he learned about IES. He also visited Rhode Island University and gathered the results of IES observations. In May, 1981 T. Rossby visited Japan on an invitation from the Institute of Oceanography and his participation in a special seminar was quite useful for recognizing the effectiveness of the IES technique in oceanographic observations (Taira, 1982). Independently, Takenouchi (Dentsu University) carried out some fundamental experiments on the use of acoustic techniques. He carried out laboratory tests using a water tank and field tests where a hydrophone is dropped from an observation vessel, etc. and evaluated the various results.

It was made abundantly clear from the results of direct measurements of the Kuroshio that changes in current velocity are related to the position of the flow axis of the Kuroshio (Taira and Teramoto, 1981). In this special project, it was proposed to determine the position of the Kuroshio flow axis by measuring the currents in the lower layer of the Kuroshio (changes in Kuroshio and effect of earth's topography). For this purpose, Administrative Department formed another committee on "Measurement Methods" and the task of IES development was entrusted to that group.

The IES Committee is headed by Prof. Teramoto and includes Prof. Mochizuki, Takenouchi (Dentsu University), Okushima, Otsuki (Tokyo Institute of Technology), Taira, Fukazawa and Kitagawa (Institute of Oceanography, Tokyo University). It was proposed to procure two IES units during fiscal 1982. The development of the acoustic part was entrusted to the Japan Petroleum Research Institute whereas the Institute of Oceanography took responsibility for developing the recording and processing section. Various aspects of design or preliminary experiments were discussed in committee meetings and as a result, the hardware was completed in that year. Mooring observations were carried out during the summer of 1983 off Hiratsuka and on November 14, 1983, both IES units were installed near the flow axis of the Kuroshio around Cape Toi. They will be recovered in March, 1984.

In the following part we shall discuss the IES developed by us and results of observations at Hiratsuka. A part of this paper will also be published in the March, 1984 issue of Kaiyo Kagaku (Ocean Science).

## Construction of IES

As shown in Fig. 1, this consists of three parts. Acoustic part: This generates sound pulses at the designated time intervals and, by detecting the reflected sound wave coming from the sea surface, measures the time of propagation. Recording part: This records the propagation time output by the acoustic section in the form of a digital signal. Both these parts are included in one pressure-resistant container and lie at the seabed. Reading and processing part: After the measurements are complete, it reads the records and calculates/analyzes them.

Development of the acoustic unit was entrusted to the Japan Petroleum Research Institute (Nichiyu Giken Kogyo K.K.). This institute has developed an acoustic release unit that works at a depth of 6,000 m (*Kaiyo Onkyo Kaiho,* Vol. 10, No. 4, 1983) and the same is used to measure currents at the lower layer of the Kuroshio under this special project. From the results, it was decided to use a sound wave of 10 kHz frequency. However, it became necessary to modify the acoustic transducer. Directional properties of the release unit are such that a 3 db bandwidth in the horizontal direction is quite wide. For use in IES, we developed a transducer that has a peak in the upward direction and 3db bandwidth of 30°. The measurement time interval can be selected anywhere from 30 sec, 1 min, 2 min, 3 min, 20 sec, 5 min, 10 min, 20 min. During one measurement 40 sound pulses will be transmitted. For time measurement, a clock with a 10 kHz frequency was used along with a 16 bit binary counter. The maximum measured was 6.5535 seconds leading to a high resolution of 7.5 cm.

The recording part was developed using a semiconductor memory developed by the Department of Physical Oceanography (Kitagawa, 1980). With the help of ten CMOS EPROM chips (Fujitsu MBM27C64), a total capacity of 40960 × 16 bits was developed. This can store records from 284 days if readings are taken every 10 minutes. The memory will hold the contents even if the battery is discharged and the records can be cleared by ultra

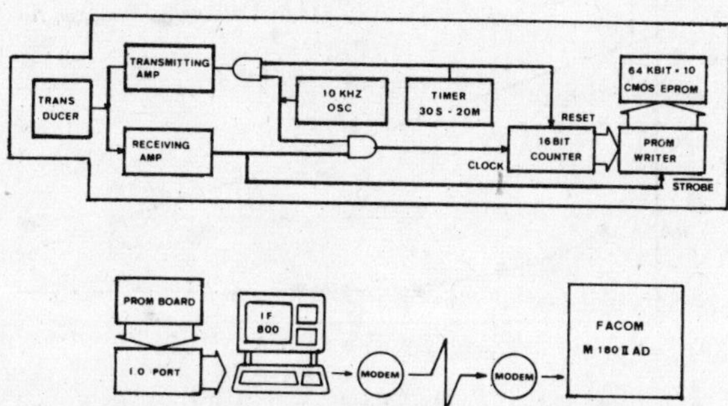


Fig. 1. Block diagram of IES.

violet rays so that the ICs can be used repetitively.

For reading and processing the data we decided to use the computer of the Institute of Oceanography. Facom-M180II AD computer had been in operation since December 1, 1982. This uses 20 modems for communication control and processing and is connected to terminals in each research room. About 150 special communication lines are used between the computer center and each research room. The terminals in the research rooms are of ten types and use a microcomputer in them. The Physical Oceanography Department uses terminals such as OKI-IF 800, TRS-80, Apple II, HP-45 Telephone Terminal (Modem), etc. for reading IES records. We decided to use an OKI-IF 800 terminal.

**Acoustic properties of IES**

The prepared IES was subjected to a wire test by suspending it with winch wire used in the observation vessel. We also obtained the records but in this section, let us introduce the analysis results of the sound signal from the tape recorder. The upper part of Fig. 2 corresponds to the one that was subjected to a high speed A/D converter whereas time progresses from left to right. The peak on the left hand side is the leakage of transmitted sound whereas the peak on the right hand side indicates reflected sound. The measured time difference between the two is $\tau$. White circles on the graph in this figure correspond to reflected sound voltage and the black circles indicate noise during the same period. The bold lines in either case show the slope as a function of $\tau$. Thus level of reflected sound corresponds to the detection level with $\tau = 3.5$ sec. This means that this IES cannot be used for depths less than 2,600 m. On the other hand, the noise level increases for smaller $\tau$. The period of 0.6 sec, immediately after transmission of the sound pulses, is used to convert the transducer to receiving equipment and during this period no measurement is made.

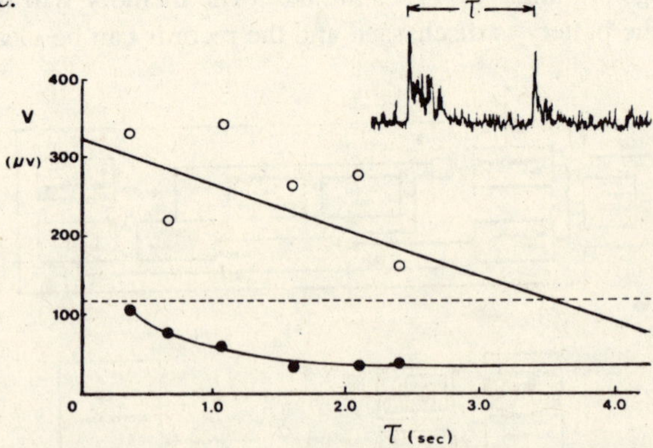


Fig. 2. Relation between the level of sound and noise reflected from the surface and propagation time.

**Mooring experiment of IES**

It was ascertained with the wire test that IES works up to (and beyond) 2,600 m. This was then moored off Hiratsuka (35°15′N, 139°20′E, 770 m deep) on August 22, 1983 and recovered on September 21. Fig. 3 shows those results. The measuring interval was 1 min and a total record of 150 hours (8,950 readings) after installation was obtained. The time axis shows each hour (Fig. 4). The working of IES was proper. Reflected sound could not be detected in 14 cases (0.16%) (time longer than 1.03 sec); whereas the counter stopped due to noise (less than 1.2 sec) in 82 cases (0.92%). This ratio is generally the same for the remaining 30,000 data points not shown here.

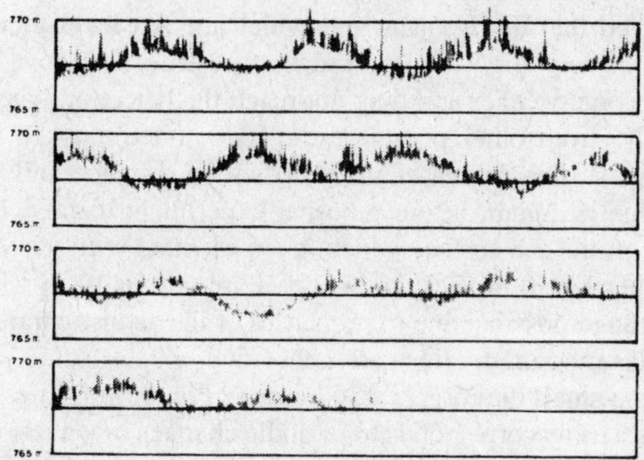


Fig. 3. Record of IES moored off Hiratsuka (35°15′N, 139°20′E, 770 m deep). For details of time axis, refer to Fig. 4.

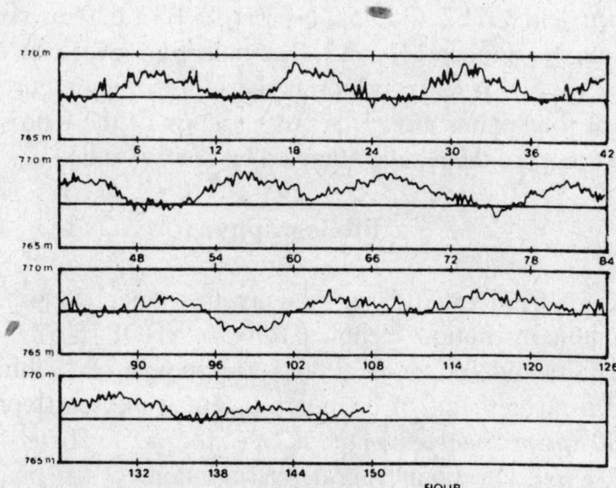


Fig. 4. IES record showing the running mean of ten points.

**Properties of IES record**

The records in Fig. 3 can be considered the IES data curve. The changes in tide levels of a semidiurnal tide are recorded as periodic variations. An important point is that each point has a random value. When a running mean of ten points is obtained, we get a smooth curve as shown in Fig. 4. It is hoped that the fluctuations in the flow axis of the Kuroshio can be expressed as a variation in sea level to a few meters using a similar low pass filter.

As some possible reasons for the irregularities in the distribution of measured values we can think of changes in the vertical distribution of sound velocity due to internal waves, etc., and the fluctuations of the sea surface due to waves. These fluctuations cause the irregular distribution of the measured value of the mean sea level. These properties are also evident in Fig. 3 and it can be observed that the frequency at which the sea level increases (arrival time longer) is very high. As the sea surface directly above IES inclines, reflections become weaker and may not reach the detection level. Under such cases, the echo from other points on the sea surface may be treated as the measuring signal. This increases the propagation distance and frequently the measured value is higher. In our mooring experiment if the echo is received from a point on the sea surface which is ten inclined with the vertical line, it will read higher by 10 m. Results of Fig. 3 are within 10 m which also conforms to the range possible due to properties of the acoustic transducer. As far as the irregular measured values are concerned, we see some periodic variation even in the small duration of 150 hours in Fig. 3. In future we would like to study whether this corresponds to periodic changes in waves, etc.

**To conclude**

IES was moored at two points off Cape Toi, CTEl (30°48.3′N, 131°41′E, 1,500 m deep) and CTE2 (30°45.2′N, 131°45′E, 1,620 m deep) using the *Keiten-maru* on November 14, 1983. The detection level was changed to 80 μV and hence we feel it will work properly. In this region, current meters are also moored at four points along the cross section of the Kuroshio. It will be interesting to compare the results obtained by these systems.

## Bibliography


1. Bitterman. 1976. Inverted Echo Sounder Instrument Report. Woods Hole Oceanographic Institution Technical Report, WHOI-76-67.
2. Kitagawa, Taira and Teramoto. 1980. Development of oil immersed current meter and measurements of boundary layers at the sea floor. *Proceedings of the 1980 Spring Symposium of the Oceanographic Society of Japan.*
3. Rossby, T. 1969. On monitoring depth variations of the main thermocline acoustically. *J. Geophys. Res.*, **74**, 5542–5546.

4. Taira and Teramoto. 1981. Velocity fluctuations of the Kuroshio near the Izu ridge and their relationship to the current path. *Deep-sea Res.*, **28,** 1187–1197.

5. Taira. 1982. Fluctuations in ocean and acoustic oceanography: Present state of US-Japan joint studies. *Report of Acoustic Oceanographic Study Association,* **9,** 6–8.

# Conclusions from the Observations of the Kuroshio Extension Current

*Jiro Fukuoka*

(Department of Fisheries, Hokkaido University)
(Received: January 23, 1984)

The special study has entered its final year and it is time to see how far we have been able to explain the mechanism of the Kuroshio circulation system which was the objective of our studies. Our group consisted of five persons and the numerical experiment subgroup within our main group has obtained significant data regarding the generation of large cold water mass or dependence of parameters on the separation of the main Kuroshio current in the Kuroshio region using the nullification phenomena. Many other results have been obtained from the analysis of other data and we propose to publish the same in the near future. In the following part I would like to report one or two features about the observations using a mooring system and also desired developments for the future.

With the beginning of this project I considered the contrast between the Kuroshio circulation system and Gulf Stream circulation system in the northwestern Pacific. The Kuroshio extension current region in the eastern part of Honshu lies along 35°N–36°N whereas the continuous bay current flows eastward and lies much south of 40°N. Another interesting point is that the water temperature in the central part of the Kuroshio extension current is much lower than the Gulf Stream region at the same latitude and this phenomenon stretches over 1,000 km in the north-south direction (Fakuoka, et al., 1983). The salinity also is much lower in the Kuroshio.

Next we considered the problem of low temperature. Fig. 1 shows the water temperature profile in the Kuroshio and Gulf Stream regions. Just a look at this figure tells us that the water temperature in the North Pacific is low, as mentioned above. However, still sufficient clarification has not been obtained about the behavior of seawater that brings such low temperatures. This is one of the topics that needs to be clarified in future. Now if we look at recent or slightly older observations, we find that there is a difference in the T-S diagram for the North Pacific and the North Atlantic oceans. This is be-

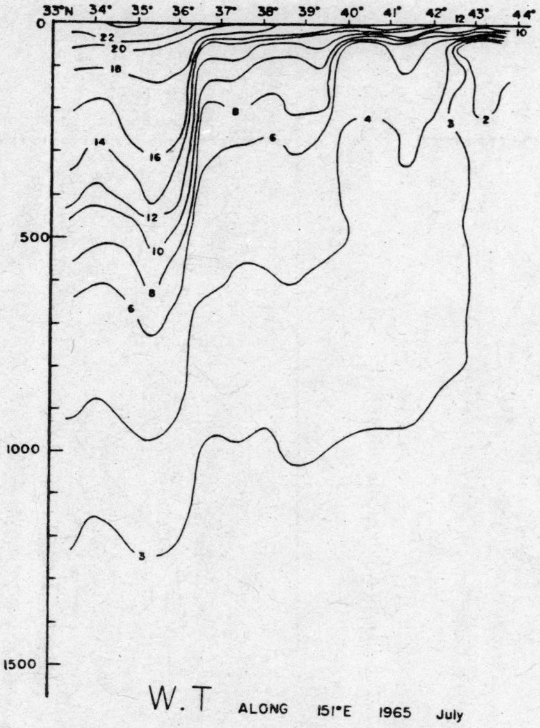

W.T ALONG 151°E 1965 July

Fig. 1a. North-south water temperature profile in the Kuroshio extension current region.

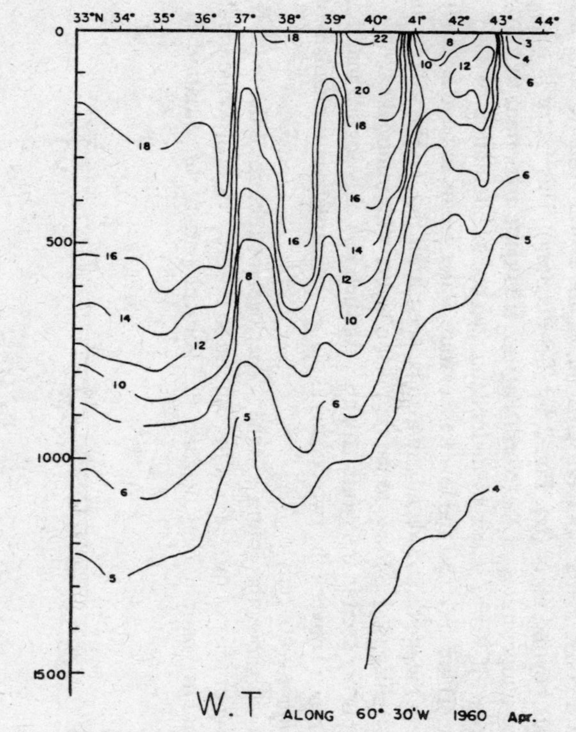


Fig. 1b. North-south water temperature profile in the Gulf Stream extension region.

cause the North Pacific has a layer that represents sudden drops in temperature to a considerable southern area but in the North Atlantic such a layer vanishes beyond 60°N (ref. Fig. 2 for T-S diagram). This factor is related to the low temperatures of the North Pacific. It is quite important to know the formation of warm water to understand the ocean circulation of the North Pacific. This is because the low temperature water seems to have an extended effect on the southern side as can also be seen from Fig. 1. The salinity in the North Pacific is also low and we find some layers with minimal salinity. This pattern is not seen in the North Atlantic. As such it is only logical that there is a difference between the two oceans as far as the formation mechanism of the water mass is concerned.

Fig. 3 shows the contrast in the winter ocean structure of the two oceans. Looking at this, we notice that in addition to differences in absolute values of temperature, salinity and density, the vertical pattern of variation is also

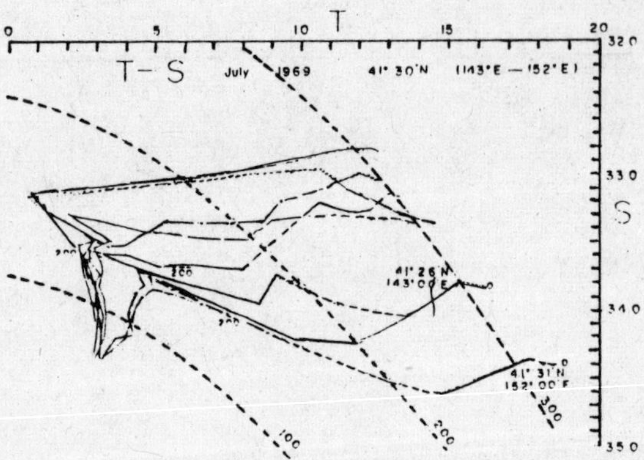


Fig. 2a. T-S diagram of North Pacific (1969).

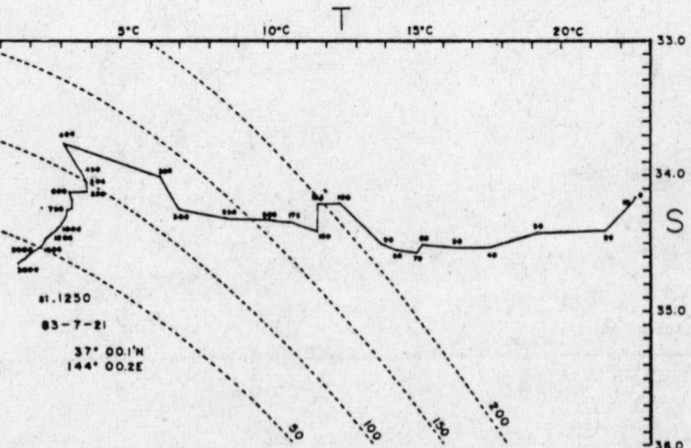


Fig. 2b. T-S diagram of North Pacific (1983).

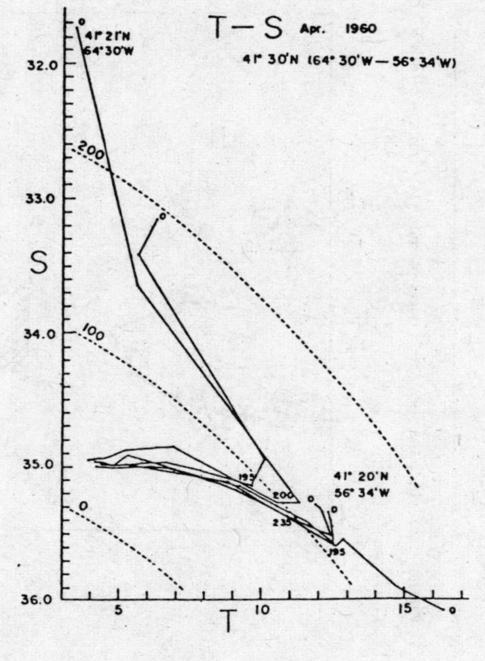


Fig. 2c. T-S diagram of North Pacific (1960).

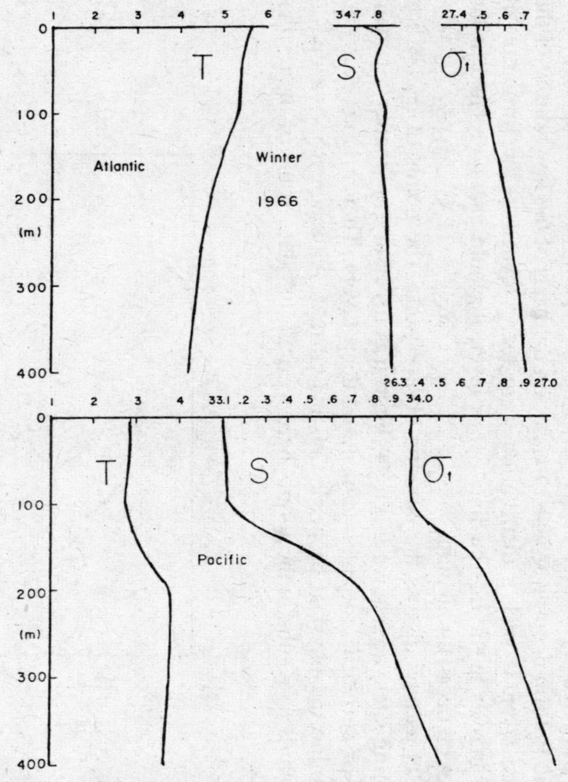


Fig. 3a. Distribution of temperature, salinity, and density in the North Pacific and North Atlantic oceans during winter (1966).

different. Recently Warren (1983) has tried to explain the failure of the North Pacific to form deep-layer water masses on the basis of the low salinity of the surface layer. If such a distribution were to be related to the formation of water masses in the intermediate ocean layer, it would also be related in the studies on ocean circulation.

Let us now discuss the mooring observations. The points of installation (mooring sites) are shown in Fig. 4. First, it is necessary to identify the main body of current in the intermediate and deep layers. Then it is necessary to determine the physical properties of this current. Although it is planned to publish these results in the future, here we will introduce some of the results.

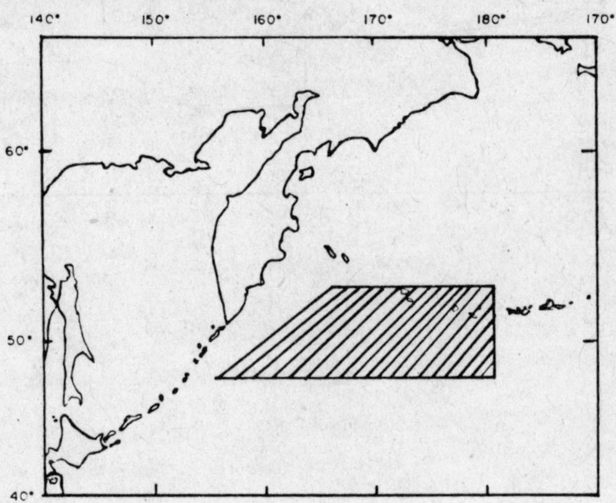


Fig. 3b. Region over which the above data was obtained (10 points in North Pacific).

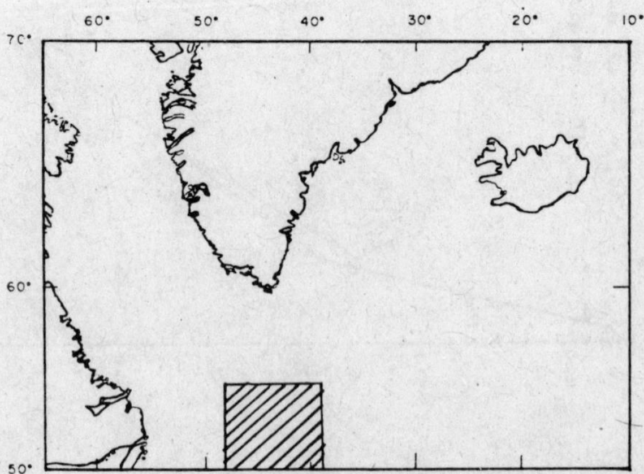


Fig. 3c. Region over which the above data was obtained (10 points in North Atlantic).

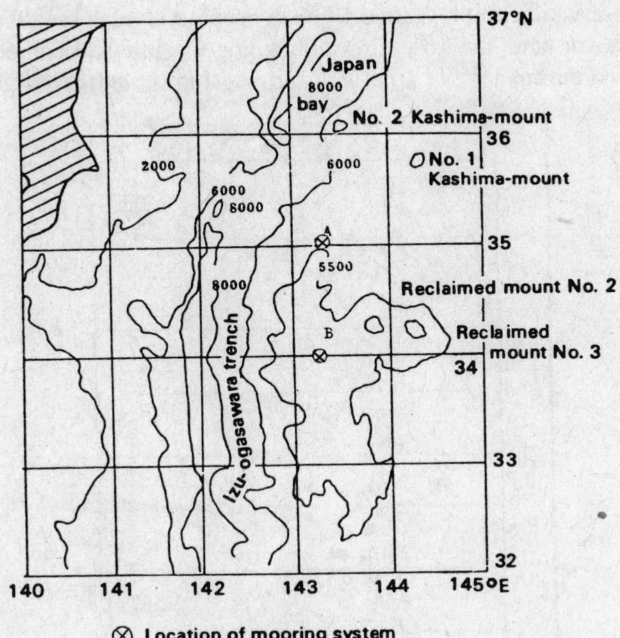


Fig. 4. Mooring points in the Kuroshio extension current region.

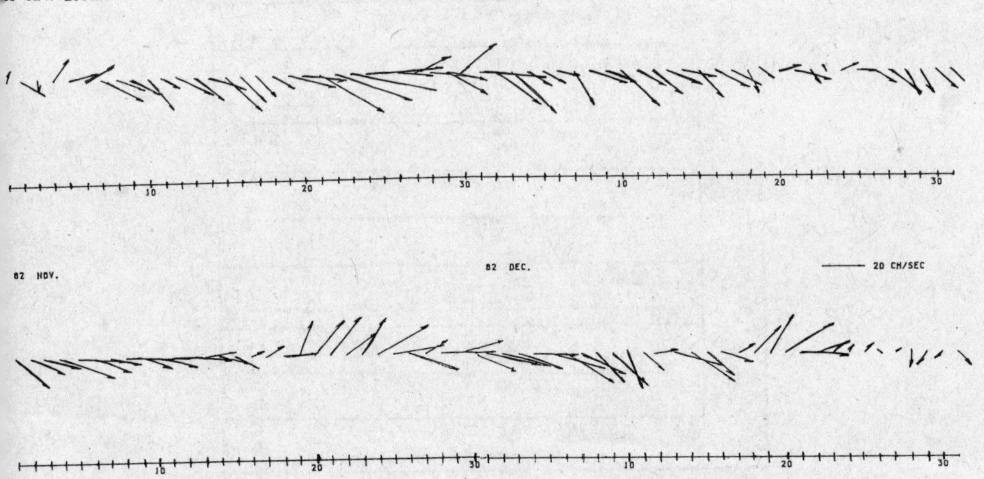


Fig. 5. Measured values of current at mooring point A (Sept.– Dec., 1982, 1,000 m deep).

Fig. 5 shows a stick diagram for the currents in the intermediate layers at northern observation points (above 1,000 m deep). Although only four months period is shown here, the same flow containing a southward component was also observed during another 10 months around June, 1983. As far as the cur-

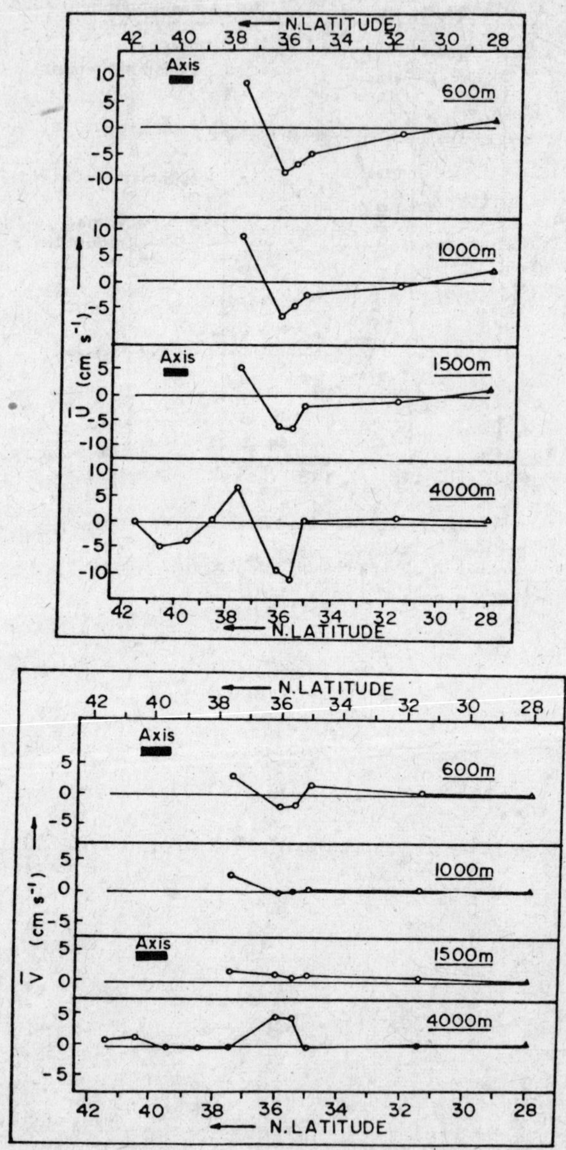


Fig. 6. Mean current velocity along 55°W (upper part shows east component and lower part shows north component; includes Schmitz's diagram).

rents 5,000 m deep are concerned, the absolute value is small but the same trends are observed. Such southward currents in the intermediate layers were also observed by Schmitz, et al. (1982) along 152°E.

Let us now turn again to the North Atlantic. It would be interesting to see the condition of currents in its intermediate layer. Although I do not have data over a wide region, the observations of Schmitz (1980) show that in the intermediate layer of the North Atlantic or deeper, southward currents are extremely rare (Fig. 6). It is quite difficult to analyze the difference in behavior of the intermediate layer currents of both these oceans. However, one explanation could be that it is an effect of Mediterranean Sea waters on the ocean structure of the North Atlantic. The high salinity waters of the intermediate layers of the North Atlantic (temperature also considerably high) maintain a proper horizontal coefficient of diffusion and hence it is possible to approximate them with the extent of the Mediterranean Sea waters (Richardson and Mooney, 1975). Worthington (1976) has also stated from the salinity distribution along the 10°C level that, Mediterranean Sea waters con-

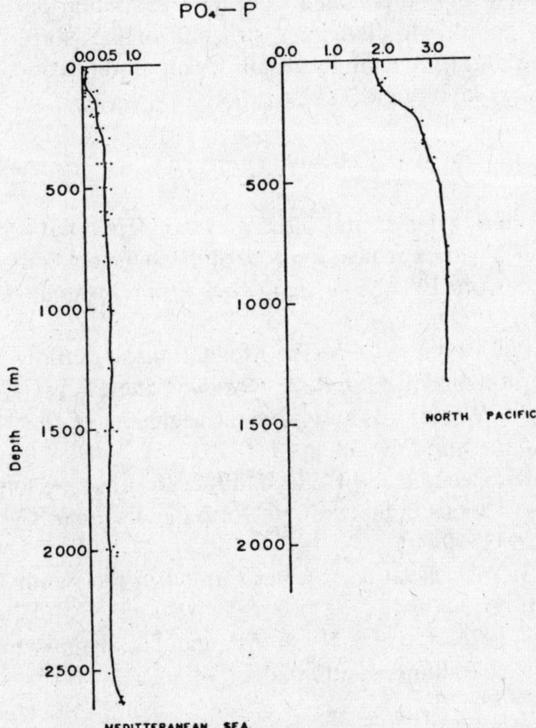


Fig. 7. Distribution of $PO_4$-P in the Mediterranean Sea and the North Pacific Ocean (1966 and 1967). Locations in the North Pacific: 39°N–41°N, 144°E–147°E.

siderably affect northwest waters. Also it is estimated that around 40°N in North Atlantic, there is a good amount of exchange with Mediterranean waters and hence a southward current is not quite clear. Naturally there is no extreme drop in salinity.

As against this, no intermediate layer is found in the North Pacific which can trace its origin to the Mediterranean Sea. Then what could be the reason for the southward current we measured? It is considered that it connects to low temperature, low salinity waters including those in subtropic regions, but the connection has not yet been established directly. Unless we continue observations and measurements of the Kuroshio currents including its northern region, it may not be possible to completely understand the circulation of the subtropical region.

Now, while considering the role played by the Mediterranean Sea in the ocean circulation of the North Atlantic, when we look at Fig. 7, showing that the water has extremely low nutrient salt, we feel that the North Pacific basically is favorable as far as the elements that affect the productivity of the ocean. Now if we compare the ocean circulation system as estimated from the above viewpoint with that obtained from the distribution of the water mass, then it may be possible to clarify the structure of the North Pacific Ocean, which has the highest productivity of all oceans in the world, even from an oceanographer's point of view.

## Bibliography


1. Fukuoka, Akiba, Miyake and Takeda. 1983. The characteristics of the oceanographical structure near the Kuroshio extension, compared with that of the Gulf Stream Extension. *Bull. Fac. Fish. Hokkaido University*, **34,** pp. 208–219.
2. Richardson and Mooney. 1975. The Mediterranean outflow—a simple advection-diffusion model, *Jour. Phys. Oceanography*, **5,** pp. 476–482.
3. Schmitz. 1980. Weekly depth-dependent segments of the North Atlantic circulation. *Jour. Mar. Res.*, **38,** pp. 111–133.
4. Schmitz, Niiler, Berstein and Holland. 1982. Recent long term moored instrument observations in the western North Pacific. *Jour. Geophy. Res.*, **87** No. C12, pp. 9425–9440.
5. Warren. 1983. Why is no deep water formed in the North Pacific? *Jour. Mar. Res.*, **41,** pp. 327–347.
6. Worthington. 1976. On the North Atlantic circulation. Johns Hopkins University Press, Baltimore and London.

# Method to Measure Drifting Deep-layer Currents Using a Relation with Loran C Measurements

*Ichitaro Sakamoto and Makoto Uchida*

(Faculty of Fisheries, Mie University)
(Received: February 16, 1984)

## Introduction

An approach to understand factors responsible for the formation of the ocean structure at the sea floor and changes therein, from the measurement of the transfer of water bodies led to binary systems in Euler type current meters. In one system, recording equipment of the shallow-water moorings which are installed mainly to observe periodic changes are courageously exposed to deep-water regions. However, it is difficult to deploy the equipment with proper spacing which can sufficiently back up the ocean structural information equipment obtained from other sources and it is also difficult to arrange the timewise structural information that supports the results of periodic measurements. Accordingly, it becomes necessary to select a few mooring sites and measuring layers as well as apply some cautious guesswork.

In another method, the double measuring system used to be carried to the deep seas to measure the currents of each layer corresponding to oceanographic observations. The equipment used here has been changed to a self-recording type. However, since under the drift current system, the standard layer (no current plane) is assumed, it is used as subdata of the geostraphic currents calculated on the basis of similar prediction.

Properties of predictions in a binary system vary and if we could eliminate the latter, it might be possible to obtain facts about deep sea currents. While the use of Aanderaa type current meters RCM (type 4, withstands pressure of 2,100 1b) has become quite common, the Loran C measuring system, where the relative accuracy is considerably increased due to automation, was studied with a possibility of one equipment system under a drift current system based on the principle of a double equipment system (does not use the assumption of no current plane).

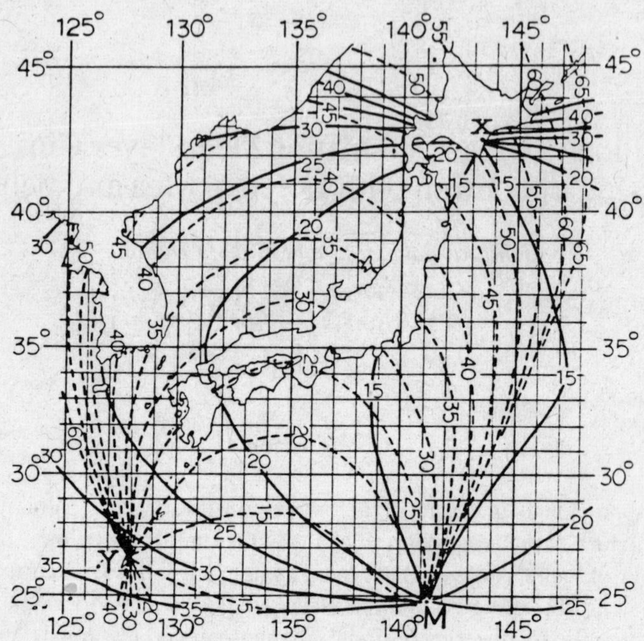


Fig. 1. Maximum error per 0.1 μS (m) using Loran C M X Y 9970 lane (solid lines: latitude direction, dotted lines: longitude direction). M: Ryuko Islands (24.8081 N, 141.3251 E), W: Nancho Islands (24.2855 N, 153.9814 E), X: Hokkaido (42.7436 N, 143.7192 E), Y: Okinawa (26.6069 N, 128. 1490 E), Z: Yap Islands (9.5461 N, 138.1652 E).

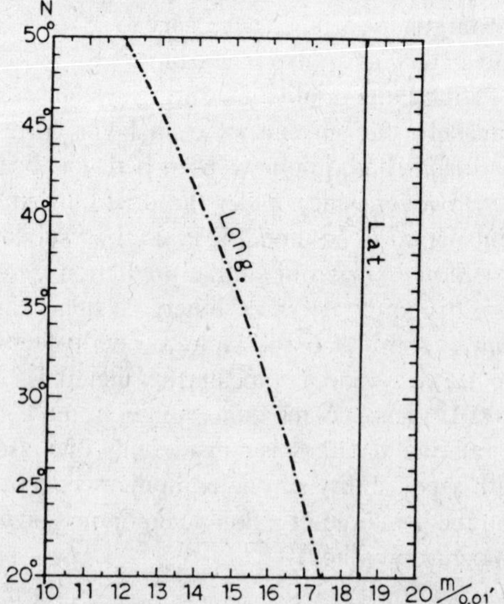


Fig. 2. Maximum error with Loran C with a selectivity of 1/100 min.

**Error in the Loran C measuring system**

In this system synchronizing signal pulses are transmitted from one main station and multiple substations. The receiving station (vessel) measures the time and phase difference between the signals arriving from each station and thereby determines its own position. As such the geographical relations between the transmitting stations and vessel affect the accuracy of such positions. In the measuring system used this time, Loran C-9970 lane was used at M: Ryuko Islands, X: Hokkaido and Y: Ground wave of Okinawa station (GW). Similarly, Yoshino Electric Co's LC-200 was used as a receiver. Latitudes were printed using a ST-70 printer (with IF-5001) through interface IF-3000 and simultaneously was fed through a plotter.

The receiver resolution was 0.1 μS whereas selectivity of the printer was 1/100 min. Fig. 1 shows the distance in latitude direction between stations MX and in the longitude direction between stations MY when the resolution is 0.1 μS. Similarly, Fig. 2 shows the distance when printer selectivity was 1/100 min. Here, the position of the ship is being determined every minute during the measurement to calculate the amount of driftage. The sum of the values in Figs. 1 and 2 therefore corresponds to the maximum error of the system. When the course of the ship is plotted with due magnification, we can see that it is not a straight line but looks like a small saw tooth form (ref. Fig. 4, upper part). At 32°N, 135°45′E which lies south of Shionomisaki, there is a possibility that each measuring position involves an error of 17.1 + 18.5 = 35.6 m in the latitude direction and 20.3 + 15.7 = 36.0 m in the longitude direction. From the deflection point of view this may correspond to an error in two ship locations and thus the maximum error will be double those values. As a result, if the measured positions of each minute are used to determine deflection, we can get a maximum error of two knots or more.

Studies indicate that if the ship position in measured every five minutes and an average of seven readings (30-minute period) is used, then the maximum error in driftage calculations in this region would be within 0.1 knot in both longitude and latitude directions.

**Error in current meter due to water head**

Frequently we encounter cases where a discontinuity in vertical structure is observed during the use of CTP. Accordingly, it is necessary to accurately determine the depth of the layer being measured. Also, while using current meter data, it is necessary to ascertain that the equipment is held at a constant depth. Fig. 3 shows the correction table for a RCM-5 type water pressure sensor which is suspended 10 m below CTP. It is quite obvious that the PSI unit should be converted into dbar units and we can see that at some stages a correction of 30–40 dbar is required. It is necessary to construct such correction tables for the equipments to be used in this method.

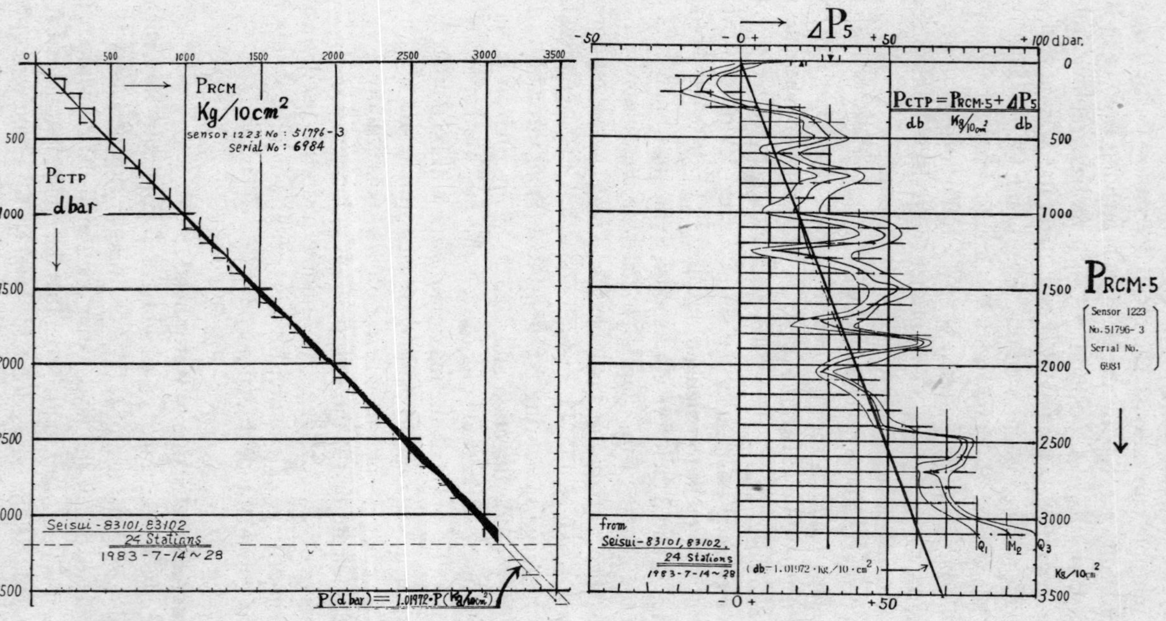


Fig. 3. Example of simultaneous corrections for multiple apparatus and pressure unit system to be applied to pressure data of RCM-5.

**Example of current measurement**

RCM can be used independently (stand-alone type) or at the end of a Nansen cast or connected to CTP, however, a good quality swivel is necessary. Fig. 4 shows an example of current measurement from the outer ridge from the Kumano trough which is 2,000 m deep to the Nankai trough which is 4,700 m deep. The RCM data collection rate is once per minute and the partition was set at 0.5. Here the water depth is calculated by normalizing the depth obtained by sonic measurements with changes in sound velocity (as a ratio with normal velocity). The RCM is connected at the end of two Nansen bottles which themselves are suspended from the point of reach of the CTP equipment (3,200 db). Drift measurements start from the instant RCM reaches its level. The position of the current meter should be about 15 m from the bottom and the winch speed is gradually reduced until it comes to a halt. The wire for correcting the inclination is collected every 500 m. (If a longer wire is used, there is a chance that the wire may reach the bottom before the measuring apparatus reaches its assigned position.) Care needs to be taken just before the apparatus reaches its position as it always has a tendency to go deeper (but shallower than the position for recovery). When RCM is connected to CTP, it is possible to monitor the sensor position continuously and this reduces measurement time. A time of 6 min (after reaching the assigned position) is used as a reference period and data during this period is rejected. If we allow about four hours for measurement to 3,000 m, it should be enough. While arranging the data, the mean displacement velocity over every 5 min, current every minute, and pressure corrections (dbar) are plotted on a time table and therefrom a period during which the apparatus is in a useful position is determined. The displacement velocity is then correlated with mean displacement. Table 1 shows the results of the sum of measured velocity and drift velocity for each layer. A corresponding ocean structure using a vertical distribution is shown in Fig. 5. A vector every 100 bar, extrapolated from the vertical distribution (every 25 db up to 200 db), was added to the immediately following current velocity measured by GEK. It can be seen that a strong WSW current, with a velocity of almost 0.6 knots, flows parallel to the ridge in the lower layer. If we take into consideration the errors in measurement, then we get the impression that a layer around 1,500 db, which is deeper than the minimum oxygen level, and a layer with minimum salinity have no currents. In many cases, a minimum salinity level represents a boundary between the upper layer and intermediate layer waters and when RCM is connected to CTP where on-line monitoring is not possible, it is not easy to install the current meters in an exact position.

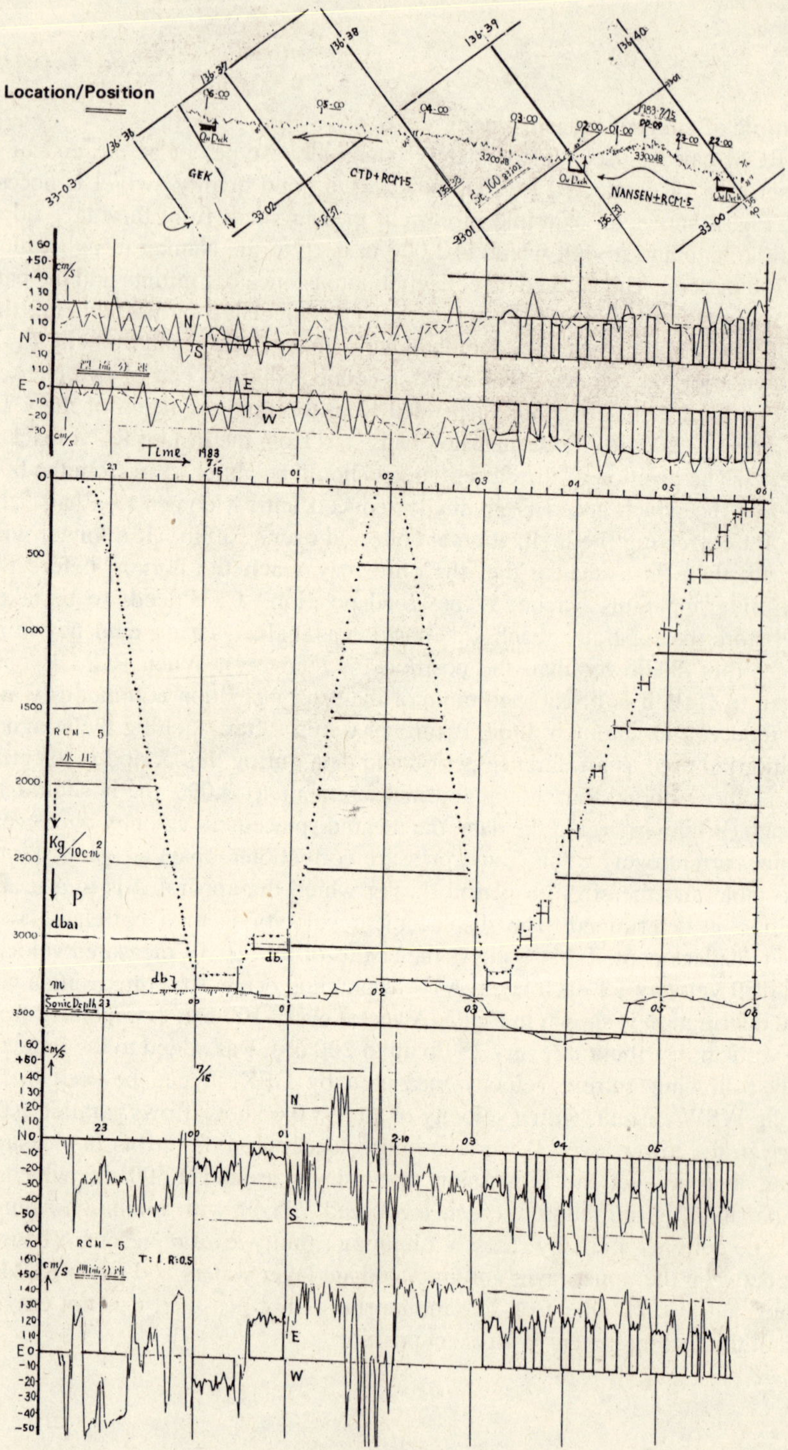


Fig. 4. Example of current observations. From top: Position determination record with Loran C, drift velocity and the average shift during a 30-min period, water-head record of RCM-5 and corresponding correction, time-based record of current velocity.

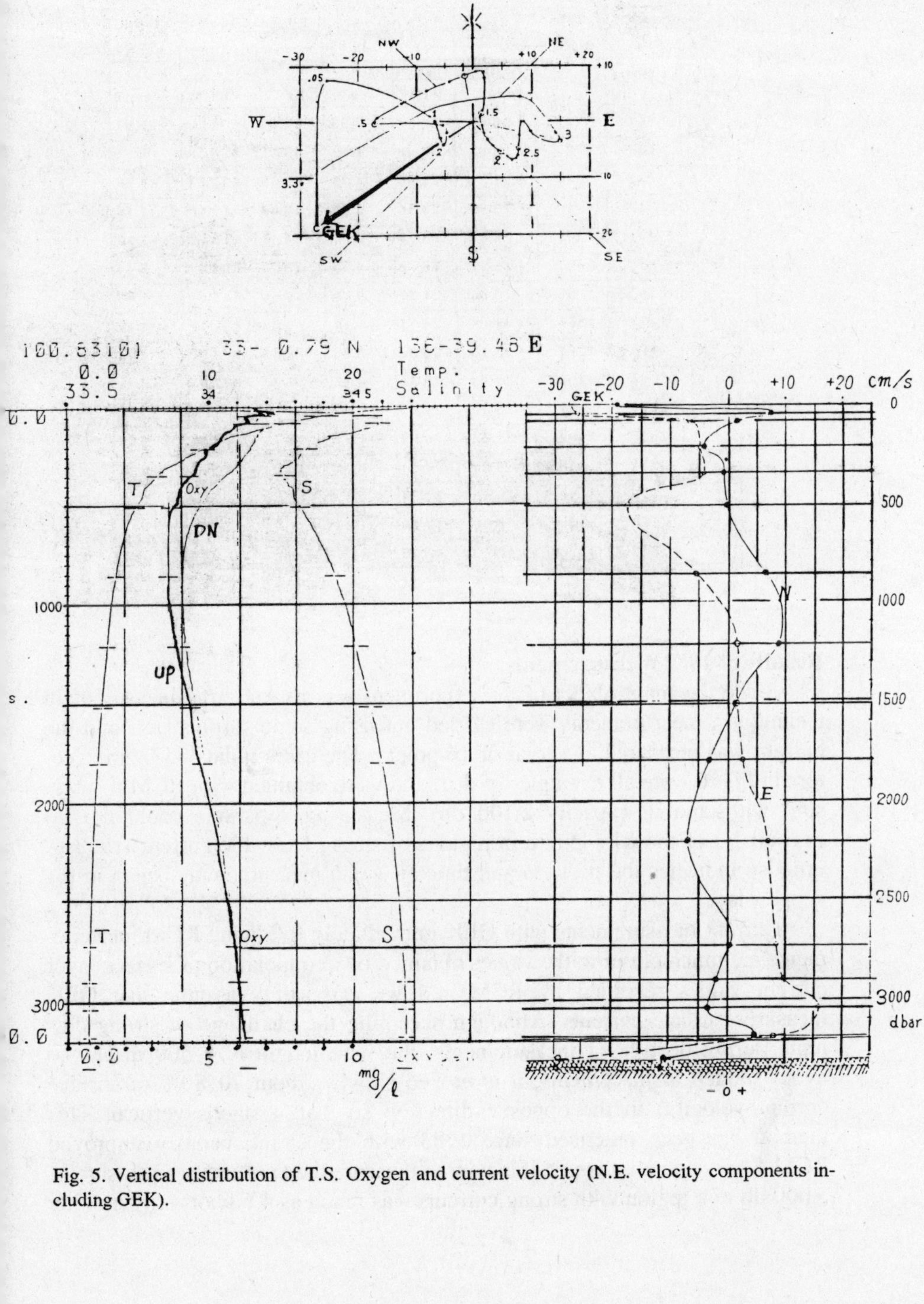

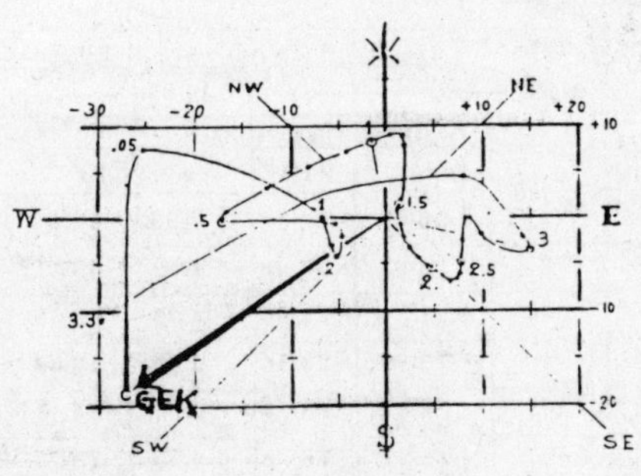


Fig. 5. Vertical distribution of T.S. Oxygen and current velocity (N.E. velocity components including GEK).

**Table 1. Results of calculations of current velocity (from Fig. 4)**

| St. 100. 83101 Jul 15 | | | Lat. 33-02.82 N | | | Long. 136-36.65 E | | |
|---|---|---|---|---|---|---|---|---|
| 06 20 | GEK | | | | -19.29 | | | -27.01 |
| Time h m~m | PRCM-5 kg/10cm² | P d bar | N Speed cm/s Drift | RCM | N cm/s | E Speed cm/s Drift | RCM | E cm/s |
| 05 48~52 | 67.5 | 48.1 | +30.87 | -23.34 | +7.53 | -32.67 | +6.80 | -25.87 |
| 05 41~45 | 109 | 85 | +28.88 | -27.28 | +1.53 | -33.21 | +26.38 | -6.83 |
| 05 31~36 | 232 | 207 | +27.27 | -31.35 | -4.08 | -32.35 | +27.37 | -4.98 |
| 05 23~27 | 342 | 346 | +23.89 | -24.76 | -0.87 | -33.64 | +29.46 | -4.18 |
| 05 13~19 | 491 | 507 | +23.57 | -23.79 | -0.22 | -34.07 | +16.94 | -17.13 |
| 04 05 57~05 | 848 | 857 | +25.31 | -18.78 | +6.53 | -33.47 | +28.03 | -5.44 |
| 04 43~48 | 1195 | 1208 | +30.36 | -21.27 | +9.09 | -29.65 | +31.47 | +1.82 |
| 04 27~35 | 1477 | 1514 | +27.68 | -26.51 | +1.17 | -29.93 | +31.40 | +1.47 |
| 04 15~20 | 1785 | 1797 | +23.16 | -26.32 | -3.16 | -28.47 | +31.07 | +2.60 |
| 03 04 59~05 | 2194 | 2208 | +22.13 | -28.86 | -6.73 | -25.88 | +32.83 | +6.95 |
| 03 42~47 | 2680 | 2737 | +25.02 | -24.65 | +0.37 | -21.45 | +29.92 | +8.47 |
| 03 29~36 | 2845 | 2903 | +25.28 | -26.99 | -1.71 | -19.41 | +29.23 | +9.82 |
| 00 37~59 | 2951 | (3040) | +3.63 | -7.14 | -3.51 | -10.04 | +26.09 | +16.05 |
| 03 08~22 | 3119 | 3195 | +25.93 | -21.34 | +4.59 | -19.50 | +28.28 | +8.18 |
| 00 00~27 | 3210 | (3300) | +6.06 | -17.30 | -11.24 | -11.68 | -17.80 | -29.48 |
| Depth | 3330m | 3310m | Lat. 33-00.75 N | | | Long. 136-39.57 E | | |

## Results of 1982 measurements

Since the main object of the experiment was to study the measurement techniques, measurements were carried out using some surplus time and the current was measured at a total of 16 points. The cases indicated by the vectors in Fig. 6 were all examples of drift and were obtained with RCM-4 (pressure withstanding capacity 2,100 db). Measuring apparatus could not go beyond 1/3 of the wire length in regions of strong Kuroshio current. The figure also indicates the position and time at which measurements began in the deepest layer.

Current measurements with GEK immediately following RCM measurements are much closer to the values obtained by extrapolation of surface layer current values recorded by RCM and we can safely assume that GEK measures surface currents. Although accepting the challenge of strong currents during deep-layer measurements, the wire length is 4–500 m and to avoid sudden displacements, it is moved slowly (about 70–80% of surface current velocity) in the opposite direction so that it stands vertical. This method has been practiced since 1983 with the simultaneously improved RCM-5 version and it is possible now to measure the currents to a pressure of 3,000 db in a region with strong currents—as much as 4.6 knots—by moving

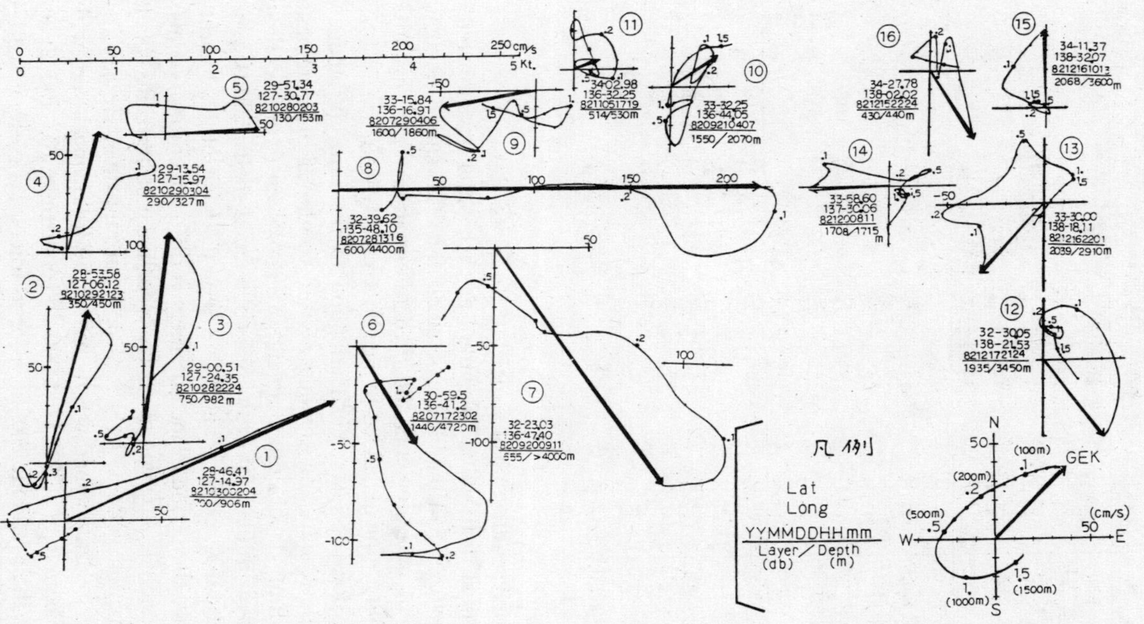


Fig. 6. Results of measurements of current velocity at 16 points in 1982.

**Table 2. Results of current measurements in specified layers**

| RCM St. No. | Time 1982 | St. No. | Nav. No. | GEK (N/E cm/s) | Sal. Max. (N/E cm/s) | Sal. Max. Layer db | Sal. Min. (N/E cm/s) | Sal. Min. Layer db | Oxy. Min. (N/E cm/s) | Oxy. Min. Layer db | 1500 db (N/E cm/s) | MaxP / Bottom P |
|---|---|---|---|---|---|---|---|---|---|---|---|---|
| 1 | 10/30<br>02~04 | 7. | 82013 | + 61<br>+ 140 | + 16<br>+ 11 | 200 | − 2<br>− 31 | 300 | − 7<br>0 | | | 700<br>906 |
| 2 | 10/29<br>21~23 | 6. | 〃 | + 79<br>+ 21 | + 14<br>+ 6 | 140 | − 2<br>0 | 350 | 0<br>0 | | | 350<br>450 |
| 3 | 10/28<br>22~24 | 4. | 〃 | + 110<br>+ 12 | + 13<br>+ 1 | 170 | + 3<br>− 19 | 475 | + 17<br>− 6 | | | 750<br>982 |
| 4 | 10/29<br>03~04 | 5. | 〃 | + 62<br>+ 15 | + 37<br>+ 26 | 130 | + 6<br>− 15 | 225 | 0<br>0 | | | 290<br>327 |
| 5 | 10/28<br>01~03 | 1. | 〃 | + 1<br>+ 48 | + 15<br>− 20 | 120 | − 2<br>− 12 | Bottom | | | | 130<br>153 |
| 6 | 7/18<br>00~02.5 | 9. | 82101 | — | | | | | | | − 11 (1400 db)<br>+ 41 | 4720 |
| 7 | 9/20<br>09~11 | 6. | 82010 | − 123<br>+ 89 | − 94<br>+ 117 | 115 | − 28<br>+ 9 | 470 | ? | 900 | ? | 655<br>>4000 |
| 8 | 7/29<br>14~16 | 26. | 82102 | 0<br>+ 216 | − 13<br>+ 224 | 120 | + 20<br>+ 32 | 480 | ? | 800 | ? | 600<br>4400 |
| 9 | 7/29<br>05~07 | 24. | 82102 | − 8<br>− 49 | − 31<br>− 30 | 75 | − 11<br>− 8 | 400 | − 20<br>+ 10 | 700 | − 9<br>− 23 | 1600<br>1860 |
| 10 | 9/21<br>04~07 | 3. | 82010 | + 16<br>+ 24 | + 21<br>+ 19 | 110 | − 25<br>− 1 | 450 | − 13<br>− 3 | 950 | + 21<br>+ 26 | 1550<br>2070 |
| 11 | 11/05<br>17~20 | 16. | 82013 | + 4<br>+ 12 | − 4<br>+ 21 | 110 | +26 / + 1; − 4 / +12 | 360; 500 | | | | 514<br>530 |
| 12 | 12/17<br>20~24 | 5. | 82017 | − 40<br>+ 31 | + 25<br>+ 17 | 100 | + 12<br>+ 10 | 350 | + 16<br>+ 1 | 700 | + 5<br>+ 7 | 1935<br>3450 |
| 13 | 12/16<br>22~25 | 7. | 〃 | − 36<br>− 33 | − 12<br>− 34 | 100 | + 25<br>− 17 | 400 | + 15<br>+ 13 | 950 | + 12<br>+ 15 | 2039<br>2910 |
| 14 | 12/20<br>08~11 | 2. | 〃 | − 1<br>− 41 | + 7<br>0 | 150 | + 8<br>+ 20 | 500 | − 4<br>+ 5 | 1000 | − 5<br>+ 9 | 1708<br>1715 |
| 15 | 12/16<br>10~13 | 9. | 〃 | + 41<br>− 1 | + 22<br>− 18 | 100 | − 2<br>+ 1 | 400 | + 2<br>− 13 | 1075 | + 2<br>− 6 | 2068<br>3600 |
| 16 | 12/15<br>22~24 | 11. | 〃 | − 37<br>+ 22 | 0<br>+ 5 | 125 | + 2<br>+ 14 | 450 | | | | 430<br>440 |
| 100 | 1983<br>7/15<br>00~06 | 0. | 83101 | − 19<br>− 27 | + 8<br>− 26 | 70 | − 1<br>− 4 | 300 | + 4<br>− 10 | 750 | + 1<br>+ 1 | 3300<br>3310 |

the wire in reverse direction at a speed of 1.6 knots.

The vertical distribution of current at each point is compared with the vertical structural distribution as obtained by CTP simultaneously or just before currrent measurements and then information regarding drift characteristics of that region are added to the above comparison.

Table 2 does not pay much attention to the time difference between various measurements and summarizes the results of measurements in the minimum salinity layer, minimum dissolved oxygen (which corresponds to the bottom layer in the East China Sea continental shelf region), and at 1,500 db along with GEK results. This is shown on the basis of the current velocity in Fig. 7. Since the period of measurement is different at these points, it is not possible to conclude anything of the cross sectional or horizontal planes as the above information pertains only to a point at any time. However, with the help of these measurements, we identified the following layers without any current; near bottom at st. 2.4.11, 1,000 db layer for st. 10, 500 db layer of st. 15 indicating their randomness. In order to forecast a current free layer, it is very necessary to get accurate readings in both a double equipment method or estimation of geostrophic current. Since 1983, this method is routine for the observation of cross currents and it is further planned to study the changes in ocean structure with a dynamic image.

One of the weak points of this method is the possibility of error in the absolute value of current velocity to the extent of about 0.1 knot which basically comes from the time accuracy of 0.1 μS and longitudinal accuracy of 1′/100 of the Loran position determination system. Some improvements may be expected in the second limitation but the first may already be the limiting value.

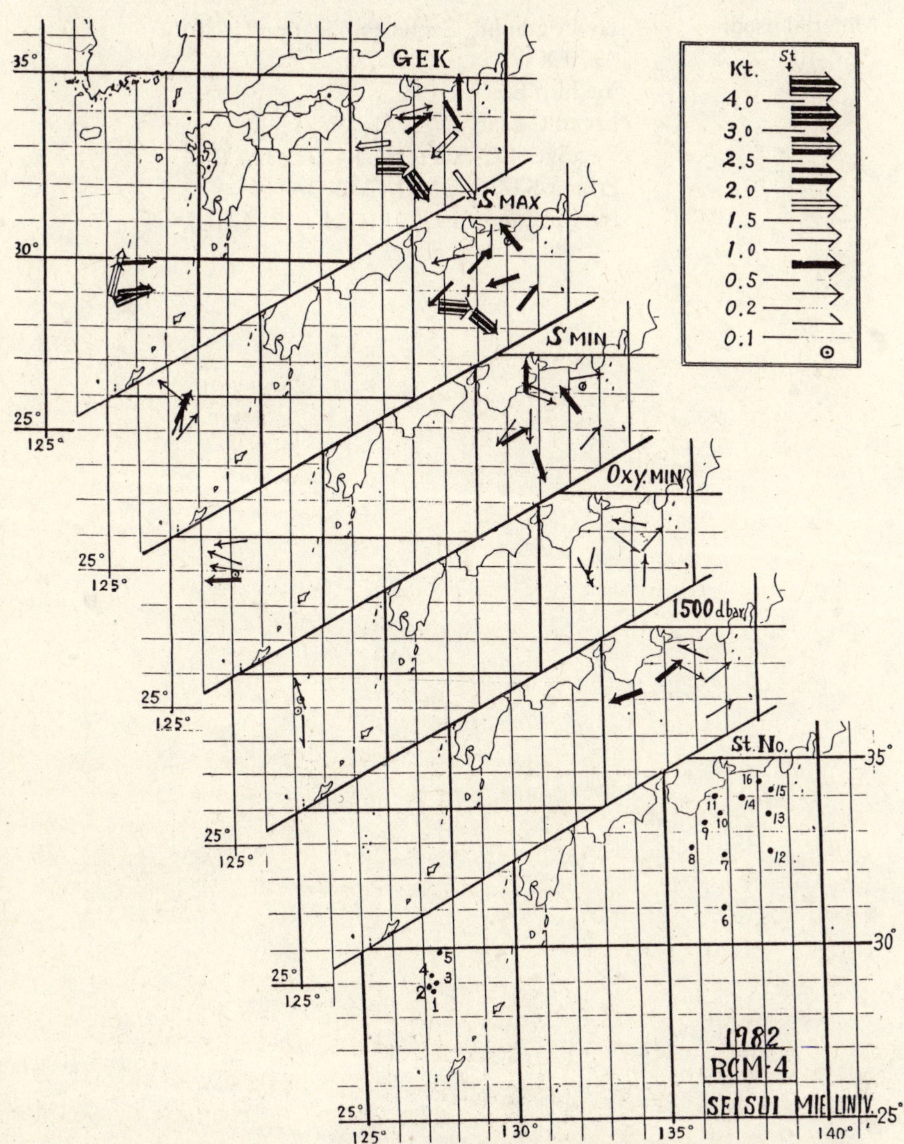


Fig. 7. Results of current measurements in specified layers (16 points measured during 1982).

As drift measurements are made repeatedly from the mooring site it may be useful to compare the readings with measured values every hour.

We would like to conclude this paper with our expression of gratitude to all members and Captain Jirro of our expedition ship the *Seisei-meru*, who helped develop this method.

Material used: Hydrographic Center Publication (1980)
9970 XY.
Yoshiro Electric Co.
Loran C Manual (1971, 1982)
Receiver LC-200 (OM-J 4233-OE)
Printer ST-70 (OM-J 4272-OA)
Interface unit IF-5001 (OM-J 4274-OB).

# Committee on Satellite Applications

*Keisuke Taira*
(Institute of Oceanography, Tokyo University)

This committee was set up as a sequel to the informal discussions in the lobby of Aoi Kaikan which followed the Managing Committee Meeting on June 19, 1981. Prof. Kajiura, Teramoto, Mochizuki, Kunishi, Toba, Aota, Taira, etc. concluded that infrared pictures with high resolution which are sent by weather satellite (NOAA series) can also be used to analyze ocean changes. During the JECSS (Japan and East China Seas Study) seminar held at Chikuha University in May, Oscar Hue (Louisiana State University) stated that the satellite is a proper equipment for the analysis of Japan and East China seas. The volume of data transmitted by satellite is quite large and it was therefore necessary to receive and analyze that data in Japan. Since Hue and his colleagues receive the data on American skies, using a tape, it was also difficult to get the same. On that day, we decided to form a committee as part of the Special Research Project and we also decided to collect information on receiving equipment.

After about a week's time, we learned that satellite signals are being received at the Satellite Center of the Meteorological Department and at Takagi Laboratory in the Production Technology Research Institute of Tokyo University. We also learned that this signal is analyzed by such organizations as Tokai University, Oceanographical Technical Center, Fisheries Information Service Center, etc. On July 3, Mr. Taira visited the Meteorological Satellite Center and obtained a copy of the magnetic tape. He also obtained copies of manuals, etc., and with great effort prepared a test analysis using a line printer. The results were scribbled under the title of "Report" and were distributed to concerned persons in three lots on June 26, July 6, and July 21.

Although the committee on "Satellite Applications" was formed under the leadership of Prof. Teramoto, the group did not even have one formal meeting. However, the information was being exchanged at various symposiums or seminars, etc. and I supported the activities of each group member as introduced below. It can therefore be assumed that the committee always extended help whenever it was necessary during the course of this special study.

Prof. Aota Masaaki (Low-Temperature Research Laboratory, Hokkaido University) has analyzed the data obtained from the Meteorological Satellite Center in cooperation with M/s Muneyama, Sasaki and Asanuma of the Oceanographical Technical Center, and M/s Tosawa and Iisaka of Japan IBM. As a result, he clarified the position of the Soya warm current or the extinction distance of intermediate cold water (dichothermal water) in combination with marine observations, mooring observations, etc. On the other hand, it was also possible to prepare a system whereby information regarding surface water temperature, as observed by about 30 ships, could be communicated by radio waves so that the satellite can also be useful in the fishing industry. Actual use of satellite signals started around October, 1983 and APT pictures have been received since November. Although it reduces the resolution somewhat, we are using the television signals being reflected from satellite for constructing a picture directly.

Prof. Toba (Faculty of Science, Tohoku University) analyzed the tape of signals received at the Meteorological Satellite Center with the help of M/s Kawai, Kawamura, Yamashita, Hanawa, etc. at the Computer Center of Tohoku University. Mr. Kawai introduced their observations in comparison with the observations off Sanrika by ferry boat, during the Autumn Symposium of the Oceanographic Society of Japan in October, 1981. His sudden demise is extremely painful. I still remember visiting Aosei Satellite Center with him. His group attained very important achievements such as the clarification of the structure of horizontal turbulent currents in the Sea of Japan, etc.

Prof. Teramoto (Institute of Oceanography, Tokyo University) continued his studies with the assistance of Taira and Fukazawa. A 6250 bpi magnetic tape unit and graphic display is essential for analyzing voluminous data and hence those units were obtained for the institute computer system which was transferred to this building in December, 1982. In August 1983, Prof. Takagi obtained the data received from December, 1982 to May, 1983 and it is being analyzed at present. If a proper analysis system becomes a reality, many researchers in our Special Research Project will be able to use that data. In November, 1983, an APT receiver was installed on the *Hakuho-maru*. When the received data can be analyzed on the vessel itself, it will be a great help to future oceanographic observations. It is planned to install an AVHRR receiving system on the Antarctic observation vessel *Shirase*.

Prof. Takagi (Biological Research Institute, Tokyo University) has proposed another special project "Use of multi-time, multi-dimensional data obtained from satellite for the analysis of physical phenomena" (March, 1983) and one of the main features of this project is oceanographical phenomena.

Prof. Teramoto is the Chairman of a subcommittee on Satellite Oceanographic Observations established in April, 1983 under the Oceanographic Society of Japan and is actively involved in getting a satellite for oceanographic studies as well as in introducing satellite observation techniques.

Most of the activities of this committee are thus not directly related to the Special Research Project but we feel they serve to create a proper background.

## ADMINISTRATIVE REPORT

Minutes of the 21st Meeting of the Executive Committee, January 13, 1984 (Friday), Institute of Oceanography, Tokyo University.

**Items reported**

National Science Foundation has decided to translate all issues of the Newsletter of this Special Research Project and publish the same.

Committee on Future Studies is now busy preparing the application to continue the Special Research Project under the new theme "Basic studies for the analysis of operations involved in deep seawater circulation."

**Items discussed**

Regarding preparation of Final Report:

Discussions are being held with a few publishing companies regarding the detailed conditions of publications.

The report will be divided into following chapters. Each group is allowed to write its report within 120 sheets of standard 400 characters manuscript including figures and references. There will be about two pages of references (50 titles) for each chapter.

**Index**

Chapter 1. Seawater circulation in the northeast Pacific and material distribution.

Chapter 2. Flow of the Kuroshio, Kuroshio extension current and topographical effects.

Chapter 3. Seawater circulation at the continental shelf region and their changes.

Chapter 4. Material circulation in ocean.

Chapter 5. Structure of the ocean floor and its Paleoenvironments.

Chapter 6. Development of measurement methods regarding seawater and material circulation.

**Progress Reports, Index, etc.**

Deadline for receipt of the first manuscript will be August. The editorial board will make changes to present a uniform style throughout the report and the final matter will be handed over to publishers by the end of September.

## Oh! Pirate Edition of Newsletter?

During the evening of December 21, 1983, in Washington D.C. the slight rain turned into snow. The hotel lobby had become extremely slippery to walk. That morning, the color of the water in the swimming pool was not so clear and when I went beyond the fence to see, I found that it had already turned into ice. I had a great Sushi dinner with hot sake after about ten days and after enjoying the same, it was difficult to return along sloppy Wisconsin avenue.

The next morning I was awakened by a phone call from Mr. Louis Brown of NSF. I was asked to come to Room No. 609, 1800 G, NW at about ten. Most of the American government buildings are without any indication. I got a taxi and walked about one block to see, but they were all big buildings with only street numbers. The upper floor of one such building is the NSF. Japanese government houses are generally tiled roof structures and look as if always on alert whereas in American style, there is a small room and the aisle becomes the typist's office. Research institutes are also the same. A closed door means a person is not in his room.

After Mr. Brown gave a demonstration of Inter-view or Tele-mail, we went to the room of Mr. C. Collins. He was ready to receive us with issues of the Journal of Oceanographic Society of Japan and Report of the cruise of *Atlantis II* in the northern Pacific Ocean. He also pointed toward a white sheet of paper showing NSF Printed Matter with a US Mail stamp and asked me are you the same Taira as mentioned here? It was the English translation of my paper "Measurement of deep layer currents on the eastern side of the Izu ridge" which appeared in Newsletter No. 1. With that in my mind, I stumbled at the unheard name of Biura Yoshijiro, and could not really make out what that paper was about. I thought maybe these are translations of Japanese papers at random and with that perspective, listened to the request of Mr. Brown for some assistance.

While we were discussing matters related to UNOLS, studies regarding material distribution, current measurements, etc., it was suddenly lunch time. I found some faces familiar to me which I had seen before when I went to see the *Atlantis* as a student. Even after being introduced, one easily forgets the family name as there were so many of them. However, during our daily life, names are quite important. I remember that one researcher at the Woods Hole Oceanographic Institution forgot his experiment notes in a neighboring room but found them one year later since his name was written on them. Now, one would ask why the note was not returned when everybody knew whom it belonged to. The answer is, nobody knew who Mr. A. Williams was. His nick-name is Sandy. In America, there is another story. All persons at NOAA have business cards, and it was a regular card exchange program just as in Japan.

Gone are the days when only businessmen used to have calling cards.

We had a nice lunch at the World Bank Employees Cafeteria with Mr. Collins and his pretty wife and he told many interesting things such as he walked barefoot in the snow last night or somebody tumbled down and was frozen to death, the schools are closed for vacation or that day was his wedding anniversary and so on and so on.

After visiting NASA, and when going to NODC, I opened the white paper in the taxi and read the title "Special studies sponsored by the Mombusho Scientific Research Fellowships, Dynamic Structure of the Ocean, Newsletter No. 1, September 1981." It was a total translation of our Newsletter. The English translation was done by Amerind Publishing Co. of New Delhi. Mr. Rene Quzon of NODC explained the Indian connection.

The English translation was quite good and generally there are no problems. Most of the problem is in names. There are such mis-spelt names like Biura Yoshijiro (Kajiura Kinjiro), Keinishi (Kunishi), Horiba (Horibe), Sisuka (Nasu), etc. Also the names of universities are correct but there are many mistakes in faculty names. For example Satellite Application is translated as Health Application. Many names have been spelled correctly such as Mochizuki or Takayanagi, etc., however, there were errors in the names mentioned above such as Kajiura Kinjiro, Kunishi, Horibe, Nasu, etc.

Mr. Louis Brown came to Japan around the beginning of January. He explained that it is proposed to translate all issues of the Newsletter and it was decided that Prof. Teramoto should send the English transcript of all typical Japanese names. Prof. Teramoto also received a letter from Mr. Prakash Kanade stating that he was translating the Newsletter for some international organization and asking for a dictionary of technical terms, etc. Prof. Teramoto had already acceded to those requests.

That was how I got an English translation of Issue No. 1. The subsequent issues will also be sent to us as they are published. Persons desirous of obtaining these issues can write directly to NSF (Dr. Louis Brown, Director of Ocean Sciences, National Science Foundation, Washington D.C. 20550) or contact the Administrative Office.

(Keisuke Taira)

**Note:**

When producing or citing material published in this Newsletter, due acknowledgment should be given. Authors' names, wherever mentioned, should also be quoted. A copy of the publication containing such a reproduction may be sent to the Editorial Section of the Administrative Office.

Published by: Administrative Office, Special Project Research "The Ocean Characteristics and Their Changes"
Published at : Administrative Office
Physical Oceanography Department,
Institute of Oceanography,
Tokyo University,
1-15-1 Minami-dai, Nakano-ku,
Tokyo 164
Tel. 03-376-1251, Ext. 251.
Editorial staff : Taeko Kitano and Noriyo Kimura